Oxygen Free Radicals in Tissue Damage

Oxygen Free Radicals in Tissue Damage

Merrill Tarr
Fred Samson
Editors

Birkhäuser
Boston · Basel · Berlin

Merrill Tarr
Dept. of Physiology
University of Kansas Medical Ctr.
Kansas City, KS 66160-7401

Fred Samson
Ralph L. Smith Research Center
University of Kansas Medical Ctr.
Kansas City, KS 66160-7401

Library of Congress Cataloging-in-Publication Data

Oxygen free radicals in tissue damage / Merrill Tarr and Fred Samson, editors.
p. cm.
Includes bibliographical references and index.
ISBN 0-8176-3609-9 (hard : alk. paper). -- ISBN 3-7463-3609-9 (hard : alk. paper)
1. Active Oxygen--Pathophysiology. 2. Free radicals (Chemistry) --Pathpophysiology. I. Tarr, Merrill, 1940- . II. Samson, Frederick E.
[DNLM: 1. Free Radicals. 2. Oxygen--metabolism. 3. Reperfusion Injury. QV 312 09558]
RB170.0984 1992 92-49815
616.07'1 --dc20 CIP

Printed on acid-free paper.

ISBN 0-8176-3609-9
ISBN 3-7643-3609-9

Typeset by ARK Publications, Inc., Newton Centre, MA
Printed and bound by Quinn-Woodbine, Woodbine, NJ
Printed in the U.S.A.

9 8 7 6 5 4 3 2 1

Table of Contents

Preface

Oxygen free radicals and other reactive oxygen species are being postulated as causal agents in an increasing number of pathological conditions. Indeed, some investigators are suggesting that highly destructive reactive oxygen species are the final common path leading to tissue damage following a wide variety of insults including trauma, hypoxia, ischemia, hyperoxia, radiation, some toxins, and even strenuous athletic pursuits. But, as Robert Floyd points out, "Proof of the importance of oxygen free radicals and the oxidative damage they initiate depend on unequivocal evidence for the presence of free radicals and a clear association of their formation with the induction of the dysfunction of pathological conditions." Since such proof does not come easily, there have been and will continue to be many controversies regarding the role played by reactive oxygen species in tissue damage.

There have been many recent reviews of the chemistry and possible role of reactive oxygen species in many types of organ dysfunctions, tissue damage, degenerative diseases, and aging. This book is not such a review. Rather it presents for a diverse audience of physical-organic chemists, biochemists, medical researchers, and other investigators of pathophysiology, discussions of a variety of issues important for understanding reactive oxygen species and their role in tissue damage. Hopefully, with the wide selection of topics that make up this volume, each written by knowledgeable scientists, the reader will gain a valuable bird's eye view, if you will, of this rapidly progressing and highly significant concept of an important mechanism underlying a variety of tissue damage.

In December, 1990 a workshop on "Oxygen Free Radicals in Tissue Damage" was held at the University of Kansas Medical Center. The intent of the workshops was to bring together experts representing various areas of research relevant to this topic for a discussion of relevant and controversial topics relative to this field. After the workshop, several participants proposed a book to be based on similar topics, with full-length, state of the art chapters. The enthusiastic willingness of the invited experts to contribute such chapters has made this book possible.

Since there have been many reviews of the chemistry, as well as of the possible role of oxygen free radicals in pathophysiology, we asked each contributor to write on a topic "close to his heart" regarding oxygen free radicals. Each contributor was encouraged to develop a theme and discuss issues relative to that theme, rather than to extensively review the literature. As with the design of the workshop, a diversity of interests was sought so as to cover important aspects of oxygen free radicals including their chemistry, their detection, their roles in tissue damage in a variety of organ systems, and methods for protecting against their damage. The chapters in this book, therefore, provide insights into current thinking regarding oxygen free radicals by experts actively involved in research on various aspects of this timely subject.

Merrill Tarr
Fred Samson
Editors

Contributors

Kenneth L. Audus, Department of Pharmaceutical Chemistry, The University of Kansas, School of Pharmacy, Lawrence, Kansas, 66047

Joseph Beckman, Department of Anesthesiology, University of Alabama at Birmingham, Birmingham, Alabama, 35294

Roberto Bolli, Section of Cardiology, Department of Medecine, Baylor College of Medecine, Houston, Texas, 77030

Donald C. Borg, Cook College, Rutgers University, New Brunswick, New Jersey, 08903-0231

Jun Chen, Department of Anesthesiology, University of Alabama at Birmingham, Birmingham, Alabama, 35294

Karl A. Conger, Department of Neurology, University of Alabama at Birmingham, Birmingham, Alabama, 35294

Robert Floyd, Molecular Toxicology Research Group, Oklahoma Medical Research Foundation, Oklahoma City, Oklahoma, 73104

Irwin Fridovich, Department of Biochemistry, School of Medecine, Duke University, Durham, North Carolina, 27710

Joshua I. Goldhaber, Division of Cardiology and UCLA Cardiovascular Research Laboratory, UCLA School of Medecine, Los Angeles, California, 90024

D. Neil Granger, Department of Physiology, Louisiana State University Medical Center, Shreveport, Louisiana, 71130-3932

Edward Hall, Central Nervous System Disease Research Group, Research Division, The Upjohn Company, Kalamazoo, Michigan, 49001

James H. Halsey, Jr., Department of Neurology, University of Alabama at Birmingham, Birmingham, Alabama, 35294

Norman R. Harris, Department of Physiology, Louisiana State University Medical Center, Shreveport, Louisiana, 71130-3932

Harry Ischiropoulos, Department of Anesthesiology, University of Alabama at Birmingham, Birmingham, Alabama, 35294

Mohamed O. Jeroudi, Section of Cardiology, Department of Medecine, Baylor College of Medecine, Houston, Texas, 77030

Sen Ji, Department of Physiology and UCLA Cardiovascular Research Laboratory, UCLA School of Medecine, Los Angeles, California, 90024

Jeffrey Kanofsky, Research Laboratories, Hines V.A. Medical Center, Hines, Illinois, 60141

Matthew E. Layton, Department of Pharmacology, Toxicology, and Therapeutics, University of Kansas Medical Center, Kansas City, Kansas, 66103

Xiao Ying Li, Section of Cardiology, Department of Medecine, Baylor College of Medecine, Houston, Texas, 77030

Thomas Pazdernik, Department of Pharmacology, Toxicology, and Therapeutics, University of Kansas Medical Center, Kansas City, Kansas, 66103

Merrill Tarr, Department of Physiology, University of Kansas Medical Center, Kansas City, Kansas, 66160-7401

Jack Uetrecht, Faculties of Pharmacy and Medecine, University of Toronto and Sunnybrook Medical Centre, Toronto, Ontario, Canada M5S 2S2

Dennis P. Valenzeno, Department of Physiology, University of Kansas Medical Center, Kansas City, Kansas, 66160-7401

James Weiss, Division of Cardiology and UCLA Cardiovascular Research Laboratory, UCLA School of Medecine, Los Angeles, California, 90024

Ling Zhu, Department of Anesthesiology, University of Alabama at Birmingham, Birmingham, Alabama, 35294

Barbara J. Zimmerman, Department of Physiology, Louisiana State University Medical Center, Shreveport, Louisiana, 71130-3932

Marcel Zughaib, Section of Cardiology, Department of Medecine, Baylor College of Medecine, Houston, Texas, 77030

Chapter 1

Getting Along With Oxygen

Irwin Fridovich

Superoxide Theory of Oxygen Toxicity

The spin restriction predisposes dioxygen to a univalent pathway of reduction. Superoxide (O_2^-), the first intermediate encountered on this univalent pathway, is consequently a commonly encountered product of dioxygen reduction. The finding that O_2^- is produced by some enzymes and is efficiently scavenged by others (McCord and Fridovich, 1968, 1969) led to the view that O_2^- is an agent of oxygen toxicity. In this view the superoxide dismutases (SODs), which catalytically scavenge O_2^-, serve a defensive role (McCord et al., 1971), much as do catalases vís a vís H_2O_2. Given the early association of O_2^- with radiation chemistry and its extensive study by pulse radiolysis (Czapski, 1971), the biological relevance of O_2^- was not easily accepted.

Subsequent work has provided abundant evidence that O_2^- is produced in aerobic living cells, that it constitutes a threat to these cells, and that the SODs provide a necessary defense. Thus, SOD is abundant in aerobes and is scarce, or lacking entirely, in sensitive obligate anaerobes (McCord et al., 1971). Moreover, exposure to dioxygen elicits an adaptive increase in the biosynthesis of SOD (Gregory and Fridovich, 1973a, b). Compounds such as paraquat, which can mediate increased intracellular production of O_2^-, exert a dioxygen-dependent toxicity and also cause an adaptive increase in cell content of SOD (Hassan and Fridovich, 1977, 1978). Mammalian cells, whose content of SOD was elevated by scrape loading, were rendered resistant to paraquat (Bagley et al., 1986). In a similar vein, increasing SOD by liposomal fusion protected against hyperoxia (Freeman et al., 1983).

Oxygen Free Radicals in Tissue Damage
Merrill Tarr and Fred Samson, Editors

The most recent and the most compelling evidence supporting the superoxide theory of oxygen toxicity has come from genetic manipulations (Touati, 1988). Thus, strains of *Escherichia coli* with insertional defects in both the sodA and sodB genes were unable to produce active SOD. Such sodA sodB strains grew as well as the parental strain in *anaerobic* minimal medium but would not grow when exposed to air. In an aerobic-rich medium these strains grew slowly and then exhibited a hypersensitivity toward paraquat and an increase in spontaneous mutagenesis. Related phenotypic deficits have been seen in both prokaryotes and in eukaryotes and they are eliminated by introduction of a plasmid bearing a functional SOD gene, even when the gene derives from an unrelated species (Bowler et al., 1989, 1990; Gruber et al., 1990; Natvig et al., 1987; Van Camp et al., 1990).

Superoxide-Imposed Auxotrophies

The ability of nutritional supplements to allow aerobic growth of the SOD-deficient strain of *E. coli* is indicative of O_2^--sensitive biosynthetic pathways. In at least one case the basis of such an O_2^--imposed auxotrophy is understood in terms of an O_2^--sensitive enzyme. Thus, the α, β-dihydroxy acid dehydratase, which catalyzes the penultimate step in the biosynthesis of branched chain amino acids, is rapidly inactivated by O_2^- (Kuo et al., 1987). There are other O_2^--imposed auxotrophies, such as those for sulfur-containing and for aromatic amino acids. It seems likely that these too reflect O_2^--sensitive biosynthetic enzymes or intermediates.

Instructive Anomalies

The foregoing leaves little doubt that O_2^- is toxic to cells and that SODs are defensive enzymes, whose role is to diminish steady-state levels of O_2^- markedly. Nevertheless, the incredible variety and complexity of living things has set the stage for apparently anomalous observations. Each of these provides an opportunity of discovery and each requires clarification by careful experimental probing. A case in point is the apparent lack of SOD in the aerotolerant *Lactobacillus plantarum* (McCord et al., 1971). How could it grow aerobically without SOD? *L. plantarum* was subsequently found to accumulate Mn(II) to an intracellular concentration of approximately 30 mM and this Mn (II) was seen to provide a functional replacement for SOD (Archibald and Fridovich, 1981a, b).

The ability to cause huge overproduction of SOD in *E. coli*, by virtue of multicopy plasmids bearing SOD genes, brought another apparent anomaly to light. Thus, previous studies had indicated that paraquat is toxic because it mediates O_2^- production and that SOD protects against the toxicity of this viologen. Yet, great overproduction of MnSOD (Bloch and Ausubel, 1986) or of FeSOD (Scott et al., 1987) was seen to sensitize *E. coli* toward the growth-inhibiting effect of paraquat. The discovery of the superoxide-inducible regulon (soxR) (Tsaneva and Weiss, 1990) provided the basis for explaining this apparent paradox.

SoxR positively regulates the biosynthesis of approximately a dozen proteins, among which are MnSOD, glucose-6-phosphate dehydrogenase, and endonuclease (IV) (Greenberg et al., 1990). These proteins provide a balanced defense against the consequences of O_2^- production, each attending to a different aspect of that defense. Thus MnSOD diminishes the steady state level of O_2^-, glucose-6-phosphate dehydrogenase provides NADPH for reduction of G-S-S-G and for the alkyl hydroperoxide reductase and endonuclease IV participates in DNA repair. The other members of this regulon remain to be identified, but they too must contribute to the totality of the defense against O_2^-.

Suppose that gross, unregulated, and selective overproduction of SOD so lowered the steady-state level of O_2^- that induction of soxR failed to occur, even in the presence of paraquat. The affected cell would then fail to produce the constellation of proteins controlled by soxR and would fail to achieve an effective, balanced defense. This has been tested by examining induction of glucose-6-phosphate dehydrogenase in wild-type and in SOD-overproducing strains, in response to paraquat (Liochev and Fridovich, 1991). As expected, the wild-type strain did show this induction, whereas the SOD overproducer did not.

Superoxide-Sensitive Enzymes

The catalog of enzymes that are inactivated by O_2^- has been growing steadily. The α,β-dihydroxy acid dehydratase, involved in the biosynthesis of branched chain amino acids, is only one member of a rather large group of related dehydratases, many of which may be susceptible to attack by O_2^-. The 6-phosphogluconate dehydratase has already been examined and is sensitive to O_2^-, exhibiting a rate constant for inactivation by this radical that has been estimated to be 2×10^8 M^{-1} s^{-1} (Gardner and Fridovich, 1991a). Aconitase, which interconverts citrate and isocitrate and so plays a key role in the citric acid cycle, is also an acid dehydratase and it too is rapidly inactivated by O_2^- (Gardner and Fridovich, 1991b). Indeed, inactivation of aconitase by O_2^- may explain why loss of the ability to respire is an early consequence of oxygen toxicity. It is possible that the great sensitivity of aconitase to O_2^- has a selective advantage for facultative organisms. Thus, in the face of increased O_2^- production, whether due to an increase in pO_2 or to exposure to viologens, quinones, or other redox cycling agents, inactivation of aconitase would interrupt the Krebs cycle and so diminish the rate of reduction of NAD^+. Since NADH feeds electrons into the pathway primarily responsible for O_2^- production (Imlay and Fridovich, 1991a), a decrease in the rate of NAD^+ reduction would decrease O_2^- production. The O_2^- sensitivity of aconitase thus provides a fuse or circuit breaker, which interrupts electron flow when the concentration of O_2^- rises and which thus protects the cell against an O_2^- overload until the inherently slower, but more effective, adaptations of the soxR regulon can be brought to bear.

There is an interesting interplay between SODs and the H_2O_2-scavenging enzymes. These enzymes are a team in the sense that the SODs convert O_2^- into $O_2 + H_2O_2$, while the catalases and peroxidases finish the job by dismuting or reducing H_2O_2 to the stable product, water. SOD, catalases, and peroxidases are also a team in the sense that they are mutually protective. Thus, H_2O_2 slowly inactivates Cu, ZnSOD, and FeSOD, whereas O_2^- more rapidly inhibits catalases and peroxidases. The catalases and peroxidases thus protect the SODs against H_2O_2 while the SODs, in return, protect catalases and peroxidases against O_2^-.

Other enzymes that have been reported to be sensitive to O_2^- include transketolase, in which case the target is the α, β-dihydroxy ethyl thiamine pyrophosphate, which is an enzyme-bound intermediate in the catalytic cycle (Asami and Akazawa, 1977), creatine kinase (McCord and Russell, 1988), and papain (Lin et al., 1978). The existence of proteases that selectively digest oxidatively modified proteins (Davies and Lin, 1988; Pacifici et al., 1989) indicates that attack by oxygen radicals is an ongoing and significant process that has been proposed to be a factor in protein turnover, aging, and oxygen toxicity (Stadtman et al., 1988).

Enhancing the Oxidative Propensity of O_2^-

The reactivity of O_2^- can be dramatically increased, and its range of susceptible targets correspondingly broadened, by association with cationic centers. In the simplest case a proton serves as the cationic ligand and yields the conjugate acid $HO_2\cdot$. The pK_a of $HO_2\cdot$ is 4.8, so even at neutral pH approximately 1% of the O_2^- is protonated. In locally acidified regions such as the interior of phagosomes and lysosomes, and the thin layer of solution immediately adjacent to polyanionic surfaces or macromolecules, this percentage of $HO_2\cdot$ will be much greater. $HO_2\cdot$ is a much more powerful oxidant than is O_2^- and it can attack polyunsaturated fatty acids with a rate constant of $\sim 1 \times 10^3\ M^{-1}\ s^{-1}$ (Gebicki and Bielski, 1981), whereas O_2^- does not do so at a measurable rate.

Other cationic centers, such as Mn(II) or $V_{(V)}$, can similarly associate with and increase the oxidizing propensity of O_2^-, such that NAD(P)H can be attacked. Since univalent oxidation of NAD (P) H yields NAD(P)· that can, in turn, reduce O_2 to O_2^-, we have the basis for a free radical chain reaction in which the cationic center acts catalytically and in which each O_2^- introduced can cause the oxidation of multiple molecules of NAD(P)H. In the case of $V_{(V)}$ the pertinent reactions are (Liochev and Fridovich, 1990):

a. $V_{(V)} + O_2^- \rightharpoonup V_{(IV)} - OO$
b. $V_{(IV)} - OO + \ NAD(P)H \rightleftharpoons V_{(\ IV)} - OOH + \ NAD(P).$
c. $V_{(IV)} - OOH + \ H^+ \rightleftharpoons \ V_{(V)} + \ H_2O_2$
d. $NAD(P) \cdot + O_2 \rightharpoonup \ NAD(P)^+ + \ O_2^-$

A similar sequence of reactions may be written for the case of Mn(II) and provides an explanation for the Mn(II)-dependent oxidation of NAD(P)H by activated neutrophils (Curnutte et al., 1976).

The interaction of O_2^- with Fe(III) can also give rise to powerful oxidants, but in this case H_2O_2 is also required. Thus:

e. $Fe(III) + O_2^- \rightleftharpoons \ Fe(II) \ + O_2$
f. $Fe(II) + \ HOOH \rightleftharpoons Fe(I)\ OOH + \ H^+$
g. $Fe(I)OOH \ \rightleftharpoons \ Fe(II)O + \ HO^-$
h. $Fe(II)O + \ H^+ \rightleftharpoons \ Fe(III)OH \ \rightleftharpoons \ Fe(III) + \ HO\cdot$

Whether the process stops with ferryl [Fe(II)O] or with hydroxyl (HO·) probably depends on conditions; this has been a source of some disagreement. However, from a biological viewpoint it hardly matters, since both FE(II)O and HO· are sufficiently reactive to react at almost diffusion-limited rates with biological molecules.

Site-Specific Oxidative Attack

Indiscriminately reactive oxidants, if generated in bulk solution, would be apt to react primarily with expendable metabolic intermediates. A large measure of selectivity can, however, be achieved by generation of the oxidants adjacent to specific targets. Thus, if the Fe(III) is bound to DNA then the Fe(II)O, or the HO·, will be generated adjacent to, and will selectively attack, that DNA (Czapski, 1984). This sequence of reactions (e–h), referred to as the metal-catalyzed Haber-Weiss reaction, has been repeatedly demonstrated *in vitro* and probably also has some relevance *in vivo*. Thus, it could explain why sodA sodB *E. coli*, which are defective in SOD but contain normal levels of catalases, should be hypersensitive toward H_2O_2 (Carlioz and Touati, 1986). It should, however, be kept in mind that living cells contain reductants, other than O_2^-, that could reduce Fe(III) to Fe(II) and thus allow production of Fe (II) O or HO· by the Fenton reaction. The role of iron in catalyzing HO· production from H_2O_2, whether the reductant is O_2^- or something else, can explain why chelating agents should protect against the toxicity of H_2O_2 (Girotti and Thomas, 1984) and why Zn(II), which can displace bound Fe(III) but which is incapable of redox cycling, should also be protective (Thomas et al., 1986).

DNA damage by the site-specific Haber-Weiss reaction is one explanation for the mutagenicity of O_2^-. This mutagenicity of O_2^- accounts for the dioxygen-dependent mutagenicity of paraquat (Hassan and Moody, 1982; Moody and Hassan, 1982) and for the mutagenicity of dioxygen in a strain of *E. coli* lacking SOD (Farr et al., 1986). Yet O_2^- does not directly attack DNA (Brawn and Fridovich, 1981) and its mutagenicity could also be the consequence of attack at remote sites, perhaps involving effects on systems responsible for repair of DNA damage. We have already seen that O_2^- directly inactivates certain enzymes, such as the dihydroxy acid dehydratases and aconitase. Suppose that one or more enzymes responsible for DNA repair were inactivated by O_2^-, or that inducible DNA repair systems failed to be induced because amino acid biosynthesis has been compromised by O_2^- attack on enzymes involved in such synthesis.

One method of plumbing the extent of oxidative damage to DNA *in vivo* depends on measurement of urinary levels of hydroxylated purines and pyrimidines, which have presumably been removed from DNA during DNA repair (Dizdaroglu and Bergtold, 1986; Kaneko and Leadon,

1986). In the case of humans, daily excretion of thymine glycols corresponds to hundreds of events per cell (Cathcart et al., 1984). Moreover, rats excreted 15 times more of the hydroxylated thymine than humans and did so even on a DNA-free diet, thus eliminating the possibility that the thymine glycol was of dietary origin. When radiolabeled thymine glycol was fed, it appeared in the feces, not in the urine, thus excluding the possibility that the urinary material originated in the gut. Levels of urinary thymine and thymidine glycols were inversely related to life span (Adelman et al., 1988). The mechanism of removal of thymine glycol from DNA has been described (Lin and Sancar, 1989). Another oxidatively modified base is 8-hydroxy deoxyguanosine. It is found in both nuclear and mitochondrial DNA of liver at levels approximating 40,000 per mitochondrial genome and 140,000 per nuclear genome (Richter et al., 1988). It is clear that oxidative damage to DNA, probably caused by endogenously generated Fe(II)O or HO·, is extensive and explains the need for ongoing DNA repair.

In the case of *E. coli*, there is reason to suspect that the cell envelope, as well as DNA and enzymes, is subject to attack by oxygen-derived radicals. The sodA sodB strain of *E. coli* exhibits dioxygen-dependent auxotrophies due to inactivation of biosynthetic enzymes by O_2^-·. It has recently been seen that nonmetabolizable osmolytes, such as salts, sucrose, or cellobiose, can partially relieve these auxotrophies (Imlay and Fridovich, 1991b). This indicates that leakage of nutrients from the sodA sodB strain under aerobic conditions compounds the consequences of damaged biosynthetic pathways. A dioxygen-dependent diminution of the structural integrity of the cell envelope, in the strain lacking SOD but not in the wild type, would explain this behavior.

Epilogue

We are left with the impression that, even for relatively simple prokaryote such as *E. coli*, living with dioxygen entails intricate adaptations to a multitude of threats and attacks by reactive entities derived from O_2. In metazoan eukaryotes, such as we, the level of complexity is vastly increased and what will suffice for one cell type is inappropriate for another. Thus, in *E. coli* reproductive fitness and therefore evolutionary selection are based on the rapidity of growth and on cell division and survival under all conditions that may be encountered. In contrast, in metazoan organisms, it is the survival and rate of reproduction of the ensemble that matters. We thus encounter phagocytic cells that engage in the suicidal production of O_2^-, H_2O_2, and OCl^-, but that in so doing foster the survival of the ensemble by destroying invading microorganisms and parasites. We also see endothelial cells that produce NO in order to relax vascular smooth muscle and thus to control blood pressure, as well as to inhibit platelet aggregation and thus to prevent inappropriate intravascular thrombosis. We may confidently anticipate that further study of the biology of oxygen radicals will provide further surprises, insights, and practical benefits. Knowledge *is* the only way out of the cages of life.

References

Adelman R, Saul RL, Ames BN (1988): Oxidative damage to DNA: relation to species metabolic rate and lifespan. *Proc Natl Acad Sci USA* 86:2706–2708.

Archibald FS, Fridovich I (1981a): Manganese and defenses against oxygen toxicity in *Lactobacillus plantarum. J Bacteriol* 145:442–451.

Archibald FS, Fridovich I (1981b): Manganese, superoxide dismutase, and oxygen tolerance in some lactic acid bacteria. *J Bacteriol* 146:928–936.

Asami S, Akazawa T (1977): Enzymic formation of glycollate in *chromatium.* Role of superoxide radical in a transketolase-type mechanism. *Biochemistry* 16:2201–2207.

Bagley AC, Krall J, Lynch RE (1986): Superoxide mediates the toxicity of paraquat for Chinese hamster ovary cells. *Proc Natl Acad Sci USA* 83:3189–3193.

Bloch CA, Ausubel FM (1986): Paraquat-mediated selection for mutations in the manganese-superoxide dismutase gene sodA. *J Bacteriol* 168:795–798.

Bowler C, Alliotte T, Van den Bulcke M, Bauw G, Vanderkerkhove J, Van Montagu M, Inze D (1989): A plant manganese superoxide dismutase is efficiently imported and correctly processed by yeast mitochondria. *Proc Nat Acad Sci USA* 86:3237–3241.

Bowler C, Van Kaer L, Van Camp W, Van Montagu M, Inze D, Dhaese P (1990): Characterization of *Bacillus stearothermophilus* manganese superoxide dismutase and its ability to complement copper/zinc superoxide dismutase deficiency in *Saccharomyces cerevisiae. J Bacteriol* 172:1539–1546.

Brawn K, Fridovich I (1981): Dan strand scission by enzymically-generated oxygen radicals. *Arch Biochem Biophys* 206:414–419.

Carlioz A, Touati D (1986): Isolation of superoxide dismutase mutants in *Escherichia coli*: is superoxide dismutase strictly necessary for aerobic life? *EMBO J* 5:623–630.

Cathcart R, Schwiers E, Saul RL, Ames BN (1984): Thymine glycol and thymidine glycol in human and rat urine: a possible assay for oxidative DNA damage. *Proc Natl Acad Sci USA* 81:5633–5637.

Curnutte JT, Kanovsky ML, Babior BM (1976): Manganese-dependent NADPH oxidation by granulocyte particles. The role of superoxide and the non-physiological nature of the manganese requirement. *J Clin Invest* 57: 1059–1067.

Czapski G (1971): Radiation chemistry of oxygenated aqueous solutions. *Annu Rev Phys Chem* 22:171–208.

Czapski G (1984): On the use of hydroxyl radical scavengers in biological systems. *Israel J Chem* 24:29–32.

Davies KJA, Lin SW (1988): Degradation of oxidatively-denatured proteins in *Escherichia coli. Free Rad Biol Med* 5:215–223.

Dizdaroglu M, Bergtold DS (1986): Characterization of free radical-induced base damage in DNA at biologically relevant levels. *Anal Biochem* 156:182–188.

Farr SB, D'Ari RD, Touati D (1986): Oxygen-dependent mutagenesis in *Escherichia coli* lacking superoxide dismutase. *Proc Natl Acad Sci USA* 83:8268–8272.

Freeman B, Young SL, Crapo J (1983): Liposome-mediated augmentation of superoxide dismutase in endothelial cells prevents oxygen injury. *J Biol Chem* 258:12534–12542.

Gardner PR, Fridovich I (1991a): Superoxide sensitivity of the *Escherichia coli* 6-phophogluconate dehydratase. *J Biol Chem* 266:1478–1483.

Gardner PR, Fridovich I (1991b): Superoxide sensitivity of the *Escherichia coli* aconitase. *J Biol Chem* 266: 19328–19333.

Gebicki JM, Bielski BHJ (1981): Comparison of the capacities of the perhydroxyl and the superoxide radicals to initiate chain oxidation of the linoleic acid. *J Am Chem Soc* 103:7020–7022.

Girotti AW, Thomas JP (1984): Damaging effects of oxygen radical on resealed erythrocyte ghosts. *Biochem Biophys Res Commun* 118:474–480.

Greenberg JT, Monach P, Chou JH, Josephy PD, Demple B (1990): Positive control of a global antioxidant defense regulon activated by superoxide-generating agents in *Escherichia coli. Proc Natl Acad Sci USA* 87:6181–6185.

Gregory EM, Fridovich I (1973a): The induction of superoxide dismutase by molecular oxygen. *J Bacteriol* 114:543–548.

Gregory EM, Fridovich I (1973b): Oxygen toxicity and the superoxide dismutase. *J Bacteriol* 114:1193–1197.

Gruber MY, Glick BR, Thompson JE (1990): Cloned manganese superoxide dismutase reduces oxidative stress in *Escherichia coli* and *Anacystis nidulans. Proc Natl Acad Sci USA* 87:2608–2612.

Hassan HM, Fridovich I (1977): Regulation of the synthesis of superoxide dismutase in *Escherichia coli.* Induction by methyl viologen. *J Biol Chem* 252:7767–7772.

Hassan HM, Fridovich I (1978): Superoxide radical and the oxygen enhancement of the toxicity of paraquat in *Escherichia coli. J Biol Chem* 253:8143–8148.

Hassan HM, Moody CS (1982): Superoxide dismutase protects against paraquat-mediated dioxygen toxicity and mutagenicity: studies in *Salmonella typhimurium. Can J Physiol Pharmacol* 60:1367–1373.

Imlay JA, Fridovich I (1991a): Assay of metabolic superoxide production in *Escherichia coli. J Biol Chem* 266:6957–6965.

Imlay JA, Fridovich I (1991b): Suppression of oxidative envelope damage by pseudoreversion of a superoxide dismutase-deficient mutant of Eschericia coli. *J Bacteriol* 174:953–961.

Kaneko M, Leadon SA (1986): Production of thymine glycols in DNA by N-hydroxy-2-naphthylamine as detected by a monoclonal antibody. *Cancer Res* 46:71–75.

Kuo CF, Mashino T, Fridovich I (1987): α, β-Dihydroxyisovalerate dehydratase: a superoxide sensitive enzyme. *J Biol Chem* 262:4724–4727.

Lin WS, Armstrong DA, Lal M (1978): Effects of SOD, dithiothreitol and formate on the inactivation of papain by hydroxyl and by superoxide radicals in aerated solutions. *Int J Radiat Biol* 33:231–243.

Lin JJ, Sancar A (1989): A new mechanism for repairing oxidative damage to DNA: (A)BC exinuclease removes AP sites and thymine glycols from DNA. *Biochemistry* 28:7979–7984.

Liochev SI, Fridovich I (1990): Vanadate-stimulated oxidation of NAD (P) H in the presence of biological membranes and other sources of O_2^-. *Arch Biochem Biophys* 279:1–7.

Liochev SI, Fridovich I (1991): Effects of overproduction of superoxide dismutase on the toxicity of paraquat towards *Escherichia coli. J Biol Chem* 266:8747–8750.

McCord JM, Fridovich I (1968): The reduction of cytochrome *c* by milk xanthine oxidase. *J Biol Chem* 243:5753–5760.

McCord JM, Fridovich I (1969): Superoxide dismutase: an enzymic function for erythrocuprein (hemocuprein). *J Biol Chem* 244:6049–6055.

McCord JM, Keele BB Jr, Fridovich I (1971): An enzyme-based theory of obligate anaerobiosis: the physiological function of superoxide dismutase. *Proc Natl Acad Sci USA* 68:1024–1027.

McCord JM, Russell WJ (1988): Superoxide inactivates creatine phosphokinase during reperfusion of ischemic heart. *UCLA Symp Mol Cell Biol* New Ser 82:27–35.

Moody CS, Hassan HM (1982): Mutagenicity of oxygen free radicals. *Proc Natl Acad Sci USA* 79:2855–2859.

Natvig DO, Imlay K, Touati O, Hallewell RA (1987): Human copper-zinc superoxide dismutase complements superoxide dismutase-deficient *Escherichia coli* mutants. *J Biol Chem* 262:14697–14701.

Pacifici RE, Salo DC, Davies KJA (1989): Macrooxyproteinase (M. O. P.): a 670 kilodalton proteinase complex that degrades oxidatively denatured proteins in red blood cells. *Free Rad Biol Med* 7:521–536.

Richter C, Park JW, Ames BN (1988): Normal oxidative damage to mitochondrial and nuclear DNA is extensive. *Proc Natl Acad Sci USA* 85:6465–6467.

Scott MD, Meshnick SR, Eaton JW (1987): Superoxide dismutase-rich bacteria. Paradoxical increase in oxidant toxicity. *J Biol Chem* 262:3640–3645.

Stadtman ER, Oliver CN, Levine RL, Fucci L, Rivatt AJ (1988): Implications of protein oxidation in protein turnover, aging and oxygen toxicity. *Basic Life Sci* 49:331–339.

Thomas JP, Bachowski GJ, Girotti AW (1986): Inhibition of cell membrane lipid peroxidation by cadmium and zinc metallothioneins. *Biochim Biophys Acta* 884:448–461.

Touati D (1988): Molecular genetics of superoxide dismutases. *Free Rad Biol Med* 5:393–402.

Tsaneva IR, Weiss B (1990): SoxR, a locus governing a superoxide response regulon in *Escherichia coli* K12. *J Bacteriol* 172:4197–4205.

Van Camp W, Bowler C, Villarroel R, Tsang EWT, Van Montagu M, Inze D (1990): Characterization of iron superoxide dismutase from plants obtained by genetic complementation in *Escherichia coli. Proc Natl Acad Sci USA* 87:9903–9907.

Chapter 2

Oxygen Free Radicals and Tissue Injury

A Reference Outline[1]

Donald C. Borg

Introduction

This introduction aims to help the reader appreciate more fully the specialized chapters in this book. In comprehensive outline form, it is designed to provide a general overview of the field and a framework to help organize what follows. It is addressed to a diverse audience, ranging from physical-organic chemists and biochemists deeply involved in free radical research to cell biologists and physicians of many specialties. Consequently, there are some oversimplifications and generalizations, especially in the definitions and discussion of terms, in order to establish an interface between free radical chemistry and major topics of current biomedical research. At the other extreme, there are also a number of chemical equations in order to set a broad perspective that may help clarify the rationales in the literature on oxygen free radicals in tissue injury.

Background

Most organic compounds have even numbers of spin-paired electrons. The reader may recall that most organic and biochemical molecules contain even numbers of electrons that fill the available energy levels in pairs. The Pauli exclusion principle forbids any two electrons in a molecule to have exactly the same energy. However, there is a very small energy difference (minuscule in comparison to the energy scale of chemical

Oxygen Free Radicals in Tissue Damage
Merrill Tarr and Fred Samson, Editors

reactions) between each member of an electron pair in a molecule, including valence electrons in covalent bonds. That energy difference results from opposite electron spins.

As a spinning charge, each electron has a magnetic moment. Quantum restrictions allow only two possible spin states, so the spins of the two electrons that make up each atomic pair or molecular bond are opposite, and their associated magnetic moments effectively cancel each other out.

Definition

A free radical (FR) is a molecule or molecular fragment with an unpaired valence electron.

Free radicals, on the other hand, are molecules, or parts of molecules, where in place of one normal covalent chemical bond consisting of a pair of valence electrons, there is an odd electron that is associated with the molecular system. Or, more exactly, a free radical is a distinct molecular species in which an unpaired valence electron of an atomic constituent does not contribute to the bonding within the molecule and is, in that sense, "free."

Abbreviations used in this chapter: **A**, a [generic] autoxidizable molecule (autoxidant), with **AH_2** its fully reduced form and **AH·** an intermediate free radical; **ATP**, adenosine triphosphate; **DMSO**, dimethyl sulfoxide; **EDRF**, endothelium-derived relaxation factor; **FR**, free radical; **GSH**, reduced glutathione; **HNE**, 4-hydroxynonenal; **H_2O_2**, hydrogen peroxide; **HO·**, hydroxyl radical; **HO_2·**, perhydroxyl radical; **L·**, lipid alkyl radical; **LH**, [poly]unsaturated lipid; **LOH**, lipid alcohol; **LOOH**, lipid hydroperoxide; **LO·**, lipid alkoxyl radical; **LOO·**, lipid peroxyl radical; **LPO**, lipid peroxidation; **M**, [generic] metal ion; **NAD^+**, nicotine adenine dinucleotide, oxidized form; **NADH**, nicotine adenine dinucleotide, reduced form; **NHE**, normal hydrogen electrode; **NO·**; nitric oxide radical; **$O_2^{\overline{\cdot}}$**, superoxide radical anion; **OFR**, oxygen free radical(s); **$OONO^-$**, peroxynitrite anion; **PRFO**, partially reduced forms of oxygen; **PUFA**, polyunsaturated fatty acid; **R**, a substrate for attack by HO·, with **RH** a representation depicting a hydrogen atom that may be abstracted; **RFO**, reactive forms of oxygen; **ROOH**, organic hydroperoxide; **RO·**, organic alkoxyl radical; **ROO·**, organic peroxyl radical; **SOD**, superoxide dismutase; **TH**, [generic] target biomolecule.

FR's are paramagnetic, with net electronic magnetic moments.

Since the spin of the unpaired electron of a FR is *not* mutually compensated by an orbital partner, the whole FR molecule or molecular fragment will carry an uncancelled electron spin magnetic moment. Such a molecule is paramagnetic, meaning it possesses some quantized value of net electronic magnetic moment.

A FR is almost always an odd-electron species, but a few paramagnetic organic species have even numbers of electrons: 1) diradicals with two separate radical centers on one macromolecule sufficiently far apart as to be energetically and magnetically uncoupled, and 2) triplet states with two unpaired electrons, each in a different orbital but interacting energetically and magnetically. Usually the latter are short lived electronic excited states, but molecular oxygen is an important exception in that its *ground* state is triplet and, therefore, paramagnetic.

Paramagnetic metal ions are not, strictly speaking, FR's.

Transition metal ions, many of which are paramagnetic, are not FR's, strictly speaking, even though they may react readily with FR's to pair up unpaired electrons. Furthermore, whereas the unpaired electron of a FR is an outer – or valence – electron, in transition metals the unpaired electrons are in unfilled inner shells.

FR's tend to undergo redox reactions to pair up the unpaired valence electrons.

In free radicals the unpaired electron tends to undergo reactions wherein it is lost (i.e., oxidation) or a partner is gained (i.e., reduction) so that the product will be a more stable, electron-paired species. Hence free radicals often are reactants or products in oxidation/reduction reactions in which there is transfer of only one electron at a time. This tendency to pair unpaired electrons also underlies the high reactivity of FR's with paramagnetic metals.

Chemical Reactivity of FR's

Reactivity often is high, but not always. As half-oxidized, half-reduced species, FR's tend to be very reactive with a transitory existence.

It is sometimes assumed by biomedical scientists that all FR's are highly reactive species. Although that is a tolerable generalization, its uncritical acceptance can lead to serious misunderstandings.

Free radicals are *physically* stable species: that is, in isolation they will not decay in some fashion. However, since each has a free valence electron that does not partake in a chemical bond and is readily lost or paired up, it is true that *usually* they are highly reactive, chemically aggressive and, therefore, short-lived.

Biomedically important examples of highly reactive FR's are legion, far too many to undertake a listing. Perhaps the archetype of reactive FR's, however, is the hydroxyl radical (HO·), an oxygen free radical that is often the most powerful oxidant in the chains of FR reactions causing tissue injury. More will be said about HO· later in this chapter, and other chapters will develop its special relevance to particular clinical and pathophysiological situations.

Some FR's, however, are quite unreactive and chemically stable.

Steric protection of the paramagnetic center by bulky groups is one way of gaining stability. Biomedically important examples include spin labels (which are mostly sterically shielded nitroxide FR's), and there is a vast biophysical literature on their application to membrane dynamics. Closely related are spin traps, often used to track reactions wherein FR intermediates can *not* be detected directly. Spin traps usually are nitrones or nitroso compounds that can form conjugates with some highly reactive FR's, giving rise to relatively stable nitroxide FR's ("spin adducts," which are very similar to spin labels).

Resonance stabilization from delocalization of the unpaired electron over many conjugated or aromatic chemical bonds is another way of lowering FR reactivity. Important examples are free radicals (often cationic or anionic FR's) of polycyclic compounds, like hemes, other porphyrins, chlorophylls, and their congeners.

Four biomedically important examples of some unreactive FR's:

1) ASCORBYL FR: In conjunction with ascorbate's reducing power, the low reactivity of this FR explains the efficacy of ascorbate as a protective antioxidant and FR scavenger *in vivo*. In other words, once formed, the ascorbyl FR is much less likely to cause further damaging reactions than are most other FR's.

2) Chromanoxyl FR's of VITAMIN E, a mixture of tocopherols: More will be said about lipid peroxidation and the role of peroxyl radicals, so it suffices now to note that the low reactivity of these FR's largely explains the effectiveness of tocopherols as antioxidants inhibiting LPO *in vivo*. Also important, however, are the lipophilicity of tocopherols, which locates them in the hydrophobic interior of membrane bilayers, and their high effectiveness in reacting with chain-carrying LOO·.

3) PHENOLIC COMPOUNDS often are good antioxidants because they readily form relatively stable phenoxyl radicals: Once again, the relative *un*reactivity of *some* of these FR's due to resonance stabilization from their aromaticity partially explains their effectiveness. Phenoxyl FR's, however, range widely in reactivity, and some are oxidants while others are reductants.

4) SUPEROXIDE radical anion ($O_2^{\bar{\cdot}}$), one of the charter members of the infamous oxygen free radical (OFR) gang, is a FR that is a weak reducing agent and an even weaker oxidizing agent in aqueous environments. However, its nucleophilicity in aprotic media leads to *non* FR behavior as a strong base, which may be important in initiating lipid oxidation within membrane bilayers. This is a challenging research problem but cannot be discussed further here.

Even among the OFR that are the main subject of this volume there is a full span of the range from very high to very low reactivity. Typical lifetimes (Pryor, 1986) vary by ten or more orders of magnitude, assuming representative reaction rate constants and local substrate concentrations:

Species	$T_{1/2}$	Substrate
HO·	10^{-9} sec	1 M Linoleate
RO·, LO·	10^{-6} sec	100 mM Linoleate
ROO·, LOO·	10 sec	1 mM Linoleate
L·	10^{-8} sec	0.02 mM O_2

with lifetimes of other OFRs in between these extremes.

The redox potentials of FR's span the range from oxidizing to reducing.

Associated with the broad range of reactivities and lifetimes of radicals noted just above, FR's exhibit a correspondingly broad range of redox potentials. At one extreme there is the almost overwhelmingly powerful oxidizing potential of HO· at about +1.9 V (vs. normal hydrogen electrode, 'NHE'). The rank order of other FR's manifests progressively weaker oxidizing power until the span of FR redox activity culminates with strongly reducing radicals, such as the paraquat cation radical, whose formal reducing potential of about −0.45 V (vs. NHE) is in the range of the most potent reducing enzymes, such as some strong reductases, xanthine oxidase, and so on.

OFR of biological interest, however, are mostly on the oxidizing side of the range, even though they are partially *reduced* forms of oxygen, and there is often a further, apparently paradoxical, requirement for strongly *reducing* reactants in order to generate the most potent oxidizing OFR's in a biological environment (see below) HO· at +1.9 V (vs. NHE) is the most potent oxidant among OFR, while $O_2^{\cdot-}$ at −0.33 V is a mild reductant. RO·'s and most ROO·'s range in between these extremes.

Representative reactions (given below) of *oxidizing* FR's (electrophilic behavior) include abstraction and addition reactions. Abstraction of hydrogen atoms, usually of aliphatic compounds, leaves a new organic FR as a reaction product. Electron transfer (i.e., abstraction of *one* electron) can occur from metal ions or metal-organic centers or from hydroperoxides. In the case of H_2O_2, electron abstraction forms hydroperoxide radical, $HO_2\cdot$, the conjugate acid of $O_2^{\cdot-}$. With ROOH (and LOOH, which represent the subset of lipid hydroperoxides) it forms peroxyl radicals, ROO· (and LOO·). Addition to unsaturated chemical bonds is another typical reaction of oxidizing FR's. Notable substrates for these additions are aromatic compounds and polyenes, including membrane phospholipids and nucleic acids.

Representative reactions of *reducing* FR's (nucleophilic behavior) include electron transfer and addition of molecular oxygen. Electron transfer (i.e., one-electron addition from FR to substrate) is most apt to occur with: (1) metal ions or metal-organic centers not in their lowest accessible valence states, (2) molecular dioxygen (O_2) to produce $O_2^{\cdot-}$, and (3) hydroperoxides. Spontaneous reactions with O_2 are called autoxidations and are important sources of $O_2^{\cdot-}$ in oxidative toxicity of xenobiotics. Reactions with hydroperoxides may involve HOOH[2] to form HO·, but this appears to involve *obligatory* metal catalysis, whereas direct reactions with organic hydroperoxides (ROOH and LOOH) form alkoxyl radicals, RO· (and LO·).

Finally, some FR's add molecular oxygen directly to their paramagnetic centers to form peroxyl radicals (ROO· and LOO·), which may continue the chain of FR reactions.

It is important to note explicitly an implication of previous remarks associating FR stability with antioxidant effectiveness. Remember that FR's, whether oxidizing or reducing in nature, typically lead to chain reactions. An odd-electron species reacting with an even-electron organic molecule will yield at least one odd-electron product, usually another, less reactive FR. FR chains will be quenched when radicals react with other radicals or with appropriate metal centers. As the chain reaction proceeds, products tend to become progressively weaker oxidants or reductants, as the case may be, and their average lifetimes tend to lengthen, with a corresponding *increase* in the likelihood that they will survive long enough to encounter and react with a FR partner or metal center and terminate the FR reaction chain.

The "take-home" lesson is this: Do *not* assume that reaction of a potentially damaging FR with so-called "scavengers" means that damage control is necessarily complete. One *must* extend the analysis to consider how the reaction chain terminates. Failure to do this may lead one astray, as emphasized below in the case of HO· reactions.

One-Electron Steps in Reduction of Molecular Oxygen to Water

Before addressing oxygen free radicals, the class of FR's thought to be of greatest importance with regard to biochemical damage and tissue injury, it is important to recall that the reduction of molecular oxygen to water can be broken down to four one-electron addition reactions directly affording two FR forms of oxygen and H_2O_2.

(1) $O_2 + e^- \rightarrow O_2^{\bar{\cdot}}\ (+\ H^+ \rightleftharpoons HO_2\cdot)$

(2) $O_2^{\bar{\cdot}} + e^- \rightarrow O_2^{=}\ (+\ 2H^+ \rightarrow H_2O_2)$

(3) $H_2O_2 + e^- \rightarrow HO\cdot + OH^-$

(4) $HO\cdot + e^- \rightarrow OH^-\ (+H^+ \rightarrow H_2O)$.

The pK_a for the equilibrium between $O_2^{\bar{\cdot}}$ and its conjugate acid, $HO_2\cdot$, is 4.75, implying that less than 1% of the total $O_2^{\bar{\cdot}}$/ $HO_2\cdot$ is present as $HO_2\cdot$ at pH ~7-7.4. However, the reaction rate constants for both the dissociation and association reactions are high, so equilibrium is established rapidly. Hence reactions with $HO_2\cdot$ that are much more rapid than the corresponding ones with $O_2^{\overline{\cdot}}$ cannot be discounted a priori as quantitatively unimportant despite the low ratios of $HO_2\cdot/O_2^{\overline{\cdot}}$ that can be expected physiologically. Furthermore, the Stern-Vollmer effect induced by the negative surface charge of most phospholipid membranes may provide a surface layer with H^+ concentrations several log units above that of the bulk extramembranous medium (i.e., several pH units lower). This could lend added emphasis to the potential physiological significance of $HO_2\cdot$.

A few one-electron enzymatic reactions have been known for many years, notably those of peroxidases in oxidizing organic substrates and those of a fair number of reductases. However, most biochemical and enzymatic redox reactions do *not* appear to occur as separate one-electron events but proceed two electrons at a time or are so sequestered within the interstices of enzymes as to be functionally two-electron reactions (e.g., catalase, tyrosinase, amine oxidases, fatty acyl CoA oxidase of peroxisomes and many other oxidases, etc.) or even an overall four-electron reaction, as in cytochrome oxidase. However, there is growing realization that the exceptions to this rule, i.e., the one-electron steps, are much more common and important in biochemistry than realized heretofore, especially with regard to reactions of oxygen itself!

OXYGEN FREE RADICALS (OFR), PARTIALLY REDUCED FORMS OF OXYGEN (PRFO) AND REACTIVE FORMS OF OXYGEN (RFO):

Often incorrectly used as synonyms.

Oxygen free radicals, partially reduced forms of oxygen, and reactive forms of oxygen often are used loosely *as though* synonymous, but really they are not. When these terms are used in the narrowest sense, which is frequently the case when they are applied to inflammation, tissue injury, etc., then usage is restricted to redox intermediates of molecular dioxygen itself. Among the partially reduced forms of oxygen, OFR refers only to the $O_2^{\overline{\cdot}}/HO_2\cdot$ equilibrium pair plus $HO\cdot$, strictly speaking, while PRFO becomes OFR plus H_2O_2, which is *not* even a FR! RFO is then PRFO plus singlet O_2, an electronically excited state, usually formed photochemically, that has strong oxidizing *potential* but is not, itself, the product of a one-electron reduction.

OFR may ≡ FR's where redox reactions occur at oxygen centers (e.g., LO· and LOO·), with RFO including hypohalides from myeloperoxidase, like HOCl.

The term OFR may be defined more completely as FR's where redox reactions occur at oxygen atom centers. Endogenous OFR, then, include alkoxyl and peroxyl radicals, notably those of lipids, as well as $O_2^{\bar{}}/HO_2\cdot$ and HO·. An even broader, less common usage of OFR would also cover semiquinones and other oxy-radical intermediates of normal metabolism, including nitroxide FR metabolites of some xenobiotics.

Correspondingly, endogenous RFO would then include oxidizing hypohalides from myeloperoxidase action on H_2O_2, such as hypochlorite. Hypohalous acids as endogenous oxidants will not be discussed in this outline, but they are of undoubted importance in determining the physiological and pathological effects of activated granulocytes.

*Ex*ogenous RFO and OFR of biomedical and toxicological importance include ozone (O_3) and the FR's nitrogen dioxide ($NO_2\cdot$) and nitric oxide (NO·), components of polluted air and, in the case of the nitrogen oxides, cigarette smoke and combustion of all kinds.

Whatever the terminology, the prominence of these species in current biomedical research is so great that some physicians and more physiologically oriented investigators have come to use the term "free radical" to mean only oxygen free radicals (OFR) or even to represent *all* PRFO and/or RFO. Others appear to believe that these various terms actually *are* synonymous, and – mistakenly – they (1) include H_2O_2 as an OFR, (2) cite catalase as a "radical scavenger," and (3) usually neglect the biochemical importance of the free radicals that are *not* OFR but which are often reactants or products of OFR reactions. This sloppy or mistaken usage too often leads to misunderstanding and wrong reasoning and, at the least, is a source of confusion in communication.

"PARADOX": PARTIALLY *REDUCED* FORMS OF OXYGEN ARE A MAJOR SOURCE OF OXIDIZING DAMAGE:

Critical role of redox-active metals (Fe [and Cu?]) as catalysts in vivo.

This relates to the crucial role of redox-active metals as catalysts for the formation of super-strong oxidants. Iron clearly is involved, and copper may be important in prokaryotes and in biochemical reactions *in vitro*, where it is often much more potent than iron in stimulating lipid peroxidation, but probably copper redox toxicity is not significant in higher animals. In animals copper appears to be tightly bound in certain enzymes and other proteins (such as ceruloplasmin in plasma) or else complexed by histidine moieties so as to be removed from sites of potential damage to sensitive biomolecules and/or inactivated with respect to redox cycling.

Whether metal catalysis is actually *obligatory* or only a common potentiating reaction is a research issue of significant therapeutic importance, because chelation therapy looks like a relatively safe and effective way to ameliorate tissue damage from OFR. Although many groups are working on this, however, there does not appear to be a comprehensive and general program to evaluate chelation therapy systematically.

Fe e^- transfers are mostly outer sphere, Cu often inner sphere (with transient organo-Cu-peroxo complexes).

Regardless of whether, in due course, examples become manifest of toxic redox-cycling of copper in animals, copper turns out to have redox properties significantly different from those of iron. In general, Cu^{2+} is a weaker oxidant than Fe^{3+}, so redox cycling of copper can indirectly support redox cycling of iron through $Cu^{+} + Fe^{3+} \rightarrow Cu^{2+} + Fe^{2+}$. Some years ago Walling (1975) noted that Fe^{3+} oxidations of organic substrates have characteristics consistent with an outer sphere electron transfer, being more selective in nature than corresponding Cu^{2+} oxidations and capable of being very fast, with correspondingly selective and rapid reductions by Fe^{2+}. Conversely, copper oxidations are much slower than the fastest ones seen with iron, and they are less discriminating, consistent with inner sphere electron transfer involving organocopper intermediates.

Critical role of HO· as the strongest oxidant in the chain of FR reactions causing tissue injury. (Also see Ferryl iron below)

What underlies the apparent paradox is the requirement for a reducing metal ion to interact with H_2O_2 to produce the most powerful oxidant in the chain of FR reactions ultimately leading to tissue injury.

Fenton-"like" reactions as source of HO·.

The reaction written by Fenton toward the end of the last century dealt with iron salts solubilized in acid:

$$Fe^{2+} + H_2O_2 + H^{+} \rightarrow Fe^{3+} + HO\cdot + H_2O.$$

"Fenton-like" refers to similar reactions of complexed or chelated iron:

$$Fe^{2+}\text{-ligand} + H_2O_2 + H^{+} \rightarrow Fe^{3+}\text{-ligand} + HO\cdot + H_2O.$$

An important research issue is to establish more definitively than at present the identity of Fenton-reactive iron *in vivo*. Some, or most, of the reactive iron may be the small pools (probably at submicromolar to micromolar concentrations) in active exchange from ferritin, the storage form of cellular iron, to transferrins or other transport forms. Local metabolic acidosis and/or the pathological local increase in $O_2^{\cdot-}$ flux in postischemic conditions may significantly increase the availability of ferritin iron for redox reactions. It is known, however, that heme-bound iron is *not* an effective Fenton reactant.

Fenton-like reactions are the main, or only, biochemical sources of HO·, but are they rapid enough to explain oxidizing tissue injury *in vivo*? This also remains an important research issue. When EDTA is the ligand binding iron, the answer is "yes, they are fast enough," but EDTA, of course, is a xenobiotic very rarely present *in vivo*. With the best physiological ligand yet identified (citrate), the Fenton rate is 7–10 times slower than with EDTA, and the answer is only "maybe." If iron is speciated as $Fe(OH)^+$ to any significant extent *in vivo*, however, then "yes" may also be appropriate, because for $Fe(OH)^+$ and H_2O_2 the reported rate constant is $k = 1.9 \times 10^6\ M^{-1}s^{-1}$ (Moffett and Zika, 1987). Nevertheless, a slower Fenton reaction rate may be reconciled with observed biological damage when HO· generation is "site specific" and very close to target molecules, as will be discussed next.

Site specificity: the reactivity of HO· approaches the diffusion limit, so HO· reacts within a few Å of where Fenton catalysts (Fe) are bound, and scavengers must compete for HO· within these reaction "cages."

Some indication of the oxidizing power of HO· is now in order: The oxidizing potential is about 1.9 V, as noted, resulting in typical reaction rate constants of about 10^9–$10^{10}\ M^{-1}\ s^{-1}$. In fact, HO· is so strongly electrophilic as to be highly *in*discriminant regarding substrate.

This leads to yet another apparent "paradox": the greater the oxidizing potential of a FR, the greater the number of potential substrates and modes of reaction with each of them; hence, the *lower* the specific reactivity of the FR with any given damaging reaction (except those requiring the full oxidizing power of the FR). Thus HO· tends to "waste" itself on functionally unimportant targets. In principle, comparable reasoning applies to strongly reducing FR's, but this is less well documented.

Another result of such extreme reactivity is a short lifetime and very restricted radius of action. This gives rise to what is now called "site specificity" in terms of direct damage by HO·. With typical reaction rate constants close to the diffusion limit, HO· usually interact with local organic molecules within one to about ten molecular collisions. Hence HO· in a cellular environment are not apt to diffuse more than a few Ångstroms from where they are born. If, then, they are commonly born at sites where iron is bound, HO· damage characteristically will be site specific, with the sites determined by the binding of iron.

One consequence of site specific HO·formation is that "scavengers" must compete for HO· with organic substrates in the local microenvironment. The "effective" concentrations of the latter can be thought of as the number of molecules within the reaction (diffusion) range of a single HO·, normalized to a molar volume. Because these effective concentrations may range up to several molar, it is no wonder that scavengers often fail to compete significantly and, therefore, may hardly affect the pathways of site-specific HO· reactions. It is *characteristic* of site-specific reactions, therefore, to be resistant to exogenous competitive substrates.

On the other hand, soluble chelators (like exogenous EDTA, DETAPAC, etc.) or freely diffusible endogenous ligands may mobilize redox-active iron to support rather homogeneous Fenton-like generation of HO·. When sensitive biomolecular targets are diffusely deployed, this redistribution of iron previously bound at certain sites may enhance cytotoxicity. However, when the targets are concentrated near the fixed sites of iron binding, mobilization of the metal will inhibit the measured damage.

In summary, then HO· either reacts directly with a functionally important biomolecule close to the site of its formation or wastes itself on unimportant targets equally close by. Or else within its small reaction radius it initiates further FR radical chains involving less strongly oxidizing radicals, which may terminate in damaging reactions with target biomolecules long after the HO· has disappeared and far distant from its birthplace. A very important "take home" lesson regarding the reactivity of HO· is that the only HO· of real concern with regard to tissue injury are those born no further than about 1–5 molecules away from important biological targets, including sites for initiation of FR chain reactions among those targets.

This lesson can be generalized to all strong oxidants: the more reactive the oxidant, the more it pays to think kinetically and to think small. To think kinetically means to consider the competition of all potential substrates and all possible fates for a chemically aggressive reactant, and to think small means to focus that consideration on the submicroscopic molecular environment where the reactant is generated, i.e., the reaction cage.

Exogenous scavengers or other xenobiotics which are much less concentrated within the reaction cage than are endogenous local reactants with comparable reaction rates cannot be expected to be very effective, and exogenous enzymes unable to reach their intended substrates will, of course, be inactive. Obvious as these declarations may be, they are all too often forgotten or ignored.

Importance of autoxidation reactions in driving Fenton-like HO· production above cellular antioxidation capacities: threshold nature of oxidizing cell death as "rechargeable" substrates like GSH and ATP are transiently exhausted.

Before turning to the importance of autoxidations or other reactions, like neutrophil activation or xanthine oxidase stimulation, that give rise to abnormally high tissue fluxes of $O_2^{\overline{\cdot}}$ and cause HO·-dependent tissue injury, attention should be paid to the threshold nature of oxidizing cell death.

In contrast to tumorigenesis, clastogenesis, eicosanoid elaboration and other sublethal cell damage that may result from oxidative stress, acute oxidative cell *killing* is frequently "all or none" in nature (i.e., cells either die within minutes to a few hours or survive functionally intact). It ensues when the oxidant stress exceeds the capacity of metabolic defenses against oxidizing challenges. A key point to remember is that the resulting threshold for cell death is a highly dynamic one! It reflects the competition between constitutive levels of antioxidants plus their metabolic regeneration versus the rate of antioxidant consumption by the total flux of oxidants.

Depending on the cell, its stage in the cell cycle, and the exact nature of the oxidative stress imposed upon it, the critical metabolite leading to irreversible cellular disequilibrium and a final common path toward death (with increased concentrations of calcium in the cytosol, blebbing of the cell membrane, discharge of creatine phosphate and ATP, etc.) may vary. Commonly, reduced glutathione (GSH) falls, because GSH is a cofactor for hydroperoxide-dependent peroxidases and for transferases that are important components of cellular antioxidative defenses, and GSH is a diffusible radical scavenger in its own right. Normally, oxidized GSH (GSSG) is rapidly recycled to GSH, and loss of GSH can be tolerated;

however, at sufficiently low levels (below ~ 5–10% of normal), irreversible depletion may occur, especially within mitochondria. In other cases sufficient H_2O_2 is present in cell nuclei to cause massive single-strand breakage of DNA, and in some cells high-fidelity, single-strand-break repair dependent upon poly(ADP-ribose)polymerase is so rapid and massive as to reduce cellular stores of NAD^+ below recoverable levels. Thus thresholds for acute oxidative cell killing will vary with cell type and may change, over time, with nutritional status, the history of recent subthreshold challenges, and so on.

Because there is a background level of oxidant stress from normal metabolism, cells have a robust — even if highly variable — antioxidative capacity. For the flux of HO· from Fenton-like reactions to represent a dangerous oxidizing challenge to tissue antioxidant defenses in the presence of only micromolar or submicromolar amounts of redox-active iron, that iron must be used catalytically: i.e., rapidly recycled. In addition, there must be a source of H_2O_2. Both of these requirements can be met when the flux of $O_2^{\bar{\cdot}}$ is markedly increased. Cellular sources of enhanced $O_2^{\bar{\cdot}}$ production will be addressed by other chapters, so this discussion will be confined to $O_2^{\bar{\cdot}}/HO_2\cdot$ damage from rapid intracellular autoxidation of a xenobiotic or one of its metabolites as described below.

Cytotoxicity of Oxygen Free Radicals Driven by Autoxidation Cycles

This *is* an important topic, but because of much ground to cover and limitations of space, only the high points will be discussed.

Generation of Autoxidizable Substrate and Redox Cycling to Form $O_2^{\bar{\cdot}}$

The autoxidizable substrate may be a FR.

The oxidized form of the autoxidant (A) is converted by NAD(P)H-dependent or other reducing enzymes to reducing free radicals (AH·). These rapidly autoxidize to form $O_2^{\bar{\cdot}}/HO_2\cdot$ and regenerate A:

$$AH\cdot + O_2 \rightleftharpoons A + O_2^{\bar{\cdot}} + H^+.$$

Since A is regenerated, cycling can occur with a stoichiometry of O_2 consumption and $O_2^{\bar{\cdot}}$ production, relative to the concentration of A, that may exceed 100.

Other compounds form FR's as they autoxidize.

An alternative path that applies when the autoxidizable compound is not a FR, for example 6-hydroxydopamine or dialuric acid (a reduced metabolite of alloxan), involves the reduced form of the autoxidant ($A'H_2$) reacting with O_2 to afford its free radical ($A'H\cdot$), which does not autoxidize so readily:

$$A'H_2 + O_2 \rightleftharpoons A'H\cdot + O_2^{\dot{-}} + H^+.$$

Ascorbate and some other reductants (DH_2) can act as pro-oxidants and increase O_2 uptake and $O_2^{\dot{-}}$ yield by nonenzymatic recycling of $A'H\cdot$ to $A'H_2$, a pro-oxidant action enhancing toxicity:

$$A'H\cdot + DH_2 \rightarrow A'H_2 + DH\cdot.$$

Generation of H_2O_2

$O_2^{\dot{-}}$ dismutates spontaneously (with $HO_2\cdot$) or enzymatically (using superoxide dismutase, SOD) to form H_2O_2 and $O_2\cdot$:

$$O_2^{\dot{-}} + HO_2\cdot + H^+ \rightarrow H_2O_2 + O_2.$$

H_2O_2 also can be formed directly by some oxidases, such as monamine oxidase, as noted previously.

Redox Cycling to Reduce H_2O_2 to $HO\cdot$

In addition to the pro-oxidant redox cycling shown above, another is redox cycling of the metal needed to reduce H_2O_2 to $HO\cdot$. H_2O_2 plus complexed, reduced nonheme iron (or copper) give rise to the Fenton-like system which generates very reactive $HO\cdot$:

1. $$H_2O_2 + [M^{n+}] \rightarrow HO\cdot + OH^- + [M^{(n+1)+}]$$

Because the brackets signify complexed or chelated metal ions, reaction 1 is comparable to the reaction discussed under the section on Fenton "like" reactions as a source of $HO\cdot$.

Although Fenton-like reactions with stoichiometric amounts of the needed reactants are sometimes convenient to run *in vitro*, the amounts of reducing metal species, $[M^{n+}]$, present *in vivo* (usually a complex of Fe^{2+} in animals, as explained above) are thought to be submicromolar (or a few micromolar, at best). Almost always, therefore, the Fenton-like reaction itself is the rate-limiting step in Fenton-dependent, $HO\cdot$-induced tissue injury. To generate sufficient $HO\cdot$ to cause significant oxidative stress, the metal must act catalytically, which requires effective recycling of the oxidized form, $[M^{(n+1)+}]$, back to $[M^{n+}]$. Hence this redox cycle can be thought of as the "engine" of HO· formation *in vivo.*

Reductants of several kinds [such as $O_2^{\overline{\cdot}}$ or AH· and DH_2] can recycle the oxidized metal.

Three variants of this redox reaction which recycles the oxidized form of the Fenton reaction's metal catalyst are:

2. $[M^{(n+1)+}] + O_2^{\overline{\cdot}} \rightarrow [M^{n+}] + O_2$, or
3. $[M^{(n+1)+}] + AH\cdot \rightarrow [M^{n+}] + A + H^+$, or
4. $[M^{(n+1)+}] + DH_2 \rightarrow [M^{n+}] + DH\cdot + H^+$, respectively.

The coupling of reaction 2 with a Fenton-like reaction often is called a metal-catalyzed Haber-Weiss reaction, and it can *be inhibited by SOD*[3].

When the H_2O_2 required for Fenton chemistry comes from dismutation of $O_2^{\overline{\cdot}}$ resulting from autoxidation according to the reaction: $O_2^{\overline{\cdot}} + HO_2\cdot + H^+ \rightarrow H_2O_2 + O_2$, the resulting HO·-dependent tissue injury will be $O_2^{\overline{\cdot}}$-dependent as well. However, because SOD expedites the dismutation of $O_2^{\overline{\cdot}}$ to H_2O_2, it would not be expected to provide protection. On the other hand, to the extent that the redox cycling of iron or other Fenton-active metal is the "engine" of HO· generation (reaction 1), reaction 2 shows how the Haber-Weiss variant depends directly upon $O_2^{\overline{\cdot}}$ itself and *is*, therefore, susceptible to quenching by SOD. $O_2^{\overline{\cdot}}$ dependence would occur even if the needed H_2O_2 were produced enzymatically.

Reductants that are not $O_2^{\overline{\cdot}}$ can drive the recycling of Fenton catalysts by way of reactions 3 and 4. When these reductants are, themselves, autoxidizable, their respective autoxidations compete with reactions 3 and 4, by removing a reactant, and in the absence of an effective Haber-Weiss cycle to regenerate $[M^{n+}]$ by reaction 2 may, therefore, be *de*toxifying! Some O_2 will be required to produce H_2O_2 via autoxidation followed by dismutation of the $O_2^{\overline{\cdot}}$ produced or via an oxidase that forms H_2O_2 directly, as explained above.

Another apparent paradox may result from this competition, however: maximum HO· production, cytotoxicity and tissue injury from oxidizing intermediates induced by oxidant stress may occur under *hypoxic* conditions rather than normoxic or hyperoxic conditions (e.g., toxicity from halothane, paraquat, anthracycline antibiotics, desferrioxamine in the presence of a strong reductase, some antimalarials, etc.)! To summarize: increased concentration of oxygen favors the detoxifying (in this situation) autoxidation of the reducing reactants needed to fuel the "engine" of HO· formation.

If a reducing FR is toxic mostly direct *electron transfer or addition reactions with critical biomolecules, autoxidation can still detoxify by competing for AH·, and microreversibility of the reaction may allow inhibition of FR effects by SOD even in the absence of primary oxidation damage.*

Some warnings are necessary at this point. The previous section showed how hypoxia can maximize cytoxicity that is dependent upon coupled redox cycles that fuel Fenton-like reactions. However, a decrease in tissue injury with increasing pO_2 *need not* signify that damage is mostly due to OFR!

If AH· in reaction 3 is toxic primarily by direct electron transfer to critical biomolecules or by addition reactions with them rather than by way of reaction 3 itself, autoxidation will still serve as a detoxification pathway by competing for the damaging AH·. Furthermore, autoxidations like $AH\cdot + O_2 \rightleftharpoons A + O_2^{\overline{\cdot}} + H^+$ or $A'H_2 + O_2 \rightleftharpoons A'H\cdot + O_2^{\overline{\cdot}} + H^+$ are actually redox equilibrium reactions usually balanced toward the right. Hence SOD will remove one of the reaction products, $O_2^{\overline{\cdot}}$ itself, and thereby draw such reactions even further to the right, resulting in accelerated depletion of AH·. In this way an autoxidizing side reaction of a damaging FR (AH·) in which $O_2^{\overline{\cdot}}$ is a not a reactant but a *product*, can give rise to toxicity that is inhibited by SOD but which remains largely *in*dependent of direct injury from OFR's. Clearly that is a lesson to remember!

Secondary interactions of Fenton reactants affect final products.

Another, earlier warning deserves expansion here. The section above that discusses the range of redox potentials spanned by FR's reached the conclusion that one *must* extend the analysis of all HO· reaction systems to consider how the entire chain of possible secondary and tertiary reactions terminates. Because HO· is such a powerful oxidant, secondary products in some HO·-induced reaction systems may still be good oxidants. With other substrates for HO· attack, however, the products are reductants, sometimes very strong ones.

In order to understand the yield and nature of the final products from HO· reactions, considerations of this kind must be systematized. Such an approach is given below.

Secondary Reactions From Fenton-Generated Hydroxyl Radicals

An acknowledgment is in order. The prescience of Cheves Walling in the account of some years ago entitled "Fenton's Reagent Revisited" (Walling, 1975) provided guidance and inspiration for this discussion, especially the sections dealing with organic substrates.

Reaction of HO· With the Metal Catalyst of the Fenton Reaction

The reaction rate constants of HO· with the Fenton-active reduced ions, Fe^{2+} and Cu^{+} (and virtually all of their chelated forms that have been assayed), are high. Hence, as the concentrations of these compounds increase, they become important competitors of other substrates for HO· oxidation. Thus, instead of increased Fenton catalyst always augmenting HO· damage, it may, at sufficient concentration, reduce damage as its role as a HO· "scavenger" begins to dominate: i.e.,

$$HO\cdot + [M^{n+}] \text{ — (fast)} \rightarrow OH^- + [M^{(n+1)+}].$$

Thus, excessive concentrations of the Fenton catalyst can inhibit HO· reactions with other substrates.

Reaction of HO· with Organic Substrates

The reactions of HO· with organic compounds fall into four categories: (1) those generating reducing free radicals (reaction 5 below), (2) those affording free radicals that dimerize readily, as in reaction 6, (3) those giving rise to oxidizing free radicals, characterized by reaction 7 and a summarizing sentence, and (4) addition of HO· to unsaturated bonds and aromatic rings, the latter producing hydroxycyclohexadienyl radicals, which often initiate complicated secondary reaction chains, as indicated by reaction 10. The important point is that so many of the products of HO· attack on organic compounds are themselves highly reactive, thereby affecting strongly the final products produced by Fenton-like generation of HO·.

5. $$HO\cdot + RH \text{ —(fast)} \rightarrow H_2O + R_{red}\cdot$$

R is a substrate-(including a reactive site on a macromolecule) that affords reducing free radicals, such as those yielding relatively stable carbonium ions (e.g., ethanol or methanol). R also may represent the ligand[s] of $[M^{n+}]$, especially when the chelator is a polydentate acid compound such as EDTA, DTPA, etc.

6. $$HO\cdot + RH \text{—(fast)}\rightarrow H_2O + R_{dim}\cdot$$

R is a substrate that affords free radicals that dimerize readily.

7. $$HO\cdot + RH \text{—(fast)} \rightarrow H_2O + R_{ox}\cdot$$

R is a substrate-including a reactive site on a macromolecule—that affords oxidizing free radicals, such as carbonyl conjugated radicals and others with relatively stable anions.

8. $$HO\cdot + RH \text{—(fast)}\rightarrow \text{HO-HR}\cdot$$

HO· readily adds to the unsaturated bonds and to aromatic rings.

9. $$\text{HO-HR}\cdot + [M^{n+}] + H^+ \rightarrow \text{HO-RH}_2 + [M^{(n+1)+}]$$

The net results of the Fenton reaction on RH is, in effect, the reduction of H_2O_2 to hydrate RH.

10. $$\text{HO-HR}\cdot + [M^{(n+1)+}] \rightarrow\rightarrow \text{HO-RH}_2 + H^+ + [M^{n+}]$$

When RH is an aromatic, HO-HR· is a hydroxy-cyclohexadienyl radical that can be further oxidized to a hydroxylated product, but the overall reactions usually are complicated.

The radicals from HO· addition may compete — depending on their own redox properties, for either the oxidized or reduced forms of the Fenton metal catalysts, as in reactions 9 and 10. The former will compete with the Fenton-like reaction by consuming more of the reactant $[M^{n+}]$. Conversely, the latter will tend to enhance Fenton. However, just as excess of the Fenton reductant competed for HO· in the reaction of HO· with the metal catalyst shown above, there will be an analogous reaction of high concentrations of reducing HO·-adducts with HO· itself, such as $HO\cdot + \text{HO-HR}\cdot \rightarrow\rightarrow H_2O + ROH$, where the consecutive arrows imply elimination, dehydration, or other complicating intermediate steps.

Reaction of Secondary Radicals

The reducing and oxidizing secondary radicals formed in reactions 5 and 7, respectively, can react with the catalytic metals in much the same way as do HO· addition radicals in 9 and 10, above.

Reaction of a reducing radical with oxidized Fenton catalyst can propagate a chain reaction of HO· production.

This occurs because redox cycling of the catalyst occurs, and the effect is like that of $O_2^{\bar{}}$, AH·, or DH_2 in reactions 2–4, namely:

$$R_{red}\cdot + [M^{(n+1)+}] \rightarrow [M^{n+}] + \text{product}$$

This regenerates $[M^{n+}]$ to propagate a redox chain reaction.

An oxidizing radical may suppress HO· damage.

This effect is double barreled: (1) the Fenton reactant, $[M^{n+}]$, is competitively consumed, suppressing further the formation of HO·, and (2) insofar as the substrate RH of the initial HO· attack in reaction 7 is regenerated, potential (i.e., irreversible) damage from the HO· causing that reaction is reversed: i.e.,

$$R_{ox}\cdot + [M^{n+}] + H^+ \rightarrow [M^{(n+1)+}] + RH\ .$$

If the original RH is regenerated, the net effect is simply the reduction of H_2O_2 to H_2O by two metal ions.

Other Competing Reactions of HO·

HO· damage to functionally or structurally important molecules can be inhibited by other competing reactions as well.

Excessive H_2O_2 can be inhibitory.

Analogous to the biphasic stimulatory/inhibitory action of the metallic Fenton reactant as a function of its concentration, excess H_2O_2 can reduce HO· damage because the reaction of HO· with H_2O_2 is rapid, and the product radical, $HO_2\cdot$, is a weaker reactant than HO· in the reaction, $HO\cdot + H_2O_2$ —(fast)→ $H_2O + HO_2\cdot$. Hence, excessive H_2O_2 can *inhibit* HO· reactions with other substrates.

Self reactions of HO· are not apt to be important.

The rate constant for 2HO· —(very fast)→H_2O_2 is nearly diffusion-limited ($k > 5 \times 10^9\ M^{-1}s^{-1}$). Nevertheless, because Fenton-like reaction 1 is usually slow, instantaneous concentrations of HO· will not cause this second order reaction to be significant.

Final Products May Differ for Fenton-Generated HO· and for Radiolytic HO·

With radiolytically-produced HO·, further reactions of oxidizing and reducing radicals will occur, of course, just as described in previous sections for Fenton chemistry. However, the potentially important reactions with metallic catalysts that can occur with Fenton-derived HO· and its immediate products will not take place in radiolysis, where these metals are absent. Because all secondary and tertiary interactions can strongly affect the overall reactant/product constitution of the reaction systems, the final product yields from Fenton-like and radiolytic production of HO· may differ markedly in some cases.

As emphasized already, there has been insufficient recognition of the importance of downstream reactions in the evaluation of Fenton-like formation of HO·. Reaction 5 notes that HO· gives rise to reducing radicals from polydentate polycarboxylic chelators like EDTA or DTPA, yet Fe-EDTA or Fe-DTPA are often used in Fenton reactions *in vitro* without considering the effects of the back reaction of HO· on them. Walling, however, did evaluate the impact of these reactions and concluded that "[a] plausible explanation of... [the findings with EDTA]... lies in the very high reactivity of EDTA ($k = 2.76 \times 10^9$) and presumably its complexes toward hydroxyl radicals.... The chief point of attack should be the ethylene bridge, to yield a very easily oxidized radical [i.e., *a strongly reducing one*]... Attack on an Fe(III)-EDTA complex would give a product containing its own oxidant, so that... some fraction of the process is actually a cage reaction between freshly generated HO· and Fe(III)-EDTA pairs. These considerations predict very complex overall kinetics...," but Walling was able to interpret them in terms of expected HO· chemistry in the two examples he analyzed in detail.

Walling noted that "the extreme sensitivity of product distributions to conditions suggests that many previous conclusions (based on such distributions) about the intermediacy or nonintermediacy of hydroxyl radicals in metal ion-induced oxidations of aromatics needs reexamination."

Interactions with metals have no counterpart in radiolysis, and the concentration of H_2O_2 is usually much lower in the latter. Hence net product yields from HO· with a given [organic] substrate can be both quantitatively and qualitatively different in the two situations, even when pH effects are accounted for.

Direct Reaction of HO· with Target Molecules

Clearly the site specificity of HO· relies on the fact that it can damage targets directly, as with ionizing radiation. However, the earlier section on the importance of autoxidation reactions in driving Fenton-like HO· production above cellular antioxidation capacities notes that oxidant toxicity has a threshold because of dynamic enzymatic antioxidant defenses, such as glutathione peroxidases, catalase, SOD, etc., which can tolerate a great deal of oxidant stress without residual damage and because of regeneration of GSH and other constitutive radical scavengers and of metabolic substrates as they are utilized.

Cytotoxicity at a Distance from the Sites of Autoxidation or HO· Formation

The matter of toxicity at a distance is important, fascinating, and frequently overlooked. Within cellular dimensions HO· can initiate chain peroxidation in polyunsaturated lipids (LH), propagating radical damage slowly (minutes, or even hours) in membranes and forming lipid hydroperoxides (LOOH). Redox-active metals can react with LOOH to start new chain reactions. (See section on *Free radical reactions in LPO* below.) Metastable $O_2^{\overline{\cdot}}$ can pass through anion channels, and its conjugate acid, $HO_2\cdot$, may be able to diffuse through some membranes, giving rise to H_2O_2 — and even to the Haber–Weiss reaction — intracellularly.

Beyond cellular dimensions H_2O_2 permeates cell membranes and is quite stable, but it will be toxic only if it "powers" redox cycling and damage from HO· to levels above cellular antioxidant defense thresholds. Toxic aldehydic products of lipid peroxidation can escape cells and migrate to distant sites, even moving through lymph and plasma to distant organs in sufficient concentrations to exert a measurable oxidizing stress. Physiological amplifying mechanisms that instigate new foci of oxidizing damage may be brought into play by oxidizing cell injury at an initial site. An example would be oxidant attack in a given locus causing complement (C5a) activation that results in neutrophil chemotaxis and activation in the lung, where the resulting levels of H_2O_2 exceed antioxidant thresholds of endothelial cells and give rise to an acute respiratory distress syndrome.

Other Aspects of Iron-Dependent and Iron-Independent Tissue Injury

Ferryl iron (FeO^{2+}; Fe(IV)-oxy) as an oxidant instead of HO·.

Ferryl iron is a poorly defined complex — usually with oxygen — wherein iron has a formal valence of +4, and it is often represented as FeO^{2+}. There is no doubt that ferryl iron can exist in hemes and related porphyrins. Indeed, physiological roles are known for ferryl compounds in large molecules with polycyclic π-electron systems, such as hemes and other porphyrins. These include Compound I and II forms of most peroxidases and probably of cytochromes of the P-450 class as well. Some of their ferryl forms have been isolated and are recognized oxidants, such as ferryl myoglobin with lipids and, probably, with LOOH. Ferryl heme iron is the likely oxidant produced by interaction of hemoglobin or myoglobin with H_2O_2.

Ferryl iron is a weaker oxidant than HO· thermodynamically, but it has been implicated by kinetic studies (pulse radiolysis, stop flow) in a few special circumstances. For example, with catalysis of Fenton-like reactions by iron-EDTA, there is some indication of an oxidant weaker than HO· (which also is present, however). Nonetheless, spectroscopic evidence for low-molecular-weight ferryl and perferryl (formally Fe(V)) species has been obtained only at high and nonphysiological pH or in other aprotic media, because ferryl species not stabilized in porphyrin-like macrocycles appear to hydrolyze rapidly in protic milieux.

It is doubtful that nonheme ferryl oxidizing intermediates are important for living organisms, and it is likely that the confusion which had led indirectly to their implication is an example of failure to follow the admonition that secondary reactions of Fenton reactants affect final products. Yields are very sensitive to conditions and usually differ from those produced by radiolytic HO·. That is, one *must* extend the analysis of all HO· reactions to consider how the *entire chain* of [possible] reactions terminates. This is the warning that is explicity emphasized and expanded from the discussion of Fenton-generated HO·.

At present, therefore, one can accept whole-heartedly that ferryl oxidation states of hemes play important biochemical and biological roles. Regarding ferryl compounds of low molecular weight, however, their participation seems very unlikely. They can be primary oxidants in a few unphysiological situations in media of very low proticity, but Fenton-like reactions *in vivo* probably always give rise to HO·.

*Metal-*independent *oxidant damage* may *occur from direct (but slower) reactions of $O_2^{\bar{\cdot}}$, a weak oxidant and reductant.*

It is shown above how oxidant stress and damage from HO· can depend on $O_2^{\bar{\cdot}}$. Because $O_2^{\bar{\cdot}}$ is known to be a weak reductant and a very weak oxidant in aqueous media, many researchers studying oxidative stress and damage have concluded that $O_2^{\bar{\cdot}}$ is noxious or toxic only by way of its participation in the production of HO· through the metal-dependent Haber-Weiss reactions. However, direct reactions of $O_2^{\bar{\cdot}}$ with some biomolecules are known to have respectable rate constants, and evidence is accumulating that these may be significant *in vivo*. (Recall, moreover, that earlier discussion pointed out that at pH's found *in vivo*, $O_2^{\bar{\cdot}}$ is in rapid equilibrium with its conjugate acid, HO_2·. Although $O_2^{\bar{\cdot}}$ is a weak oxidant, HO_2· is slightly stronger thermodynamically and, often, kinetically.)

Compounds known to be oxidized by $O_2^{\bar{\cdot}}$ include catecholamines, oxyhemoglobin, ascorbate, some hydroquinones, and some protein thiols. Ferricytochrome *c*, methemoglobin, NADH, nitroblue tetrazolium, and some quinones can be reduced, probably without tissue injury. Because $O_2^{\bar{\cdot}}$ concentrations are low *in vivo*, especially where SOD is present, it remains unclear when or whether the net oxidation of these substrates might become consequential.

Nitric oxide (NO·) from endothelium or EDRF, neurons, etc. + $O_2^{\bar{\cdot}}$ give rise to peroxynitrite ($OONO^-$), which can undergo acid-catalyzed, metal-independent homolytic scission to HO·, apparently a minor path.

In the past few years it has become apparent that NO· is an important and widespread cytokine. A well documented example is the endothelium-derived relaxing factor (EDRF), which is either NO· or an immediate precursor of NO· that may be induced to release it given appropriate stimulation.

Even more recent research has identified a rapid reaction of NO· with $O_2^{\bar{\cdot}}$ that gives rise to the reactive peroxynitrite anion ($OONO^-$). $OONO^-$ can undergo spontaneous intramolecular scission (favored at $pH < 7$) to afford HO· by a metal-independent route. The ability of some known HO· scavengers to inhibit certain oxidations mediated by $OONO^-$ has led to the conclusion the this may be a significant alternative source of HO· damage *in vivo*. On the other hand, because so much oxidative stress and damage attributed to HO· is inhibited by appropriate chelators, Fenton-like reactions appear to be the dominant pathway to nonradiolytic HO· *in vivo*, if not the only significant one. The latter conclusion is supported by current research on direct reactions of $OONO^-$ with known substrates of HO· *in vitro* which indicate that even at pH < 7, $OONO^-$ homolysis to produce HO· is either slow or a minor path.

As a good oxidant, $OONO^-$ may react directly with some HO· scavengers, thus explaining some of the HO·-like behavior reported. In other experiments, contaminating metals may have induced formation of NO_2^+ or a related reactive secondary product. (Metal-dependent formation of a potent nitrating intermediate, probably the nitronium ion, NO_2^+, also can occur). The competitive importance of these reactions in causing oxidative tissue injury is not yet clear.

Free Radical Chain Reactions in Lipid Peroxidation (LPO)

LPO is a very important manifestation of oxy radical cytotoxicity, and the cardinal features of LPO that relate to tissue injury are outlined below. Because of this focus, the secondary and tertiary reactions that occur in the lipids themselves and give rise to an extensive set of final products are omitted.

Initiation of Oxidation in Polyunsaturated Lipid (LH)

It has long been recognized that oxidative hydrogen abstraction from lipids is markedly facilitated by unsaturated carbon-carbon bonds and that the higher the degree of unsaturation of the polyene (commonly denoted as PUFA [for **P**oly**U**nsaturated **F**atty **A**cid]), the greater the susceptibility to oxidation. Considering its importance, however, surprisingly little has been documented regarding the kinds of reactions that initiate LPO in biological systems. Nonetheless, it is customary to write a generalized and nonspecific reaction like, reaction 11:

11. $$\mathrm{LH} \xrightarrow{(X)} \mathrm{L}\cdot,$$

where L· is a lipid alkyl FR and X = gamma or ultraviolet irradiation, certain metals and free radicals, etc. When X=HO·, reaction 11 becomes

11a. $$\mathrm{LH} + \mathrm{HO}\cdot \rightarrow \mathrm{L}\cdot + \mathrm{H_2O}.$$

Chain Propagation

LPO is characterized by free radical chain reactions and requires molecular dioxygen. The "main" reaction cycle that propagates LPO is short:

two reactions. In the first, the lipid alkyl FR produced by initiation reacts very rapidly with O_2, which adds on to give a lipid peroxyl radical, LOO·:

12. $$L\cdot + O_2 \text{ —(very fast)→ } LOO\cdot.$$

In the second, the peroxyl radical abstracts a hydrogen atom from *another* polyunsaturated lipid molecule to give a lipid hydroperoxide, LOOH, while regenerating a lipid alkyl radical, L′·:

13. $$LOO\cdot + L\cdot H \text{ —(very slow)→ } LOOH + L'\cdot,$$

(where the prime symbol indicates that the lipid substrate may be different from the parent of LO·). L′· can then serve as a reactant reaction 11 to propagate the chain reaction. Note that no metals are involved in the chain propagation cycle of these reactions.

Characteristically the rate constants for the second reaction are so low that the lifetime of a LOO· in a biological membrane or other phospholipid bilayer is very long in chemical terms: 10 or more seconds, depending on the particular lipid and membrane. Most lipid molecules in a membrane bilayer are highly mobile in the plane of bilayer (but not *across* the plane) and typically can migrate from one pole of a lipid vesicle to the opposite pole in about a second when the vesicle is of the size of cells or organelles. Thus the reaction can occur far away (*in cellular dimensions!*) from the preceding reaction. This generalization underlies the proclivity of LPO to induce action at a distance.

In principle, LPO can proceed until either lipid substrate or available O_2 are exhausted, and this might require minutes — or even hours — in a very pure lipid. However, chain termination reactions do exist even in pure lipids, some of which are noted in below, and different terminations are possible when other reactants are present.

Secondary Initiation from Lipid Hydroperoxides (LOOH)

Secondary initiation can be thought of as a kind of auxiliary cycle of LPO in the presence of appropriate redox-active metals. Lipid hydroperoxide, LOOH, a product of each cycle of reaction 13, is metastable. It can give rise to reactive alkoxyl radicals, LO·, and [usually] less reactive peroxyl radicals, LOO·, by way of reactions 14 and 15, respectively.:

14. $$LOOH + [M^{n+}] \text{—(fast)→} LO\cdot + OH^- + [M^{(n+1)+}]$$

15. $$LOOH + [M^{(n+1)+}] \rightarrow LOO\cdot + H^+ + [M^{n+}].$$

In turn, LO· can react with PUFA to reinitiate LPO by way of reaction

16,

16. $$LO\cdot + L''H\text{—(fast)}\rightarrow LOH + L''\cdot,$$

and LOO· can carry out the slow component of the primary LPO chain reaction cycle, reaction 13.

Reaction of LOOH with $[M^{n+}]$ (reaction 14) is analogous to Fenton-like reactions in many ways, as can be visualized readily by substituting H for L in LOOH. Nevertheless, it may differ from Fenton redox cycling in effectively utilizing ferrous heme iron.

The corresponding oxidation of LOOH by metals, reaction 15 usually is slower than the reduction of LOOH by $[M^{n+}]$, but it should not be neglected. Indeed, it was pointed out earlier that ferryl myoglobin may be an effective oxidant of lipids and possibly of LOOH.

Characteristically LPO *in vivo* and in many biological preparations is strongly inhibited by metal chelators that can gain access to the sites of active LPO. This indicates that secondary initiation of LPO, sometimes referred to as chain branching (i.e., branching of the chain reactions of LPO), is a very important part of LPO *in vivo* and may even be the dominant component. It is consistent with the fact that for most low-molecular-weight complexes of Fe^{2+} that have been studied, the rate constants for reactions with LOOH are much higher than the rate constants for the corresponding Fenton-like reactions with H_2O_2. Hence, if LOOH and H_2O_2 were equally accessible to a hypothetical, redox-active Fe^{2+} complex, reaction with the former (i.e., secondary initiation of LPO) would be more likely than reaction with the latter (i.e., a Fenton-like generation of HO·, reaction 1).

Lipid alkoxyl radicals, LO·, are strong oxidants, so reaction 16 is important in reinitiating LPO in a fashion analogous to reaction 11a. Although rapid, reaction 16 is typically about a hundred to a thousand times *slower* than the oxidation of PUFA by HO· (reaction 11a). On the other hand, it is on the order of a million to ten million times *faster* than the oxidation by peroxyl radicals, reaction 13.

Chain Termination

In LPO of neat lipids, chain-terminating interactions of reaction intermediates and products are important, and some of the major reactions are these:

17. $$2L\cdot \rightarrow L\text{-}L$$

18. $2LO\cdot \rightarrow LOOL$ or other nonradical products

19. $2LOO\cdot \rightarrow LOOL + O_2$, etc.

The likelihood of reaction 18 increases with respect to 17 as pO_2 rises due to the production of LO·, and 19 dominates when pO_2 exceeds 1 atmosphere. Cross reactions between L·, LO·, and LOO· also occur. *In vivo*, however, there are usually sufficient potential reactants of other kinds that chain termination is apt to be of the kinds addressed in the next section.

Interception

Reactions with target molecules, TH, in contact with the peroxidizing lipids may allow the reactive intermediates of metal-independent chain propagation or of metal-dependent secondary initiation to be intercepted before they initiate another cycle of LPO. This is a form of chain termination often referred to as co-oxidation, and it is important as a way to effect the action at a distance cited earlier and discussed below. A generalized co-oxidation reaction is the following:

20. $$\left.\begin{array}{c} L\cdot \\ LO\cdot \\ LOO\cdot \end{array}\right\} + TH \rightarrow \left.\begin{array}{c} LH \\ LOH \\ LOOH \end{array}\right\} + T\cdot\text{—}(O_2) \longrightarrow TO\cdot \ldots.$$

There is much supporting evidence that LPO plays a causal role in oxidative stress and damage, at least sometimes and in some tissues or cells. To the extent it does, it is likely that damage to functionally important membrane structures and embedded enzymes is of greater consequence than is consumption of lipid by the process, because membrane lipids normally are in constant flux and may be removed in bulk by pinocystosis. Since co-oxidation of lipids with nucleic acids and their components produces free radicals in the latter and causes both strand breaks and many kinds of chemical reactions in nucleic acid constituents, it is even possible that LPO in the nuclear membrane may be a source of oxidative damage to chromatin *in vivo*.

Inhibition

When tissue injury or damage to an important biomolecule is preceded by a long propagative chain reaction (and LPO in cell membranes may have average chain lengths of $\sim$ 15–30), partial inhibition of a chain-carrying reaction may shorten the average chain length by many times and results in more efficient antioxidant protection than would be provided by comparable inhibition of initiation. Thus the most potent endogenous lipid antioxidants (tocopherols) are chain breakers that compete very effectively for the LOO· required for the slow step of LPO chain, reaction 13.

Metal Dependence

Intrinsically LPO does not depend on metal catalysis. However, in many situations the initiation reaction is caused by a free radical whose generation is metal-dependent, such as HO· from a Haber-Weiss or other Fenton-like reaction. In those cases appropiate chelation treatment may suppress LPO.

Even when initiation is metal-independent, LPO can be *amplified* by metals able to induce secondary initiation, and the discussion of secondary initiation from lipid hydroperoxides by reaction 14 and 15 concluded that this may be the most important component of LPO-induced tissue injury *in vivo*. Appropriate chelation therapy may then effect partial inhibition of LPO propagation reactions or of the biochemical damage or cell injury caused by them.

Chelation therapy has another advantage in that oxidative damage caused by Fenton-like HO· reactions also can be modulated by sequestering the required metal catalysts. Hence, as a first approximation, chelation might be useful in prevention or treatment of oxidative tissue injury regardless of whether or not LPO is important in the process. More careful consideration of the role of LPO, however, might lead to the choice of different chelating agents with different lipophilicities than would be favored if HO·-dependent damage were known to dominate.

Action at a Distance

The discussion of chain propagation in LPO explained why LPO is intrinsically a highly delocalized process, even at the scale of subcellular and cellular dimensions. This generalization would be expected to hold even if metal-dependent secondary initiation (chain branching) were to dominate because reactions like LO· formation by metal reduction of LOOH (reaction 14) and LOO· formation by LOOH oxidation (reaction 15) feed back into the main cycle of LPO chain propagation through the rapid attack of LO· on unreacted lipid followed by the addition of O_2 and the slower attack by LOO·, respectively. Moreover, even the strongly oxidizing LO· radicals from reaction 14, which are so effective in causing co-oxidation damage, have far greater diffusion radii than do HO·. This diffuse nature of LPO and the great separation (on the molecular scale) of structurally or functionally important molecular damage from the loci of catalytic metal complexes is in sharp contradistinction to the site-specific nature of HO· damage stressed previously in the section on site specificity.

Because reaction 13 is slow (seconds or more), the chain lengths of LPO usually are long (~1–30), and secondary initiation is common, co-oxidation damage typically occurs far removed in time and space from the site(s) of primary- or even secondary-initiation. In addition, reactive carbonyls from LPO, especially 4-hydroxynoneanal and the less reactive malondialdehyde, can induce oxidative stress in distant tissues and organs.

The reactions leading to final breakdown products of lipids themselves in the process of LPO have not been dealt with at all in this overview, with the occasional exception of a reaction such as 16 or those in the section on interception which show lipid alcohols, reconstituted lipids, and lipid hydroperoxides among the reaction products. Many other products also are known, including rearrangement products, like epoxides which complex with sulfhydryl groups on proteins, and those from various bimolecular and unimolecular scission reactions on the parent lipids, such as gaseous alkanes and alkenes and a broad spectrum of reactive carbonyls. Among the latter are several aldehydes, prominent among them and most studied being 4-hydroxynonenal (HNE). HNE and its congeners are released to the blood and lymph from some tissues undergoing LPO, and there is evidence of oxidative stress distant organs, such as temporarily reduced hepatic GSH levels. Whether such distant transport of LPO products can become a significant source of oxidative *damage* is uncertain at present.

LPO also can follow *cell damage from other sources.*

Despite the many ways whereby LPO may effect tissue injury, it is important to point out that LPO can be either a cause or a nonspecific consequence of cytotoxicity. LPO as a result of cell damage from other causes can occur as metabolic antioxidative capacity becomes deficient, and so on. Hence another "take home" lesson of this introductory chapter is that without further documentation of damage mechanisms, one should not attribute a causal role to LPO upon the mere demonstration of its abnormal presence or amount.

MEASUREMENT OR OTHER DETERMINATION OF FR'S, ESPECIALLY OFR, *IN VIVO*:

Detection of FR's usually is done by the use of substances which form "characteristic" reaction products. Measurement may be by electron paramagnetic (spin) resonance or by the competitive kinetic effects of scavengers that compete with precursors, reactive species, or reactive products. Detection and measurement of OFR's deserves serious consideration, because often it is difficult and highly subject to errors of false attribution, on the one hand, and insensitivity on the other hand, especially *in vivo*. However, for the purposes of this overview, the topic will be left to speak for itself for the most part. A few sentences summarizing very briefly the main problems with scavengers are warranted to alert the reader, because scavengers are used so frequently, and the analyses of results are so often misinterpreted or oversimplified.

Nonspecificity can give false positive results.

A scavenger used because of a known and potentially useful competitive reaction with a reactive intermediate of oxidative stress can have other metabolic effects than inhibition of the particular reaction of interest, and very few, if any, radical scavengers have only one possible fate *in vivo*. For example, many so-called FR scavengers are reducing agents which react readily with electrophiles or oxidants other than the FR under study. Hence they may produce effects *un*related to FR scavenging. Conversely, many inhibitors of other oxidants, such as quenchers of singlet oxygen, are equally nonspecific and often serve as good substrates for HO· and other oxidizing radicals. The frequency with which consideration of scavenger nonspecificity is overlooked in biomedical reporting is alarming.

Limited access can give false negative results.

False negative results can be obtained because of the inability of a putative scavenger to reach the site[s] of FR reactions in competitive concentrations or for other reasons. Three examples will be cited here: (1) HO· or another ultimate oxidant is being formed in a highly site-specific way at the molecular level so that there is competition only with local substrates within the reaction cage, some of which may be effectively in the mutilmolar range, as described earlier; (2) there is intracellular compartmentalization that precludes effective contact between the scavenger and its intended substrate; and (3) the reactivity of a secondary FR produced from the inhibitor by the scavenging reaction is sufficient to produce toxicity in its own right.

An example of the last case is the apparent failure of dimethyl sulfoxide (DMSO), a good "scavenger" of HO·, to inhibit LPO actually being initiated by HO·. This can occur because one major product of HO· attack on DMSO is the methyl FR, $H_3C\cdot$, which rapidly reacts with O_2 to form the methyl peroxyl FR, $H_3COO\cdot$. $H_3COO\cdot$, in turn, is an effective initiator of LPO, so although HO· may, in fact, have been "scavenged" by the experimenter's intervention with DMSO, LPO was not quenched.

Concluding Remarks

Fenton-like chemistry that requires metal catalysis and affords HO· as a primary oxidant appears to underlie much of the tissue injury that can be attributed to oxygen free radicals. In animals the catalytic metal is redox-active, nonheme, complexed iron, either always or almost always.

Lipid peroxidation also is common and may make another significant contribution to oxidative stress and damage *in vivo*, but LPO can be a result of both oxidative and nonoxidative cytotoxicity as well as a possible cause of the former. Intrinsically LPO is independent of HO· and of metal catalysis. However, HO· can initiate LPO in homogeneous systems *in vitro* and may be able to do so *in vivo*, and augmentation of LPO by branching of the lipid chain reaction scheme catalyzed by metals (iron, in particular) seems to be important in LPO-associated cell and tissue damage.

Hence the critical reactions underlying tissue injury from OFR are those of Fenton-like production of HO· and those supporting the chain reactions of LPO. Because both of these reaction sets, in turn, are straightforward, the foundation chemistry of oxidative stress and damage is fundamentally simple. Often overlooked, however, is that the powerful reactants involved (the most notable being HO·) can give rise to a rich interplay of secondary interactions (and even tertiary and higher-order downstream reactions) among themselves and with other substrates.

Two basic lessons to be learned by students of oxidative damage derive from this high reactivity of the chemical intermediates involved: (1)

the direct reactions of HO· are intrinsically site specific and, hence, resistent to competitive interception ("scavenging"), and (2) the final products to be expected cannot be foretold reliably without due consideration of *all* derivative reactions. The first lesson is developed largely in the section on site specificity of this outline, whereas the discussion on action at a distance contrasts this site specificity of primary HO· reactions with the intrinsically diffuse nature and long duration of LPO, leading more naturally to "action at a distance" for the latter. The section on the wide range of redox potentials, of different FR's discusses some implications of the second lesson, concluding that scavenging of a primary reactant is not necessarily equivalent to damage control, and the discussion on secondary reactions from Fenton-generated hydroxyl radicals focuses upon the lesson's special relevance to reactions initiated by HO·.

The critical roles played by metal (iron) catalysis in both HO·-induced reactions and LPO commends consideration of preventive and therapeutic chelation therapy to a more systematic study than it has received heretofore. The importance to oxidative stress and tissue damage of certain other, metal-independent reactions of oxygen free radicals remains to be clarified and may not be negligible, as summarized in the section on metal-independent oxidant damage from direct but slow reactions of $O_2^{\cdot-}$, a weak oxidant and reductant.

Finally, it is important to remember that not all important cytotoxic reactions of reactive forms of oxygen (RFO) are mediated by oxygen free radicals (OFR), the main subject of this chapter. Hypohalides produced by myeloperoxidases and singlet oxygen formed photochemically are notable exceptions.

Notes

Tables 1–4 provide a detailed outline of the chapter contents. Section headings correspond closely with those in the text.

[1] This chapter is based on the plenary lecture of an unpublished symposium on oxygen radicals and tissue injury organized by Prof. B. Luchessi at the University of Michigan in September, 1986.

[2] Written in this way rather than the conventional H_2O_2 to emphasize the homology with ROOH and LOOH.

[3] Strictly speaking, the Haber–Weiss appellation applies only when the metal catalyst, M, is iron.

References

Moffett JW, Zika RG (1987): Reaction-kinetics of hydrogen peroxide with copper and iron in sea water. *Environ Sci Technol* 21:804–810.

Pryor WA (1986): Oxy-radicals and related species: their formation, lifetimes, and reactions. *Annu Rev Physiol* 48:657–667.

Walling C (1975): Fenton's reagent revisited. *Acc Chem Res* 8:125–131.

Table 1. Oxygen free radicals and tissue injury

1. **Background**: Most organic compounds have even numbers of spin-paired electrons.
2. **Definition:** A free radical (FR) is a molecule or molecular fragment with an unpaired valence electron.
   - 2.1. FR's are paramagnetic, with net electronic magnetic moments.
     - 2.1.1. Paramagnetic metal ions are not, strictly speaking, FR's.
   - 2.2. FR's tend to undergo redox reactions to pair up the unpaired valence electrons.
3. **Chemical Reactivity of FR's**: Often high, but not always.
   - 3.1. As half-oxidized, half reduced species, FR's tend to be very reactive with transitory existence.
   - 3.2. Some FR's, however, are quite unreactive and chemically stable.
   - 3.3. The redox potentials of FR's span the range from oxidizing to reducing.
4. **One-electron steps in reduction of molecular oxygen to water:**
   - 4.1. $O_2 + e^- \rightarrow O_2^- \cdot (+H^+ \rightleftharpoons HO_2\cdot)$
   - 4.2. $O_2^- \cdot + e^- \rightarrow O_2^{=} (+2H^+ \rightarrow H_2O_2)$
   - 4.3. $H_2O_2 + e^- \rightarrow HO\cdot + OH^-$
   - 4.4. $HO\cdot + e^- \rightarrow (+ H^+ \rightarrow H_2O)$
5. **Oxygen free radicals (OFR), partially reduced forms of oxygen (PRFO), and reactive forms of oxygen (RFO)**: Often incorrectly used as synonyms.
   - 5.1. OFR strictly confined to $O_2^- \cdot / HO_2\cdot$ and $HO\cdot$
   - 5.2. PRFO$\equiv$ OFR + H_2O_2, which is *not* an FR!
   - 5.3. RFO $\equiv$ PRFO + singlet O_2
   - 5.4. OFR may $\equiv$ FR's where redox reactions occur at oxygen centers (e.g., LO· & LOO·), with RFO including hypophalides from myeloperoxidase, like HOCl.
6. **"Paradox": partially *reduced* forms of oxygen are a major source of oxidizing damage**:
   - 6.1. Critical role of redox-active metals (Fe and ?Cu) as catalysts *in vivo*.

6.2. Critical role of HO· as the strongest oxidant in the chain of FR reactions causing tissue injury.

6.2.1. Fenton-like reaction as source of HO·:

Fe^{2+}-ligand + $H_2O_2 \rightarrow Fe^{3+}$-ligand + HO· + OH^- (see *Cytotoxicity Outline*, Table 2)

6.2.2. Site specificity: the reactivity of HO· approaches the diffusion limit, so HO· reacts within a few Å of where Fenton catalysts (Fe) are bound, and scavengers must compete for HO· within these reaction "cages."

6.3. Importance of autoxidation reaction in driving Fenton-like HO· production above cellular antioxidation capacities: threshold nature of oxidizing cell death as "rechargeable" substrates like GSH and ATP are transiently exhausted. (see outline *Cytotoxicity of OFR driven by autoxidation cycles*, Table 2)

6.4. Ferryl iron (FeO^{2+}; Fe(IV)-oxy) as an oxidant instead of HO·

6.4.1. Important in hemes and aprotic media (including high pH) but *not* in small aqueous Fe complexes at pH $< \sim 9$.

6.5. Metal-*in*dependent oxidant damage *may* occur from direct (but slower) reactions of O_2^-·, a *weak* oxidant and reductant.

6.5.1. Catecholamines, some protein thiols can be oxidized (also oxyhemoglobin, ascorbate, some hydroquinones).

6.5.2. Ferricytochrome *c*, methemoglobin, NADH, notroblue tetrazolium, some quinones can be reduced, probably without tissue injury.

6.5.3. Nitric oxide (NO·) from endothelium or EDRF, neurons, etc. + O_2^-· $\rightarrow$ peroxynitrite ($OONO^-$) which can undergo acid-catalyzed, metal-independent homolytic scission $\rightarrow$ HO·, apparently a minor path.

(Metal-dependent formation of a potent nitrating intermediate, probably the nitronium ion, NO_2^+, also can occur.)

6.5.4. The competitive importance of these reaction in causing oxidative tissue injury is not yet clear.

7. **Profound biological importance of lipid peroxidation** (LPO), Table 4

(See outline *Free radical chain reactions in* LPO)

7.1. LPO also can *follow* cell damage from other sources.

8. **Measurement or other determination of FR's, especially OFR**, *in vivo*:

8.1. Scavengers.

8.1.1. Competition with precursors, reactive species, reactive products, and reactants in equilibrium.

8.1.2. Nonspecificity can give false positive results.

8.1.3. Limited access can give false negative results.

8.2. Detection of "characteristic" reaction intermediates and products.

8.3. Electron paramagnetic (spin) resonance.

8.3.1. Spin trapping.

Table 2. Cytotoxicity of oxygen free radicals driven by autoxidation cycles

1. **Generation of autoxidizable substrate and redox cycling to form** $O_2^-\cdot$:

   1.1. The oxidized form of the autoxidant (A) is converted by NAD(P)H-dependent or other reducing enzymes to reducing free radicals (AH·). These rapidly autoxidize to form $O_2^- \cdot / HO_2\cdot$ and regenerate A:

$$AH\cdot + O_2 \rightleftharpoons A'H\cdot + O_2^- \cdot + H^+$$

OR, ALTERNATIVELY

   1.2. With 6-OHDA, alloxan, etc. the reduced form of the autoxidant ($A'H_2$) reacts rapidly with O_2 to afford its free radical ($A'H\cdot$), which does not autoxidize so readily:

$$A'H_2 + O_2 \rightleftharpoons A'H\cdot + O_2^-\cdot + H^+$$

   1.2.1. Ascorbate and some other reductants (DH_2) may increase O_2 uptake and $O_2^-\cdot$ yield by nonenzymatic recycling of $A'H\cdot$ to $A'H_2$, a pro-oxidant action *enhancing* toxicity:

$$A'H\cdot + DH_2 \rightarrow A'H_2 + DH\cdot$$

2. **Generation of** H_2O_2:

   $O_2^-\cdot$ dismutates spontaneously (with $HO_2\cdot$) or enzymatically (SOD) to form $H_2O_2 + O_2$:

$$O_2^- \cdot + HO_2 + H^+ \rightarrow H_2O_2 + O_2$$

3. **Redox cycling to reduce** H_2O_2 **to** $HO\cdot$:

   H_2O_2 + complexed, reduced nonheme iron (or copper) = a Fenton-like system forming very reactive hydroxyl radicals (HO·):

$$H_2O_2 + [M^{n+}] \rightarrow HO\cdot + OH^- + [M^{(n+1)+}]$$

   3.1. Reductants of several kinds can then recycle the oxidized metal, such as $O_2^-\cdot$ formed in 1.1. or AH· and DH_2, the reducing agents in 1.1. and 1.2.1., respectively:

   3.1.1. $[M^{(n+1)+}] + O_2^-\cdot \rightarrow [M^{n+}] + O_2$, or

   3.1.2. $[M^{(n+1)+}] + AH\cdot \rightarrow [M^{n+}] + A + H^+$, or

   3.1.3. $[M^{(n+1)+}] + DH_2 \rightarrow [M^{n+}] + DH\cdot + H^+$, respectively.

   3.2. The coupling of 3.1.1. with a Fenton-like reaction (as in 3.) often is called a metal-catalyzed Harber-Weiss reaction, and it can be inhibited by SOD. When AH· or DH_2 are readily autoxidizable, 3.1.2. and 3.1.3. compete with 1.1. and 1.2. for reducing reactants. Some O_2 may be required to produce H_2O_2 via 1. and 2., but maximum HO· production and tissue injury often occur under *hypoxic* conditions.

3.2.1. *If* a reducing FR is toxic mostly via *direct* electron transfer or addition reactions with critical biomolecules, autoxidation via 1.1. can still detoxify by competing for AH·, and microreversibility of 1.1. may allow inhibition by SOD even in the absence of primary oxidation damage.

3.3. Secondary interactions of Fenton reactants affect final products. Yields are very sensitive to conditions and usually differ from those produced by radiolytic HO·. (See *Secondary reactions from Fenton-generated* HO·, Table 3)

4. **Direct reaction of HO· with target molecules**:
HO· can damage targets directly, as with ionizing radiation. However, acute toxicity has a threshold, occurring only when dynamic enzymic antioxidant defenses (especially glutathione peroxidases, catalase, and SOD) are overwhelmed and regeneration of metabolic substrates like ATP and of GSH, tocopherols, and other constitutive radical scavengers becomes deficient.

5. **Cytotoxicity at a distance from the sites of autoxidation or HO· formation**:

5.1. **Within cellular dimensions**:

5.1.1. HO· can initiate chain peroxidation in polyunsaturated lipids (LH), propagating radical damage slowly (minutes, or even hours) in membranes and forming lipid hydroperoxides (LOOH). Redox-active metals can react with LOOH to start new chain reactions. (See *Free radical chain reactions in* LPO, Table 4.)

5.1.2. Metastable $O_2^-\cdot$ can pass through anion channels, and its conjugate acid, $HO_2\cdot$, may be able to diffuse through some membranes to initiate Step 2.

5.2. **Beyond cellular dimensions** (but diluting with distance to subthreshold levels):

5.2.1. H_2O_2 permeates cell membranes and is quite stable, but it will be toxic if it "powers" Steps 3 and 4 to levels above cellular antioxidant defense thresholds.

5.2.2. Toxic aldehydic products of lipid peroxidation can escape cells and migrate to distant sites, even moving through lymph and plasma to distant organs in sufficient concentrations to exert a measurable oxidizing stress.

5.2.3. Physiological amplifying mechanisms that instigate new foci of oxidizing damage may be brought into play by oxidizing cell injury at an initial site. An example would be oxidant attack in a given locus causing complement (C5a) activation that results in neutrophil chemotaxis and activation in the lung, where the resulting levels of H_2O_2 exceed antioxidant thresholds of endothelial cells and give rise to an acute respiratory distress syndrome.

Table 3. Secondary reactions from Fenton-generated hydroxyl radicals

1. **Production of HO· by Fenton-like reactions:**

$$H_2O_2 + [M^{n+}] \rightarrow HO\cdot + OH^- + [M^{(n+1)+}]$$

Generalized form of "Fenton-like" reaction from *Cytotoxicity Outline*, Table 2.

2. **Reaction of HO· with metal catalyst of the Fenton reaction:**

$$HO\cdot + [M^{n+}] \text{ — (fast)} \rightarrow OH^- + [M^{(n+1)+}]$$

Excessive concentrations of the Fenton catalyst can *inhibit* HO· reactions with other substrates.

3. **Reaction of HO· with organic substrates:**

3.1. $HO\cdot + RH$ —(very fast)→ $H_2O + R_{red}\cdot$

R is a substrate—including a reactive site on a macromolecule—that affords reducing free radicals, such as those yielding relatively stable carbonium ions e.g., (ethanol or methanol). R also may represent the ligand[s] of $[M^{n+}]$, especially when the chelator is a polydentate polycarboxylic acid compound such as EDTA, DTPA, etc.

3.2. $HO\cdot + RH$ —(very fast)→ $H_2O + R_{dim}\cdot$

R is a substrate that affords free radicals that dimerize readily.

3.3. $HO\cdot + RH$ —(very fast)→ $H_2O + R_{ox}\cdot$

R is a substrate—including a reactive site on a macromolecule—that affords oxidizing free radicals, such as carbonyl conjugated radicals and others with relatively stable anions.

3.4. $HO\cdot + RH$ —(very fast)→ HO-HR·

3.4.1. $HO\text{-}HR\cdot + [M^{n+}] + H^+ \rightarrow HO\text{-}RH_2 + [M^{(n+1)+}]$

The net result of the Fenton reaction on RH is, in effect, the reduction of H_2O_2 to hydrate RH.

3.4.2. $HO\text{-}HR\cdot + [M^{(n+1)+}] \rightarrow \rightarrow ROH + H^+ + [M^{n+}]$

When RH is an aromatic, HO-HR· is a hydroxycylohexadienyl radical that can be further oxidized to a hydroxylated product, but the overall reactions usually are complicated.

4. **Reaction of secondary radicals:**[1]

4.1. $R_{Red}\cdot + [M^{(n+1)+}] \rightarrow [M^{n+}]$ + product

This regenerates $[M^{n+}]$ to propagate a redox chain reaction.

4.1.1. $R_{Red}\cdot + HO\cdot \rightarrow H_2O$ + product

4.2. $2R_{dim}\cdot \rightarrow$ product (dimer)

4.3. $R_{ox}\cdot + [M^{n+}] + H^+ \rightarrow [M^{(n+1)+}]$ + (RH or R′H)

If RH from 3.3. is regenerated, the net effect is the reduction of H_2O_2 to H_2O by two metal ions.

5. **Other competing reactions of** HO·:

5.1. $HO\cdot + H_2O_2$ —(fast)→ $H_2O + HO_2\cdot$

Excessive H_2O_2 can *inhibit* HO· reactions with other substrates.

5.2. 2 HO· —(very fast)→ H_2O_2

Because Step 1, the Fenton-like reaction, usually is slow, the instantaneous concentration of HO· is not apt to cause this reaction to be significant despite its ∼ diffusion-limited rate ($k > 5\times 10\ M^{-1}s^{-1}$).

6. **Final products may differ for Fenton-generated HO· and for radiolytic HO·:**

Interactions with metals in Step 2, reactions 3.4.1/3.4.2, and reactions 4.1/4.3 have no counterpoint in radiolysis, and the concentration of H_2O_2 is usually much lower in the latter. Hence net product yields from HO· reaction with a given (organic) substrate can be both quantitatively and qualitatively different in the two situations, even when pH effects are accounted for.

---

[1] Derived from the prescient review by Walling: Fenton's reagent revisited. *Acc. Chem. Res.* **8**: 125–131, (1975).

Table 4. Free radical chain reactions in lipid peroxidation (LPO)

1. **Initiation of oxidation in polyunsaturated lipid (LH):**
   - 1.1. LH —(X)→ L·,
     where X = gamma or ultraviolet irradiation, certain metals and free radicals, etc.
     When X = HO·, then
   - 1.2. LH + HP· → L·+ H_2O.
2. **Chain propagation:**
   - 2.1. L·+ O_2 —(very fast)→ LOO·
   - 2.2. LOO·+ L′H —(very slow)→ LOOH+ L′,
     whereupon L′· can serve as a reactant in 2.1. to propagate the chain reaction.
3. **Secondary initiation from lipid hydroperoxides (LOOH):**
   LOOH, a product of 2.2., is metastable and can give rise to reactive alkoxyl, LO·, and (usually) less reactive peroxyl radicals, LOO·, by way of 3.1.-3.4.
   - 3.1. LOOH + M^{n+} —(fast)→ LO·+ OH^-+$M^{(n+1)+}$
   - 3.2. LOOH + HO_2· —(slow)→ (X·) — (indirect)→ LO·+ H_2O
   - 3.3. LOOH + $M^{(n+1)+}$ → LOO·+ H^++ M^{n+}
   - 3.4. 2 LOOH → LO·+ LOO·+ H_2O ,
     where 3.4. becomes significant only when the concentration of LOOH approaches 1%.
     LO·, in its turn, can reinitiate LPO via 3.5., and reinitiation can also be brought about via 2.2. by LOO· from 3.3. or 3.4.
   - 3.5. LO·+ L″H — (fast)→ LOH +L″.
4. **Chain termination:**
   - 4.1. 2 L· → L-L
   - 4.2. 2 LO· → LOOL or other nonradical products
   - 4.3. 2 LOO· → LOOL + O_2, etc.,
     where the likelihood of 4.2. increases with respect to 4.1. as pO_2 rises, and 4.3. dominates when pO_2 exceeds 1 atmosphere. Cross reactions between the reactants of 4.1.-4.3. also occur.

5. **Interception**:
A special case of termination involves reactions with target molecules, TH, in contact with the peroxidizing lipid. This is often referred to as co-oxidation:

$$5.1.\quad \left.\begin{matrix} L\cdot \\ LO\cdot \\ LOO\cdot \end{matrix}\right\} + TH \longrightarrow \left.\begin{matrix} LH \\ LOH \\ LOOH \end{matrix}\right\} + TR\cdot - - - (O_2) \rightarrow TOO\cdot \ldots$$

6. **Inhibition**:
When tissue injury or any response being measured is preceded by a long propagative chain of LPO, partial inhibition of chain-carrying reaction 2.2. may shorten the average chain length by many times and provide more efficient antioxidant protection than comparable inhibition of initiation.

7. **Metal dependence**:
Intrinsically LPO is not metal-dependent, although generation of an initiating free radical, such as HO· (reaction 1.2.), *may* depend on metal catalysis. In any case, metals able to induce secondary initiation via 3.1. or 3.3., including heme iron, can induce explosive potentiation of LPO.

8. **Action at a distance**:
Because reaction 2.2. is slow (seconds or more), the chain lengths of LPO usually are long (~ 15–30), and secondary initiation (section 3.) is common, cooxidation damage from 5.1. *typically* occurs far removed in time and space from the site(s) of primary—or even secondary—initiation. In addition, reactive carbonyls from LPO, especially 4-hydroxynonenal and the less reactive malondialdehyde, can induce oxidative stress in distant tissues and organs.

Chapter 3

Photosensitizers as Model Systems to Study Reactive Oxygen Effects in Biological Preparations

Dennis P. Valenzeno and Merrill Tarr

The choice of an appropriate model system to generate reactive intermediates for the study of free radical/reactive oxygen effects in biological systems is not trivial. At present, experts in the field have not determined which intermediates are the primary species responsible for the effects observed in cells and tissues exposed to conditions known to produce free radicals/reactive oxygen. Neither is it clear that one species can be introduced into a biological preparation without initiating production of others that may ultimately be the most damaging (Cadenas, 1989; Valenzeno and Tarr, 1991b). Thus, until the damaging reactions are better understood, it is likely that a variety of model systems will be, and should be, employed by different investigators. Comparisons among these systems are likely to provide clues regarding free radical/reactive oxygen mechanisms of action. In this chapter we present the relative merits and demerits of the use of photosensitizers as a means to generate reactive intermediates. In addition, we provide a brief introduction to the theory, terminology, and special techniques necessary for a novice in photobiology.

Choice of a Model System

The investigator faced with the choice of a model system is likely to select one of two broad approaches. Either he will try to choose a system that

Oxygen Free Radicals in Tissue Damage
Merrill Tarr and Fred Samson, Editors

mimics the *in vivo* generation of reactive intermediates, or he will select a system that initially generates a single (or limited number of) defined species. In studies with direct clinical application the former may be indicated, but in studies aimed at an understanding of mechanism the latter approach has distinct advantages.

Selection of a specific model system will be based on many factors including ease of use, toxicity to the biological preparation, reactive species generated, control of timing and concentration of reactive species generated, and localization of the generating system in the biological preparation. The latter factor cannot be overemphasized. Most reactive oxygen species have limited diffusion distances because of their short lifetimes. For example, singlet oxygen can diffuse no more than 0.06 to 0.16 μm within its lifetime in biological preparations (Valenzeno, 1987). Diffusion distances for hydroxyl radical are considerably shorter: < 2nm (Girotti, 1990). For these transient species reaction is most likely to occur with nearby molecules. Thus, a generating system in the extracellular medium may produce no effects within a cell, or even at the inner surface of the plasma membrane, whereas a different generating system that can penetrate the cell membrane may produce entirely different results. In general, biological preparations present themselves to reactive species as a forest of susceptible reactive groups that are difficult to bypass. If a particular target is not nearby, it will never be reached. Localization of the free radical/reactive oxygen generator thus becomes a critical consideration.

Model Systems Currently Available

It is not our intent to survey all model systems currently available, for there are many. Rather, the available systems can be divided into two classes, chemical generating systems and light-activated systems. Both have strengths and weaknesses. Whereas chemical systems have been exploited extensively, light-activated systems have been used much less frequently. Yet light-activated systems offer some distinct advantages that not only allow some studies to be performed better, but in some cases they allow studies that could not be accomplished using chemical systems.

Chemical systems for generation of reactive intermediates include the well-known xanthine–xanthine oxidase, hydrogen peroxide, hydroperoxides, and the use of iron in various combinations, among others.

In general, the agents can be introduced relatively simply into biological preparations *in vitro*, and offer some degree of certainty regarding the initial species promoting the observed effect. They offer the advantage that they are thought, in some cases, to be involved in *in vivo* generation of reactive intermediates that produce the effects being studied.

On the other hand, chemical systems frequently suffer from localization problems. For example, xanthine oxidase is such a large molecule that it will not cross cell membranes and thus will be confined to the extracellular space when introduced exogenously. As discussed above, localization may well be critical with short-lived intermediates and the effects produced may differ significantly if reactive intermediates are produced inside or outside a cell (or in different parts of a cell). Another problem with chemical generators is their slow action. It is not unusual for many minutes to pass before the monitored effects are observed (Valenzeno and Tarr, 1991b). Certainly, in some cases this is due to the process being monitored, but even in ideal cases there are diffusion limitations to the onset of the effect. Termination of the process is even more difficult to define for similar reasons.

Photosensitizers offer an alternative that overcomes some of the problems with chemical generating systems. There are a variety of photosensitizers available providing a choice of properties relating to localization, wavelength of activating light, reaction mechanism, and others. For example, water-soluble sensitizers are easily introduced into biological systems in aqueous solution. Water-insoluble sensitizers are easily confined to liposomes for targeting to specific tissues. Sensitizers can be chosen that are either membrane permeable or impermeable, primarily dependent on lipid solubility. Thus, localization can be manipulated. For example, fluorescein derivatives are largely membrane impermeable. Acridine orange, on the other hand, penetrates cell membranes quite easily and is well known to intercalate into DNA. Localization and interaction with the biological preparation can also be controlled by selecting either a cationic or anionic sensitizer, or by using a photosensitizer attached to long lipid chains or to polystyrene beads. The lipid-bound forms are expected to bind to membranes; those bound to beads are confined to the extracellular medium. Both forms are available commercially. Finally, different photosensitizers produce different reactive intermediates, so Rose Bengal is frequently chosen as a singlet oxygen generator (Lamberts and Neckers, 1985; but see below), whereas menadione (VanLier, 1991) or the flavins (Jernigan, 1985; Cusanovich, 1991) are much more likely to produce radical species in homogenous solution.

All photosensitizers offer easy control of the rate of generation of reactive intermediates by adjusting the intensity of light that impinges on them. As a result, they do not suffer from the diffusion limitations of the chemical systems. A photosensitizer can be introduced in the dark (or in light of wavelengths that it *cannot* absorb) and can be allowed to diffuse to equilibrium before activating it with light. Thus, the reaction can be initiated at a defined time and can be terminated at a defined time by ending illumination.

Similarly, the rate of generation of reactive intermediates by photosensitizers can be controlled by changing the sensitizer concentration, as for example from one experiment to the next. However, even during a single experiment, the rate can be changed simply by altering the intensity of illumination (e.g., Valenzeno and Tarr, 1991a). Together, the control over both sensitizer concentration and illumination intensity permit an extremely wide range of control over the rate of generation.

There are drawbacks to the use of photosensitizers of course. They require that precautions be taken to shield the preparation from activating light until the desired time. The preparation must also be in a position to be illuminated with light of appropriate wavelengths when generation of reactive intermediates is desired. In most cases these are not severe restrictions. Of somewhat greater concern is the question of the species of reactive intermediate produced. This varies from photosensitizer to photosensitizer and can take the form of either excited state singlet molecular oxygen or radical species (see below). Whereas the yields of each can be determined in homogenous solution, the yields in compartmentalized, complex biological preparations, wherein the sensitizer may localize in different microenvironments, is less clear. However, it may well be that interconversion of reactive oxygen species in biological systems is much more prevalent than heretofore believed (see Chapter 14 in this volume). If this is the case, the identity of the initiating species may not be a critical factor in the fate of the cell. Other potential pitfalls, which can usually be avoided by good experimental technique, are addressed at the end of the chapter.

As an example of the use of photosensitizers, and to demonstrate a study that could not have been performed using a chemical generating system, consider the results shown in Fig. 1. The figure is a plot of the electrical current that flows across the plasma membrane of a cell from the atrium of a frog heart when the membrane potential is suddenly

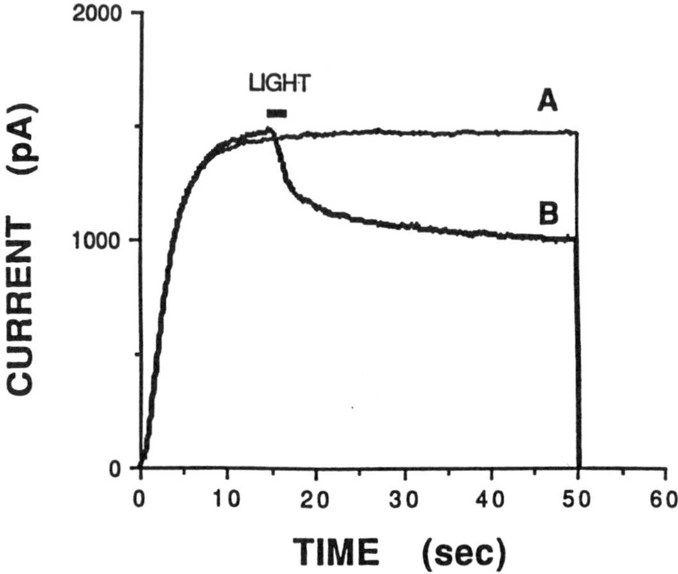

Figure 1. Block of potassium current by Rose Bengal and light. The figure plots current flowing through the surface membrane of an isolated frog atrial cell as a function of time after suddenly switching the membrane potential from −70 mV to 0 mV. The bathing medium contained 0.5 μM Rose Bengal, and pharmacological agents to block sodium and calcium currents. **A**: Control current with no illumination. **B**: Current recorded when the cell was illuminated for 2 s at the time indicated by the dark bar. Note the rapid onset of the current decrease with illumination, as well as the change in slope of the current decrease when illumination ends.

changed and held at 0 mV after being kept at a value near the normal resting membrane potential of the cell, −70 mV. Under control conditions (trace A) the current increases with time, reaching a steady-state value after about 10 s (see figure legend for experimental details). It is well established that this current is carried by potassium ions through voltage-gated channels that are selective for this ion. Trace B shows the effect on the measured potassium current of a brief (2s) exposure to reactive oxygen species generated by illuminating the photosensitizer Rose Bengal contained in the cell suspension medium. There is a *delay* of about 1 s before any change is seen, and then the current begins to decrease during light. After light and after the brief lifetime of most reactive intermediates, the current continues to decrease, so that there is a *persistence* of the effect. Both delay and persistence are significant in elucidating the mechanism of the impairment of potassium channel activity. Although they cannot be fully explored in this chapter, they suggest a long-lived intermediate state of potassium channels that, while still conductive to potassium ions, is committed to becoming blocked. Neither persistence nor delay could have been discovered using chemical generating systems with their diffusion limitations. The control of the timing of activation of the photosensitizer makes this study possible.

With this introduction to the utility of photosensitizers, let us next consider how photosensitizers function. To understand better how photosensitizers work and how to use them to study reactive oxygen effects, we must first consider what light does to photosensitizer molecules.

A Primer on Photosensitizer Mechanisms of Action

Photosensitizers (often referred to simply as sensitizers) can be thought of as transducers. They convert the energy of photons into the energy needed to form and/or break chemical bonds. In biological systems the ultimate result of a photosensitizer-mediated modification is usually oxidation of some biomolecule. In general terms, photosensitizers accomplish this energy transduction and ultimate substrate oxidation by one of two broad mechanisms (Foote, 1991). Either the sensitizer reacts directly with the substrate, which is subsequently oxidized by ground state molecular oxygen (Type I reaction), or the photosensitizer interacts directly with an oxygen molecule that then oxidizes a biomolecule, the substrate (Type II reaction).

A photosensitized oxidation reaction of a biomolecule can be thought of as occurring in three steps: light absorption, initial photophysics, and reaction (Type I or II) and substrate oxidation. We will consider each step in turn.

Light Absorption. The first law of photochemistry (enunciated by Grotthus and Draper in 1818) states that no photochemical reaction can occur unless light is first absorbed by some component of the system (Grossweiner, 1989b; Pottier and Russell, 1991). Thus, light that can be absorbed by the photosensitizer must illuminate the system. Conversely, illumination of a photosensitizer-containing system with light that cannot be absorbed allows the experimenter to manipulate the system before initiating the photosensitization reaction. Precise timing of the beginning and end of the photosensitization reaction can be accomplished by simply illuminating with the appropriate wavelengths of light.

Initial photophysics of the photosensitizer. Photochemically, photon absorption promotes an electron of the photosensitizer to a higher energy orbital, an excited (singlet) state. The photon's energy is thus converted to the potential energy of the excited state of the photosensitizer. Thermodynamically speaking, the photosensitizer is not in its lowest energy configuration: thus, it wants to release the excess energy of the excited state. It can do this in a variety of ways that compete with one another (Fig. 2A).

Most light-absorbing molecules that are not photosensitizers release absorbed energy as heat, i.e., vibrational energy (Fig. 2A, path 1). In this process, termed internal conversion, these molecules are able to return to their ground state. Alternately, the excess energy may be released by emission of a photon (Fig. 2A, path 2). This process is known as fluorescence. The fluorescence emission will be at a longer wavelength (lower energy) than the absorbed light, since some energy is usually lost as heat before fluorescence occurs. With the energy loss by fluorescence a fluorescent molecule also returns to its ground state energy level. Fluorescence and internal conversion are competing processes. Most molecules do not fluoresce appreciably because intersystem crossing is so efficient that the energy of excitation is lost by this pathway, depleting excited states before fluorescence can occur.

High efficiency of a third competing pathway for energy loss is the hallmark of effective photosensitizers. Such sensitizers can readily transfer their energy of excitation to a long-lived excited state called a triplet state (Fig. 2A, path 3). Excited triplet states have long lifetimes compared to the initial excited state produced by light absorption (a singlet state). The lifetime is sufficiently long to allow the triplet state photosensitizer to react with other molecules in its environment. Thus, the triplet state is usually the starting point for photosensitization reactions. It is important to appreciate that the triplet excited state of the photosensitizer can have significantly different properties from the ground state of the same molecule (Glass, 1961). For example, bond lengths are longer, reactivity is typically greater, and acid–base status may be altered.

Although for photosensitization reactions we are interested in reaction of the triplet state with other molecules, there are several other possible fates of triplet states. Like the initial excited singlet state, there are competing pathways for loss of energy from triplet states. These again include loss of heat energy (Fig. 2A, path A) and emission of light, termed phosphorescence when it occurs from a triplet excited state (Fig. 2A, path

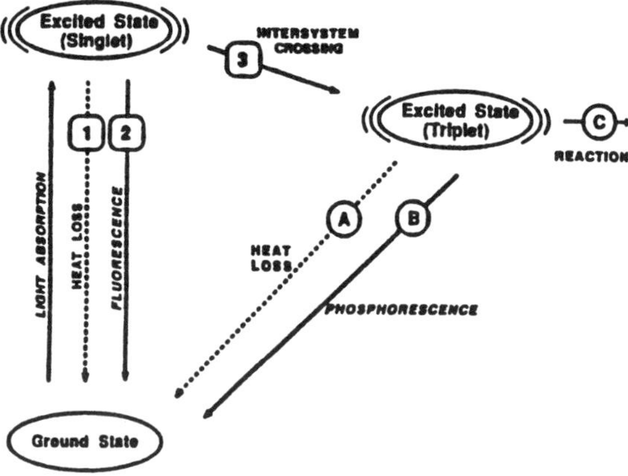

Figure 2.

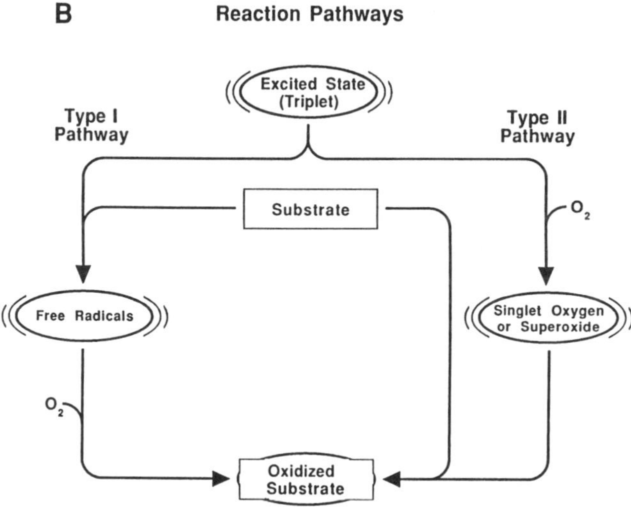

Figure 2. Photosensitization reaction pathways. **A** describes the possible deexcitation pathways for photosensitizer that absorbs light energy putting it in an excited singlet state. Paths 1, 2, and 3 are competing pathways for the excited singlet state to release some or all of the absorbed energy. If pathway 3 is followed to the excited triplet state, some of the energy of excitation remains. It can be released via paths A, B, or C. In most cases only molecules that follow paths 3 and then C produce photosensitization reactions. **B** shows the customary division of possible reactions of the excited triplet state. If the initial reaction is with oxygen the reaction is termed a Type II reaction. If the initial reaction is with another substrate molecule it is termed a Type I reaction, even though both may ultimately lead to an oxidized final product.

B). Phosphorescence emission is normally at a longer wavelength than fluorescence, since the triplet state is usually at a lower energy level than the original singlet excited state. Reaction with another molecule, in which either energy or an electron is transferred (Fig. 2A, path C), is the pathway that leads to measurable photomodification reactions.

With all the competing pathways for energy loss, it would seem that photosensitized modification would be an unlikely process, and indeed that is the case. The vast majority of light-absorbing molecules in our environment lose energy absorbed from photons without reacting with neighboring molecules. However, some molecules are efficient at channeling absorbed light energy to react with nearby molecules. A sampling of the classes of molecules that have this ability are given in Table 1. The photosensitizers listed in Table 1 are grouped according to their ability to produce singlet oxygen versus the superoxide radical. Representative references are cited.

Reaction and substrate oxidation. A favorite target of photosensitizer triplet states is molecular oxygen (Spikes, 1989). The result is a reactive oxygen species, and the photomodification process is referred to as a Type II mechanism (Fig. 2B). The reactive oxygen species formed is not unique. If only the energy of excitation is transferred to the oxygen molecule, singlet molecular oxygen can be formed. Singlet oxygen is an excited state of ground state molecular oxygen that has a relatively long excited state lifetime (a few μs in water), long enough to react with other molecules. On the other hand, an electron may be transferred resulting in superoxide radical anion and a radical species of the sensitizer. Both reactive oxygen species can oxidize substrate molecules in biological preparations, and both species may be formed from the same photosensitizer. For example, illumination of the photosensitizer Rose Bengal in aqueous solution produces singlet oxygen from 75% and superoxide from 20% of the excited states formed (Lee and Rodgers, 1987). However, the initial species produced do not necessarily indicate the species producing modification in any given case. In another chapter in this book the complexity and possible interconversion of these reactive oxygen species is discussed further. In addition, it is important to be aware that a low yield of a particular reactive species does not necessarily exclude it as a candidate in a reaction mechanism. There are numerous examples in photobiology of intermediates produced with very low efficiency (low quantum yield = low number of states produced per absorbed photon) that are responsible for observed reactions.

Photosensitizer triplet states can react by electron transfer with biological substrates other than oxygen (Fig. 2B). These Type I reactions result in radical species of the substrate and sensitizer (Spikes, 1989). Binding of the photosensitizer, as often occurs in biological preparations, is thought to promote Type I photochemistry over Type II, owing to the close proximity of the excited state photosensitizer and the potential target to which it is bound.

It bears noting that there has been some confusion in the literature regarding the nomenclature of photooxidation. In some cases Type II has been used to denote only the singlet oxygen pathway. However, as recently emphasized (Foote, 1991), Type II refers to any mechanism in which the initial reaction of the excited state sensitizer is with oxygen.

Technical Considerations for the Use of Photosensitizers

What are the technical considerations in attempting to use photosensitizers to study reactive oxygen effects? The major discriminating features of photosensitization studies center on light. These features can be divided into five areas: illumination sources, measurement of light intensity, background light, controls, and quantitative expression of data per absorbed photon.

Illumination sources. The choice of an illumination source (Grossweiner, 1989a; Oriel Cat., 1989) depends on the requirements of light intensity, spectral distribution (wavelength), area to be illuminated, source stability, source lifetime, and expense. No single source is best in all applications. Whereas the high intensity of a laser may be optimal in one setting, it cannot easily illuminate large areas as can fluorescent lamps. Arc lamps can provide high intensity, but are more expensive than tungsten–halogen sources.

Fluorescent lamps. These lamps are low pressure mercury arc lamps with a phosphor coating the envelope. The phosphor is excited by the sharp emission peaks of the arc and then emits fluorescent light over a much broader spectrum of wavelengths. The major attributes of these lamps are their ability to illuminate large areas with relatively uniform intensity, their long lifetime, and their economy. Fluorescent lamps tend to fade over time rather than burn out, as do other sources. The intensity should be checked regularly to allow comparison of data gathered at

Table 1. Selected Photosensitizers

Mechanism	Photosensitizer (% dominant path)	Class	Peak[a] absorption	Reference
$^1O_2 >> O_2^-$				
	Methylene blue (80%)	Thiazine	664	Chacon, Jamieson & Sinclair, *Chem Phys Lipids* 43:81 (1987)
	Rose Bengal (75%)	Xanthene	548	Lee & Rodgers, *Photochem Photobiol* 43:79 (1987)
	Erythrosin B (75%)	Xanthene	523	Chacon, Jamieson & Sinclair *Chem Phys Lipids* 43:81 (1987)
	Hematoporphyrin (72%)	Porphyrin	399, 505 540, 565 615	Chacon, Jamieson & Sinclair, *Chem Phys Lipids* 43:81 (1987); Kessel, *Cancer Res* 41:1318 (1981)
	Deuterophyrin	Porphyrin	[b]	Cannistraro, Jori & Van de Vorst, *Photobiochem Photobiophys* 3:353 (1982)
	Uroporphyrin	Porphyrin	[b]	Pottier & Truscott, *Int J Radiat Biol* 50:421 (1986)
	Merocyanine 540	Cyanine	510, 535	Easton, Valinsky & Reich, *Cell*, 13, 475 (1978)
	Eosin Y	Xanthene	516	Gandin, Lion & Van de Vorst, *Photochem Photobiol* 37:271 (1983)
	Toluidene blue	Thiazine	636	Ito *Photochem Photobiol* 25:47 (1977)

Table 1. Selected Photosensitizers

Mechanism	Photosensitizer (% dominant path)	Class	Peak[a] absorption	Reference
$^1O_2 \approx O_2^-$				
	Riboflavin ($\leq$ 50% $^1O_2$)	Flavin	224, 260, 373, 445	Chacon, Jamieson & Sinclair, *Chem Phys Lipids* 43:81 (1987); Krishna, Uppuluri, Riesz, Aigler & Balasubramanian, *Photochem Photobiol* 54:51 (1991); Joshi & Pathak, *BBRC* 112:638 (1983)
	Psoralen	Furocoumarin	[c]	Joshi and Pathak, *BBRC* 112:638 (1983)
	8-Methoxypsoralen	Furocoumarin	212, 245, 295, 330	Joshi and Pathak, *BBRC* 112:638 (1983) Wolf & Hönigsmann, *Pharmac Ther* 12:381 (1981)
$O_2^- > {}^1O_2$				
	4,5′ 8-Trimethylpsoralen	Furocoumarin	[c]	Joshi and Pathak, *BBRC* 112:638 (1983)
	Anthracene	Anthracene		Joshi and Pathak, *BBRC* 112:638(1983)
	Many sensitizers + reducing agents			Hartman, Dixon, Dahl & Daub, *Photochem Photobiol* 47:699 (1988); Buettner & Need, *Cancer Lett* 25:297 (1985); Girotti *Photochem Photobiol* 51:497 (1990); Ito & Ito, *Photochem Photobiol* 35:501 (1982)

[a] In aqueous solution.
[b] Similar to hematopophyrin.
[c] Similar to 8-Methoxypsoralen.

different times. Fluorescent lamps are available with output in specific spectral regions, including the UV. Their drawbacks include relatively low intensity, even for so-called very high output lamps, the fading mentioned above, and sharply peaked output. The output spectrum shows sharp peaks, corresponding to the excitation peaks of the mercury arc. If the effects of different photosensitizers are to be compared, the sensitizers may absorb different amounts of light depending on whether or not their absorption spectrum overlaps one of these peaks.

Incandescent sources. These sources produce light by heating a filament until it begins to emit light as a black body radiator. The higher the filament temperature, the shorter the wavelength of peak emission. Typically, these lamps have a tungsten filament, a quartz envelope, and contain a halogen gas. Tungsten–halogen lamps are useful sources at visible and infrared wavelengths, with a maximum output typically in the long wavelength visible. They can illuminate large areas, but plane surfaces cannot be uniformly illuminated as easily as with fluorescent sources. They provide moderate output intensity at fairly low cost. Although lamp lifetime may be short, replacement cost is low.

Arc lamps. Arc lamps produce light by generating an electrical arc between two electrodes. They emit either a sharply peaked spectrum of light or a relatively flat spectrum depending on the gas in the envelope and the internal pressure. For example, a low pressure mercury vapor arc lamp emits a spectrum of light that is characteristic of the sharp emission peaks of mercury (the mercury lines). It is an ideal source of 254 nm radiation, since 90% of its output is at this wavelength. On the other hand, a high pressure xenon arc lamp emits nearly constant, high intensity light at all visible wavelengths. Such high pressure arc lamps can be excellent general purpose sources for photobiological experiments that do not require illumination of large areas. Different power lamps (75–5000W) can provide a wide range of intensities. They are relatively expensive since they require regulated power supplies with a special starting circuit, especially the higher power models. Replacement costs can be significant ($\approx$ \$750 for a 1000W lamp at time of writing), although lamp lifetime is fairly long ($\approx$1000 hr).

Lasers. These sources are becoming more common in photobiological studies as prices decrease and their variety increases. Lasers can provide a broad range of light intensities, including intense short-duration pulses. They are excellent sources of monochromatic radiation at defined wavelengths. The wavelengths available were limited, but have now been considerably expanded by dye lasers and by devices known as frequency doublers, triplers, and so on. Lasers are best used to illuminate restricted areas, and can even be focused on subcellular structures microscopically. Many models are still expensive and require expert personnel. However, this situation is improving and many lasers are relatively inexpensive and easily run (Parrish and Wilson, 1991).

Light intensity measurements. Everything has its price, and the price of the increased control over reactive intermediate generation using photosensitizers is the necessity of accurately measuring the intensity of light reaching the preparation. To discuss light measurement intelligently, we must first be familiar with the terminology and units of optical radiation (Grossweiner, 1989a).

Radiation units are referred to as radiometric units, and are applicable to any kind of electromagnetic radiation. Table 2 and Fig. 3 list radiation units and describe their significance. As indicated in Table 2, radiation is described in terms of the radiant energy (measured in Joules) or the radiant power (measured in Watts = Joules/s). Radiant power is often termed radiant flux.

The output of a source is often given as the *intensity* of the source. As shown in Table 2 and as depicted in Fig. 3, this is the flux in a given direction (into a unit solid angle). A related quantity is the *radiance*, which is similar to the intensity, but is normalized per unit area at the surface of the source. Radiance is a particularly useful measure in considering sources such as fluorescent lamps that emit secondarily from an extended surface area.

In choosing a source you may deal with radiance and intensity. However, it is the amount of light received by the experimental preparation that is of paramount importance. Light incident on a surface, such as that falling on an experimental preparation, is characterized by its *irradiance*, the radiant flux falling on a unit surface area. Alternately, the total energy falling on a unit surface area, denoted as *exposure* or *fluence*, is often used. Irradiance may also be termed *fluence rate*. Fortunately, some manufacturers provide information on the irradiance of their sources.

A special set of units, known as photometric units, have developed to describe visible radiation. Those photometric units are based on the response of a "standard" human eye to visible radiation. Although they are in common use in the lighting and photographic industries and may be appropriate in specialized areas of scientific research such as visual science, they introduce unnecessary complications in other instances and should be avoided. For example, consider two laser beams that have the same energy or power, one at 300 nm and the other 500 nm. Although they would both have the same radiance in radiometric units, the 300-nm laser would have 0 luminance in photometric units because the eye cannot perceive 300 nm radiation. Even a 450-nm laser would have a photometric luminance more than 8-fold lower than the 500-nm source because of the reduced sensitivity of the eye at wavelengths below 550 nm.

We should also note in passing that light is typically defined as *visible* electromagnetic radiation. Therefore, properly speaking, ultraviolet (and infrared) radiation should not be referred to as light.

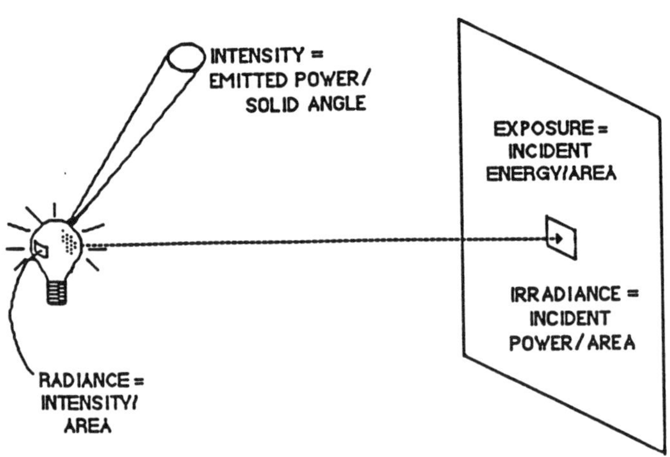

Figure 3. Units used to quantify radiation. The radiation emitted by the source at the left of the figure can be expressed as the *intensity*, which is the emitted power per unit solid angle, or as the *radiance*, which is intensity normalized to a unit surface area of the source. The surface at the right of the figure depicts the units used to express incident radiation. The *exposure* is the energy incident on a unit area, whereas the *irradiance* is the power per unit area.

Table 2. Units of measure of light and other electromagnetic radiation

Unit class	Description of quantity measured	Radiometric description	SI units (recommended)	Luminous units (not recommended)	Other units (not recommended)
Basic units	Amount of radiation	Energy	Joule(J)	lumen-sec (lm-s)	
	Flux (time rate of energy transfer)	Power	Watt(W)	lumen (lm) or candela-steradian (cd·sr)	1 lm = (1/680) lightwatt
	Flux per unit area leaving a small plane surface (by emission or reflection)	Excitance	Watt/meter^2 (W/m^2)	lumen/meter^2 (lm/m^2)	
Source units	Flux emitted into a unit solid angle	Intensity	Watt/steradian (W/sr)	Candla (cd) or lumen steradian (lm/sr)	1 cd =1 candlepower
	Flux in a given direction per unit solid angle and per unit area (normal to direction of propagation)	Radiance	Watt/meter^2· steradian (W/m^2· sr)	Luminance or candela/meter^2 (cd/m^2)	1 cd/m^2 =1 nit 1 cd/cm^2 =1 stilb
Target Incidence units	Flux per unit area incident on a small plane surface	Irradiance (also called fluence rate)	Watt/meter^2 (W/m^2)	Illuminance lux (lx) or lumen/meter^2 (lm/m^2)	1 lux = 1lm/m^2 1 lm/ft^2 = 1 foot · candle 1 lm/cm^2 =1 phot
	Energy per unit area incident on one side of a small plane surface	Exposure (also called fluence)	Joule/meter^2 (J/m^2)	lumen/meter^2· second (lm/m^2·s)	1 lm/m^2· s =1 lx · s

The discussion of units thus far has not distinguished between different wavelengths of light. When considering the wavelength dependence of emitted or incident light, the term "spectral" usually precedes the units discussed. Thus, it is the spectral irradiance of the preparation that concerns us. As discussed above, only those wavelengths of light absorbed by the photosensitizer will produce any photochemical effects. In choosing a source and in designing an optical system to deliver the light to the preparation, the aim is to maximize the irradiance of the wavelengths in the vicinity of the peak of the absorption spectrum of the photosensitizer.

With these aims and principles in mind, consider the problem of measurement of light. From the above it follows that an ideal light measurement system should provide information on the spectral irradiance of the experimental system. In other words, how much light of what wavelengths hits the target? There are several ways to approach this question using devices that range from simple, inexpensive pieces of equipment to fairly involved, moderately expensive systems. Both extremes have their utility in the photobiology laboratory, as will be explained below. First let us consider the strengths and weaknesses of the various devices available.

Thermopiles. These are commonly used, rugged, and inexpensive devices that respond to the radiant power in the incident light. They respond to a broad spectral range extending from the ultraviolet (λ = 200 nm) through the visible to the infrared (λ = 50 μm). This makes them versatile, but they respond slowly since they rely on heating of the detector. They give no indication of the wavelengths of the incident light. Some spectral information can be gained by placing narrow bandpass filters in the incident light path, but the resolution and accuracy is limited.

Photomultipliers. In many applications photomultipliers are being replaced by solid state photodetectors. However, photomultipliers are still the most sensitive detectors in the visible and long wavelength ultraviolet, and they have rapid response time. Like thermopiles they provide no spectral information. They are more expensive and less durable than thermopiles, and require high operating voltages.

Solid state photodetectors. These include photodiodes, phototransistors, and the like. They compete effectively with photomultipliers because they provide rapid response time and high sensitivity in the visible and long wavelength ultraviolet, although less than photomultipliers. On the other hand, they are inexpensive and do not require the high operating voltages used for photomultipliers. Again no spectral information is available directly.

Spectroradiometer systems. To gain information on the spectral content of the incident light, the general approach is to interpose a dispersive device in the light path before the detector. A dispersive device separates the different wavelengths of incident light in space. The simplest example is a prism. Monochromators are the modern day counterpart of the prism, and together with a photodetector form the essential components of a system capable of measuring irradiance as a function of wavelength (i.e., a spectroradiometer system).

When properly calibrated, spectroradiometer systems provide measurements of the irradiance at each wavelength. Their sensitivity is determined by the sensitivity of the detector used. Modern spectroradiometers frequently use linear arrays of photodiodes as detectors, allowing measurement of many wavelengths simultaneously. The disadvantages include the high cost and complexity of the system.

Accurate calibration of any detector system requires a standard lamp. This is a lamp whose irradiance has been determined at a variety of wavelengths against a standard that ultimately has been compared to a standard lamp at the National Institute of Standards and Technology (NIST; formerly the National Bureau of Standards). It is run by a highly regulated power supply. A new spectroradiometer system can be calibrated by using it to measure the irradiance of the standard lamp. The measured values are then compared to the correct values supplied with the lamp in order to generate a set of correction factors. Many software programs that accompany spectroradiometer systems do this automatically. Once calibrated, measurement with the new system are also "traceable" to the NIST standard lamp.

Practical everyday light measurement. In a practical sense, few investigators can take the time to measure the spectral irradiance for each experiment, and this is not necessary provided certain precautions are observed. First, make sure that your light source is powered from a regulated supply and the line voltage is free from variations. Lamp intensity varies as a high power of supply voltage, so minor voltage fluctuations can produce major changes in intensity. They will also alter the spectral content of the light. Next, when it is necessary to adjust the irradiance of the experimental preparation, do so by placing neutral density filters in the light path. Do *not* alter the current to the lamp. Good neutral density filters, as the name implies, reduce the irradiance at all wavelengths to the same degree. Changing the lamp power changes the intensity of the source differently at different wavelengths. Make sure all optical elements in the light path are securely mounted, and avoid moving them once they are in place. As much as possible keep them covered to avoid accumulations of dust and dirt, which can affect optical performance.

With reasonable care the spectral character of an optical system should not vary significantly with time. Therefore, a simple photodetector (solid state, photomultiplier, or even thermopile) can be used to determine the overall light intensity for each experiment. Keep a record of these. Periodically, once or twice a year, or whenever an optical component is changed, the detector output should be compared to new spectral measurements of the same source with the same optical filters, etc. using a spectroradiometer (James and Matchette, 1988). Although day-to-day variations in irradiance can be tracked by measuring the photodetector output, publications should always include information on the spectral content of the light used, not just the total intensity determined with a photodetector.

Since spectroradiometry is expensive, time consuming, and sometimes complex, there is now at least one company that will make spectroradiometric measurements in the customer's laboratory (Rapid Precision Testing Laboratories, P. O. Box 1342, Cordova, TN, 38018-1342. Phone: (901) 386-0175).

"Dim laboratory lighting" — how dim is dim? When working with light-activated systems, extraneous light must be kept to a minimun to avoid unintended activation. Thus, laboratory lighting must usually be reduced, but how much? There is no absolute answer to this question. This is because there are at least two variables in any photosensitization reaction: light and photosensitizer concentration. With high sensitizer concentrations even low ambient light levels may generate photosensitized changes in the preparation. If the sensitizer concentration is much lower, a significant amount of ambient light may be tolerated. Nonetheless, it is prudent to keep ambient light levels as low as possible or to shield the preparation from ambient light. A workable compromise is to filter laboratory light that can be absorbed by the photosensitizer in use.

Darkroom filters will work in some cases, depending on the absorption spectrum of the photosensitizer in use. In all cases it is a good idea to measure the percent transmission of the filter to be used. This can be done using a spectrophotometer. The filter should have zero transmission at all wavelengths absorbed by the photosensitizer in use.

The same strategy can be used to block the light that will ultimately be used to activate the sensitizer. For example, we customarily illuminate cells throughout an experiment in the presence of Rose Bengal, which absorbs at wavelengths shorter than 600 nm. Initially we have a 600-nm long pass filter in the light path, which allows only wavelengths above 600 nm to reach the cells. When we wish to generate reactive oxygen species, we simply remove the filter, exposing the cells to the full spectrum of light from the source. In this way we use only one source to provide light to manipulate the cells and to activate the photosensitizer.

Controls. Photomodification studies require slightly more extensive controls than those that do not involve light. It is necessary to demonstrate that the effects observed require both light and photosensitizer. This requires that separate controls be run without photosensitizer and without illumination. A complete set of controls also includes the absence of both photosensitizer and light.

Quantitation. When describing chemical reactions, the preferred means to express reaction rates is per mole of reactant. Similarly, the preferred means to express photoreactions is per mole of absorbed photons (1 mole of photons = 1 Einstein) or simply per absorbed photon. In photochemistry this is termed "quantum yield" and is the ratio of the number of molecules of photoproduct produced per photon absorbed. It is critical to consider only absorbed photons, since photons outside the absorption spectrum of the photosensitizer cannot be absorbed and hence cannot produce any photosensitized reactions. This issue becomes especially pertinent when the actions of two more photosensitizers with different absorption spectra are being compared. Consider UV-absorbing and visible-absorbing sensitizers. Since many standard visible sources provide much less energy in the UV, the UV-absorbing sensitizer will produce less effect when it receives what appears to be the same intensity of illumination. However, calculation of the number of photons absorbed would reveal that the UV absorber is receiving few photons that it can absorb to produce excited state chemistry. In fact, since each UV photon has more energy than each visible photon ($E=h\nu$) the ratio of photon fluxes in the UV to visible is even less than the ratio of energy fluxes in the two ranges. To avoid these confusions, reaction rates should be standardized to the number of absorbed photons wherever possible (Pooler and Valenzeno, 1982).

In simple photochemical systems, with some effort, quantum yields can usually be determined (Pottier and Russell, 1991). In biological systems the same is not always true. To determine the rate of photon absorption by the photosensitizer, the concentration and absorption spectrum of the photosensitizer must be known. In a structured, compartmentalized system like even a simple cell, where partitioning of the sensitizer occurs, these quantities are seldom known with certainty. Although there is no single resolution to this problem, several strategies have been adopted by previous investigators. Partitioning and absorption properties of sensitizers in membranes have been estimated using bulk phase partitioning and absorption properties in hydrocarbon solvents, which are thought to mimic the environment in the interior of cell membranes. In other instances it has been possible to measure uptake to cells or subcellular organelles to estimate photosensitizer concentrations in them. Likewise, absorption spectra have been measured in cellular preparations (although this is not without difficulty). Even in instances where these approaches are difficult, it is worthwhile to estimate the sensitizer concentration and to calculate reaction rates per absorbed photon. Even for seasoned photobiologists, this sometimes produces some surprising results.

Finally, it should be mentioned that in some instances, although not incorrect, it may not be most appropriate to express results per absorbed photon. When reactions to sunlight in the environment are concerned, it can be argued that reactions are most appropriately expressed per minute of exposure to a standard dose of solar radiation. However, even in this sphere, mechanistic studies will ultimately have to dissect such responses to identify and quantify the responses to specific wavelengths of radiation.

Summary

Photochemical systems offer an attractive alternative to conventional chemical systems as generators of reactive intermediates. The day-to-day experimental protocols required for photosensitizer use are not much more involved than those required for a conventional chemical system.

Their major strength is the ability to control the timing of production of the intermediates as well as their concentrations. In addition, the availability of a variety of photosensitizers allows the investigator to exercise some control over the type of reactive intermediate generated and the subcellular generation site. In some cases photosensitizers allow investigations that would be impossible with conventional chemical systems for generation of reactive intermediates.

Acknowledgments. Supported by NIH grant R01 HL43008 and by an American Heart Association Grant-in-Aid (890714).

References

Cadenas E (1991): Biochemistry of oxygen toxicity. *Annu Rev Biochem* 58: 79–110.

Cusanovich MA (1991): Photochemical initiation of electron transfer reactions. *Photochem Photobiol* 53:845–857.

Foote CF (1991): Definition of Type I and Type II photosensitized oxidation. *Photochem Photobiol* 54:659.

Girotti AW (1990): Photodynamic lipid peroxidation in biological systems. *Photochem Photobiol* 51:497–509.

Glass B (1961): Summary. In: *A Symposium on Light and Life*, McElroy WD, Glass B, eds., Baltimore: Johns Hopkins Press, pp. 817–831.

Grossweiner LI (1989a): Photophysics. In: *The Science of Photobiology*, Smith KC, ed., New York: Plenum Publishing Co, pp. 1–45.

Grossweiner LI (1989b): Photochemistry. In: *The Science of Photobiology*, Smith KC, ed, New York: Plenum Publishing Co, pp. 47–78.

James RH, Matchette SL (1988): Accurate spectral measurements of laboratory light sources. *Photochem Photobiol* 47:32S.

Jernigan HM (1985): Role of hydrogen peroxide in riboflavin-sensitized photodynamic damage to cultured rat lenses. *Exp Eye Res* 41:121–129.

Lamberts JJM, Neckers DC (1985): Rose bengal derivatives as singlet oxygen sensitizers. *Tetrahedron* 41:2183–2190.

Lee PCC, Rodgers MAJ (1987): Laser flash photokinetic studies of rose bengal sensitized photodynamic interactions of nucleotides and DNA. *Photochem Photobiol* 45:79–86.

Oriel Corporation (1989): Oriel Catalog — Light Sources, monochromators, detection systems. Oriel Corporation, Stratford, CT.

Parrish JA, Wilson BC (1991): Current and future trends in laser medecine. *Photochem Photobiol* 53:731–738.

Pooler JP, Valenzeno DP (1982): A method to quantify the potency of photosensitizers that modify cell membranes. *J Nat Cancer Inst* 69:211–215.

Pottier RH, Russell DA (1991): Quantum yield of a photochemical reaction. In: *Photobiological Techniques*, Valenzeno DP, Pottier RH, Mathis P, Douglas RH, eds. New York: Plenum Publishing Co, pp. 45–57.

Spikes JD (1989): Photosensitization. In: *The Science of Photobiology*, Smith KC, ed., New York: Plenum Publishing Co, pp. 79–110.

Valenzeno DP (1989): Photomodification of biological membranes with emphasis on singlet oxygen mechanisms. *Photochem Photobiol* 46:147–160.

Valenzeno DP, Tarr M (1991a): Membrane photomodification of cardiac myocytes: potassium and leakage currents. *Photochem Photobiol* 53:195–201.

Valenzeno DP, Tarr M (1991b): Membrane photomodification and its use to study reactive oxygen effects. In: *Photochemistry and Photophysics*, Rabek, JF, ed. Boca Raton, FL: CRC Press Inc., pp. 137–191.

Chapter 4

Singlet Oxygen in Biological Systems: A Comparison of Biochemical and Photochemical Mechanisms for Singlet Oxygen Generation

Jeffrey R. Kanofsky

Oxygen in the $^1\Delta_g$ state (called singlet oxygen in this chapter) is a potent biological toxin. It oxidatively damages a variety of biological molecules including DNA (Foote, 1981; Straight and Spikes, 1985; Wefers et al., 1987), inactivates viruses (Wefers et al., 1987), and kills both bacteria and mammalian cells (Dahl et al., 1987; Dahl, 1991). Singlet oxygen is felt to be a common mediator of photochemical damage of both plants and animals. In contrast, the existence of significant biochemical routes of production of singlet oxygen, in the absence of light, remains to be established (Kanofsky, 1989b, 1990).

This chapter reviews the mechanisms responsible for both the photochemical and biochemical generation of singlet oxygen. The technical difficulties of measuring singlet oxygen generation within living cells will also be discussed.

Detection of Singlet Oxygen Generation in Biological Systems

Several tests have been developed to identify singlet oxygen in complex systems. These include singlet-oxygen quenchers, chemical traps, the use of deuterated solvents, and the direct measurement of singlet-oxygen phosphorescence. The manner in which some of these tests perturb the

Oxygen Free Radicals in Tissue Damage
Merrill Tarr and Fred Samson, Editors

system being studied is shown in Fig. 1. All of these methods work best in systems containing low concentrations of reactants in water or in organic solvents. In these systems, the lifetime of singlet oxygen is relatively long. In contrast, the lifetime of singlet oxygen within the living cell is very short, of the order of 0.1 μs (Moan et al., 1979; Firey et al., 1988; Patterson et al., 1990; Moan and Berg, 1991; Baker and Kanofsky, 1992), making measurement of singlet oxygen extremely difficult. The manner in which the short singlet-oxygen lifetime limits the effectiveness of the various detection methods is discussed in the next few paragraphs.

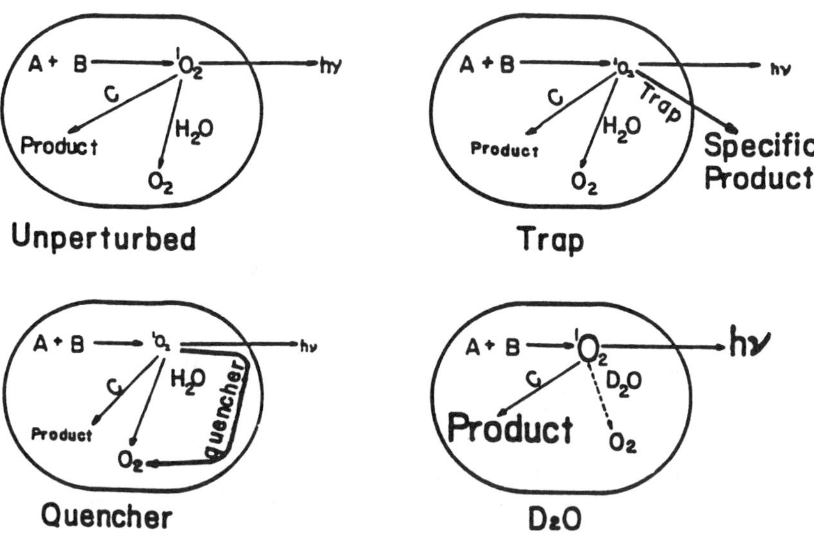

Figure 1. Schematic of perturbations of a complex singlet-oxygen generating system caused by a singlet-oxygen quencher, by a singlet-oxygen trap, and by deuterium oxide solvent. The size of the symbols is proportional to the steady-state concentration of a particular species or to the intensity of the light emission.

As shown in Fig. 1, singlet-oxygen quenchers physically or chemically inactivate singlet oxygen, consequently limiting the oxidation of biological molecules by singlet oxygen. To reduce the oxidation of biological molecules by 50%, it is necessary to reduce the singlet-oxygen lifetime in the system by 50%. The singlet-oxygen lifetime is given by the Stern-Volmer equation:

$$\frac{1}{\tau} = \frac{1}{\tau_0} + k[C] \qquad (1)$$

Here τ is the singlet-oxygen lifetime of the system containing the quencher, τ_0 is the singlet-oxygen lifetime without the quencher, k is the singlet-oxygen quenching constant of the quencher, and C is the concentration of the quencher.

Histidine is a commonly used singlet oxygen quencher and has a quenching constant of $4.4 \times 10^7 M^{-1}s^{-1}$ (Rougee and Benasson, 1986). Using Equation 1, the concentration of histidine needed to lower singlet-oxygen lifetime of a dilute aqueous solution from 3.3 μs (Egorov et al., 1989) to 1.65 μs is 7 mM. In contrast, within a cell, a 230-mM concentration of histidine would be required to reduce the singlet-oxygen lifetime from 0.1 μs to 0.05 μs. It is often not feasible to use quenchers in this high a concentration.

A similar problem occurs with the use of chemical traps. Chemical traps intercept a fraction of the singlet oxygen in the system being studied to produce a unique product made only from singlet oxygen (Fig. 1). The concentration of a chemical trap that is needed to trap singlet oxygen within a cell is roughly 30 times the concentration required to trap singlet oxygen in dilute aqueous solutions. Even if the trap were soluble at these high concentrations, the large concentration of trap would likely perturb the delicate biochemistry within the cell in an unpredictable manner.

For experiments in dilute aqueous solutions, replacement of water with deuterium oxide greatly increases the oxidation of target molecules by singlet oxygen (Fig. 1). In dilute aqueous solutions, almost all the singlet oxygen generated is quenched by solvent molecules. The singlet oxygen lifetime is 3.3 μs in water (Egorov et al., 1989), but 65 μs in deuterium oxide (Ogilby and Foote, 1982; Parker and Stanbro, 1984a). Thus, replacement of water solvent with deuterium oxide can result in up to a factor of 20 increase in the oxidation of target molecules by singlet oxygen. In contrast, deuterium oxide buffers are not useful for documenting singlet-oxygen generation within living cells. Inside the cell, the singlet-oxygen lifetime is of the order of 0.1 μs. Here almost all of the singlet oxygen will be quenched by biological molecules. Only 3% of the singlet oxygen will be quenched by water molecules. Thus, replacement of water with deuterium oxide will increase the oxidation of target molecules by a maximum of 3%.

Singlet oxygen can undergo a transition to the ground state with release of a photon. This transition is forbidden, but the development of highly sensitive near-infrared detectors has made measurements of the weak singlet-oxygen emission at 1270 nm one of the most useful techniques for detecting singlet oxygen in complex systems. Singlet-oxygen emission at 1270 nm has been detected from biochemical systems (Chemiluminescence) and from photochemical systems (phosphorescence) using either continuous or pulsed-laser excitation. The short singlet-oxygen lifetime within the living cell, however, severely limits the ability to detect singlet oxygen generated within the cell. For biochemical singlet-oxygen generation and for photochemical singlet-oxygen generation using a steady-state illumination, the intensity of the singlet-oxygen emission is proportional to the lifetime of singlet oxygen in the system being studied (Firey et al., 1988; Rodgers, 1988). Consequently, singlet-oxygen emission from cells will be roughly a factor of 30 less than that produced in dilute aqueous systems. This is generally below the detection limits of existing near-infrared spectrometers. An additional problem in photochemical systems is photosensitizer fluorescence, which may obscure the weak singlet-oxygen emission.

The short singlet-oxygen lifetime in cells also limits singlet-oxygen detection using laser excitation and time-resolved measurements of the near infrared emission. Depending on the oxygen concentration used, the singlet oxygen emission will either have a small intensity with a duration determined by the photosensitizer triplet lifetime or else have an extremely short duration so that it cannot be easily separated from the fluorescence of the photosensitizer (Parker and Stanbro, 1984b; Rodgers, 1988; Baker and Kanofsky, 1991).

Measurement of singlet oxygen generated at or near the cell surface is a much easier experimental problem than the measurement of singlet oxygen generated deep within the cell. In a cell suspension in buffer, a small fraction of the singlet oxygen generated near the surface of the cell is able to diffuse out into the buffer where the singlet-oxygen lifetime is long (Kanofsky, 1991; Baker and Kanofsky, 1991). For time-resolved studies, the long lifetime of the singlet-oxygen emission in the buffer is easily separated from photosensitizer fluorescence. The identification of singlet-oxygen emission from the buffer is also made easier because the emission is decreased by singlet-oxygen quenchers and increased with deuterium oxide solvent (Baker and Kanofsky, 1991). Kanofsky (1991) has detected 1270-nm emission from singlet oxygen that was generated in red-cell ghost membranes and subsequently diffused out into deuterium oxide buffer. Baker and Kanofsky (1991) have detected emission from singlet oxygen generated in the membrane of L1210 leukemia cells and subsequently diffusing out into the buffer.

Mathematical analysis of the data from cell surfaces is complex when compared to simple systems, however. It is necessary to set up and then solve the differential equations describing singlet-oxygen diffusion and singlet-oxygen quenching in a two-phase system (buffer and cell). The simplest form of these differential equations involves one linear dimension:

$$\frac{\partial[^1O_2]}{\partial t} = D_1 \frac{\partial^2[^1O_2]}{\partial x^2} - \frac{[^1O_2]}{\tau_1} \tag{2}$$

$$\frac{\partial[^1O_2]}{\partial t} = D_2 \frac{\partial^2[^1O_2]}{\partial x^2} - \frac{[^1O_2]}{\tau_2} \tag{3}$$

Here D_1 is the diffusion coefficient for singlet oxygen in the cell, D_2 is the diffusion coefficient for singlet oxygen in the buffer, τ_1 is the singlet-oxygen lifetime within the cell, and τ_2 is the singlet-oxygen lifetime in the buffer. Kanofsky (1991) found reasonable agreement between the kinetics of the singlet-oxygen emission from photosensitizer-labeled red cell ghosts and the numerical solutions to Equations 2 and 3.

Biochemical Production of Singlet Oxygen in Biological Systems

Production of singlet oxygen in reasonable yields by several model enzymatic systems is now well established (Kanofsky, 1983; Khan et al., 1983; Kanofsky and Axelrod, 1986; Kanofsky, 1988). Singlet oxygen is not a direct product of the enzymatic reactions, but rather a secondary reaction product. The two most carefully studied mechanisms for the production of singlet oxygen from model enzymatic systems are the reaction of hydrogen peroxide with hypohalous acid:

$$H_2O_2 + HOX \longrightarrow H^+ + X^- + H_2O + O_2(^1\Delta_g) \tag{4}$$

(where X is chlorine or bromine) and the reaction between two peroxyl radicals:

$$2\,RR'HCOO. \longrightarrow RR'CO + RR'CHOH + O_2(^1\Delta_g) \tag{5}$$

In both of these mechanisms, two highly reactive species must come together to make singlet oxygen. With the appropriate selection of conditions, some peroxidase/hydrogen peroxide/halide model systems can generate singlet oxygen in near stoichiometric yield (Kanofsky, 1984a, b; Khan, 1984; Kanofsky et al., 1984, 1988). The high yields of singlet oxygen in these systems is due to the lack of competing reactions. Within living cells, however, the singlet-oxygen precursors, hypohalous acid and hydrogen peroxide have many targets due to the extremely large concentration of biological molecules present. The same problem occurs for singlet oxygen generated from peroxyl radicals. Whereas model enzymatic systems may have singlet oxygen yields of the order of 10% (Kanofsky and Axelrod, 1986; Kanofsky, 1988) within a living cell, almost all of the peroxyl radicals generated would be expected to react with various biological molecules. Thus, it seems unlikely that the reaction of hypohalous acid with hydrogen peroxide or the reaction between two peroxyl radicals can compete with the other reaction consuming these singlet-oxygen precursors.

Indeed, only one type of living cell, the human eosinophil, produces detectable quantities of singlet oxygen (Kanofsky et al., 1988). Eosinophils are phagocytic cells that are felt to be important for the defense against the larva of parasitic worms and to be important in the pathogenesis of some allergies. Fig. 2 shows the time course of singlet-oxygen emission from human eosinophils, stimulated with phorbol myristate acetate. Equations 6 through 9 give the mechanism for the singlet-oxygen generation:

$$2O_2 + \mathrm{NADPH} \stackrel{NADPHoxidase}{\longrightarrow} \mathrm{NADP}^+ + H^+ + 2O_2{\cdot}^- \qquad (6)$$

$$2H^+ + 2O_2{\cdot}^- \stackrel{superoxide\ dismutase}{\longrightarrow} H_2O_2 + O_2 \qquad (7)$$

$$H_2O_2 + H^+ + Br^- \stackrel{eosinophil\ peroxidase}{\longrightarrow} \mathrm{HOBr} + H_2O \qquad (8)$$

$$H_2O_2 + \mathrm{HOBr} \longrightarrow H^+ + Br^- + H_2O + O_2(^1\Delta_g) \qquad (9)$$

With human eosinophils, several unusual factors contribute to the efficiency of singlet-oxygen generation. First, these cells have an extremely active respiratory burst, generating 20 fmol min^{-1} cell^{-1} of superoxide anion (Kanofsky et al., 1988). Second, eosinophils generate hypobromous acid rather than hypochlorous acid (Weiss et al., 1986). The reaction of hydrogen peroxide with hypobromous acid is much faster

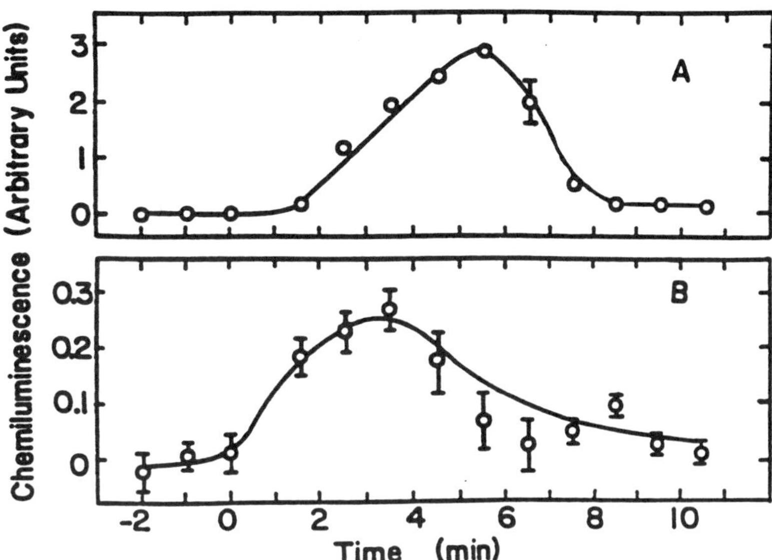

Figure 2. Time course of 1268-nm chemiluminescence from human eosinophils. Eosinophils, 10^7/ml, were suspended in Dulbecco's phosphate-buffered saline made with deuterium oxide in place of light water. Phorbol myristate acetate, 10 μg/ml, as added at time 0. Acetonitrile (0.5%, v/v) was present. Error limits are shown only when they exceed the size of the symbol. **A**: Sodium bromide replacing sodium chloride in Dulbecco's buffer. Average of three experiments. **B**: Sodium bromide, 100 μM, added to Dulbecco's buffer. Average of five experiments. Reprinted with permission of the American Society for Biochemistry and Molecular Biology from Kanofsky JR, et al. (1988): Singlet oxygen production by human eosinophils. *J Biol Chem* 263:9692–9696.

than the reaction of hydrogen peroxide with hypochlorous acid. Thus, the hydrogen peroxide–hypobromous acid reaction is better able to compete with other reactions consuming hypobromous acid or hydrogen peroxide. Third, while hypobromous acid reacts rapidly with amino acids, the products of these reactions retain the ability to react with hydrogen peroxide to form singlet oxygen (Kanofsky, 1989a). Finally, the human eosinophils secrete the hydrogen peroxide and the hypobromous acid into the extracellular space or into phagocytic vacuoles. In these microscopic environments, the concentration of biological molecules is much lower than in the cytoplasm. Despite all of these factors favoring the formation of singlet oxygen by human eosinophils, the singlet oxygen generated by human eosinophils is only 0.4% of the available oxygen (Kanofsky et al.,

1988). In view of the much larger concentrations of hydrogen peroxide, superoxide anion, and hypobromous acid generated by the eosinophil, it is not clear that singlet oxygen makes a significant quantitative contribution to the toxic oxidants produced by the eosinophil.

Recently, Kanofsky and Sima (1991) found that ozone reacts with certain biological molecules to generate singlet oxygen. Consistent with earlier studies using simple organic compounds, biological molecules containing either a sulfhydryl group or a nonaromatic carbon–carbon double bond in a ring structure generated singlet oxygen in high yield. Figure 3 shows the near-infrared singlet-oxygen emission from the reaction of ozone with cysteine and with NADPH. Uric acid, ascorbic acid, NADH, and NADPH each contain a nonaromatic carbon–carbon double bond in a ring structure and were proposed to react with ozone by the following mechanism:

$$\mathrm{R_1R_2C = CR_3R_4 + O_3 \longrightarrow R_1R_2\overset{\overset{\Large O}{/\quad\backslash}}{C - C}R_3R_4 + O_2(^1\Delta_g)} \tag{10}$$

Cysteine, glutathione, and albumin each contain a sulfhydryl group and were proposed to react with ozone by the following mechanism:

$$\mathrm{RSH + O_3 \longrightarrow \overset{\overset{O}{\|}}{RSH} + O_2(^1\Delta_g)} \tag{11}$$

The sulfhydryl-containing amino acids, peptides, and proteins are quantitatively the most important ozone targets generating singlet oxygen because their concentration within cells is high and because ozone reacts with sulfhydryl compounds at near diffusion-controlled rates (Pryor et al., 1984).

The mechanisms for singlet-oxygen generation from the ozone-biomolecule reactions differ in a fundamental way from the other biochemical mechanism for singlet-oxygen generation discussed earlier. For the ozone reactions only one highly reactive species is required by the elementary reaction generating singlet oxygen. Thus, the amount of singlet oxygen produced within the cell should vary linearly with the ozone concentration. Significant amounts of singlet oxygen may be generated in pulmonary tissue even with the low concentrations of ozone present in the urban atmosphere. In contrast, the amount of singlet oxygen generated in cells from the hydrogen peroxide–hypohalous acid reaction and

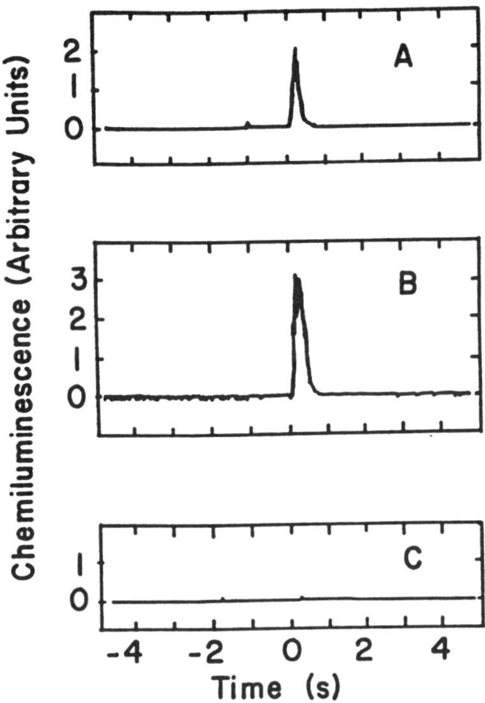

Figure 3. Chemiluminescence at 1270 nm from the reaction of ozone with cysteine and with NADPH. Conditions were 200 μM ozone, pH 7.0, 50 mM sodium phosphate, 0.5 mM sodium perchlorate. **A**: Cysteine, 200 μM, average of nine experiments. **B**: NADPH, 200 μM, average of five experiments. **C**: Control, no organic material added, average of four experiments. Reprinted with permission of the American Society for Biochemistry and Molecular Biology from Kanofsky JR, Sima P (1991): Singlet oxygen production from the reactions of ozone with biological molecules. *J Biol Chem* 266:9039–9042.

from the reactions of peroxyl radicals will have a quadratic dependence on the reactant concentrations. At the very low concentrations of reactants present in cells, the amount of singlet oxygen generated will be insignificant.

Much additional work is still required to establish a role for singlet oxygen as a mediator of ozone toxicity, however. Not all biomolecules react with ozone to generate singlet oxygen. Experiments with intact plants and animals and using ozone concentrations of one part per million or less will be needed to establish the amount of singlet oxygen actually generated *in vivo*.

Photochemical Production of Singlet Oxygen in Biological Systems

Singlet oxygen is believed to be the major mediator of biological damage from a number of photosensitizers, including porphyrins, phthalocyanines, and halogenated fluoresceins. Singlet oxygen is produced by a reaction sequence called a Type II mechanism:

$$\mathrm{P} + \mathrm{h}\nu \longrightarrow \mathrm{P}^* \tag{12}$$

$$^1\mathrm{P}^* \longrightarrow {}^3\mathrm{P}^* \tag{13}$$

$$^3\mathrm{P}^* + \mathrm{O}_2(\Sigma_g^-) \longrightarrow \mathrm{P} + \mathrm{O}_2(^1\Delta_g) \tag{14}$$

Here P is ground state photosensitizer, $^1P^*$ is photosensitizer in an excited singlet state, and $^3P^*$ is photosensitizer in an excited triplet state.

Most of the evidence for singlet-oxygen generation from photosensitizers comes from studies using pure solvents, micelles, or liposomes (all model systems in which the lifetime of singlet oxygen is relatively long). In many cases, production of a long-lived triplet state in high yield has been documented by transient absorption spectroscopy (Bonnett et al., 1983; Firey and Rodgers, 1987; Keir et al., 1987; Land et al., 1988). Photosensitizer in these long-lived triplet states has frequently been shown to react with ground state oxygen at near diffusion-controlled rates to generate singlet oxygen (Bonnett et al., 1983; Firey and Rodgers, 1987, Land et al., 1988).

The Type II mechanism is always in competition with the Type I mechanism (Foote, 1968, 1987). In the Type I mechanism, excited singlet and triplet states of the photosensitizer react directly with biological molecules to produce radicals or radical ions. Hydrogen abstraction and electron transfer are common mechanisms:

$$\mathrm{P}^* + \mathrm{HB} \longrightarrow \mathrm{PH}\cdot + \mathrm{B}\cdot \tag{15}$$

$$\mathrm{P}^* + \mathrm{HB} \longrightarrow \mathrm{P}\cdot^+ + \mathrm{HB}^-\cdot \tag{16}$$

$$\mathrm{P}^* + \mathrm{HB} \longrightarrow \mathrm{P}\cdot^- + \mathrm{HB}\cdot^+ \tag{17}$$

Here HB is a protonated biological compound, B is the biological molecule after abstraction of a hydrogen atom, P* is the photosensitizer in an excited singlet or triplet state, and P is the ground state photosensitizer. From these competing mechanisms, it can be deduced that high oxygen concentrations will favor a Type II mechanism whereas high concentrations of biological molecules would favor a Type I mechanism. *In vivo* there is an enormous concentration of biological molecules, but only a moderate concentration of oxygen. Thus, one cannot conclude that high singlet-oxygen yields will occur *in vivo* just because high singlet-oxygen yields occur in simple model systems.

Some confusion in this area also results from this misinterpretation of experimental results. First, lack of an oxygen requirement for biological damage rules out a Type II mechanism. However, an oxygen requirement for biological damage does not establish a Type II mechanism. The radicals generated in the Type I mechanism may react with oxygen as part of the mechanism of photochemical damage:

$$R\cdot + \mathrm{O_2} \longrightarrow \mathrm{RO_2}\cdot \tag{18}$$

Thus, oxygen may enhance cytotoxicity for some Type I mechanisms. Second, as discussed earlier, no deuterium oxide solvent effect is expected for singlet oxygen generated within the cell. The use of deuterium oxide solvent, however, would be expected to perturb cell biochemistry in a complex manner and may enhance cell cytotoxicity by mechanisms other than the direct prolongation of the lifetime of singlet oxygen within the cell.

There are a few studies that do support the production of singlet oxygen from photosensitizers in cell suspensions. Truscott et al. (1988) studied the photosensitization of BHK21 fibroblasts labeled with hematoporphyrin. These investigators were able to detect the excited triplet state of hematoporphyrin derivative within the BHK21 cells by absorption spectroscopy following pulsed laser excitation. In nitrogen-flushed cell suspensions, the lifetime of the triplet state was greater than 17 μs. This means that, despite the large concentration of biological molecules, the triplet state lifetime was sufficiently long to react with ground state oxygen. Indeed, in air-saturated cell suspensions, the triplet lifetime decreased to 8 μs. The authors concluded that 50% to 60% of triplet hematoporphyrin derivative reacted with oxygen in air-saturated cell suspensions and that the rate constant for the reaction between oxygen and triplet hematoporphyrin was 4×10^8 $\mathrm{M^{-1}s^{-1}}$. These results strongly favor a Type II photosensitization mechanism in this cell system. Truscott et al. were unable to detect a 1270-nm singlet-oxygen emission from their cell suspension, but this was almost certainly due to technical limitations of their near-infrared spectrometer.

Firey et al. (1988) obtained a similar result. These investigators studied the photoexcitation of zinc phthalocyanine in mouse myeloma cells. Following pulsed laser excitation, they were able to detect the triplet zinc phthalocyanine within the cells using absorption spectroscopy, but were not able to detect any singlet-oxygen emission at 1270 nm.

Another method to potentially measure singlet-oxygen emission from living cells is to excite the photosensitizer with light modulated at high frequency and then search for the expected phase-shifted singlet-oxygen emission using a synchronous detection system (lock-in amplifier). This approach has had conflicting results. Parker (1987, 1990) detected weak emission at 1260 nm from a tumor-bearing mouse, but Patterson et al. (1990), using a similar apparatus, were unable to detect any singlet-oxygen emission from solid tumors or cell suspensions.

Recently, Baker and Kanofsky (1991) were able to detect singlet oxygen emission at 1270 nm from L1210 leukemia cells labeled with purified hematoporphyrin derivative. These investigators used pulsed-laser excitation, an ultrasensitive near-infrared spectrometer, and selected conditions that favored photosensitizer accumulation in the cell membrane rather than the cytoplasm. A small amount of the singlet oxygen generated in the cell membrane was able to diffuse out of the cell membrane into the deuterium-oxide buffer used for these experiments. The long lifetime of the singlet-oxygen emission from the deuterium-oxide buffer distinguished it from the fluorescence of the hematoporphyrin derivative. Figure 4 shows the decay of the 1270-nm emission from the L1210 leukemia cells. The detection of singlet-oxygen emission from membrane-labeled cells is a much easier technical problem than the detection of singlet-oxygen emission from cells with photosensitizer in the cytoplasm.

Conclusions

Significant biochemical production of singlet oxygen is uncommon. Preliminary studies suggest that singlet oxygen may be an important intermediate for ozone toxicity in plants and animals (Kanofsky and Sima, 1991), but additional work using intact plants and animals will be necessary before a definite role for singlet oxygen is established. Photochemical generation of singlet oxygen from photosensitizers within living cells appears to be much more common, although the experimental evidence using intact living cells is limited. Recent studies have shown the excited triplet states of a few photosensitizers react only slowly with biological

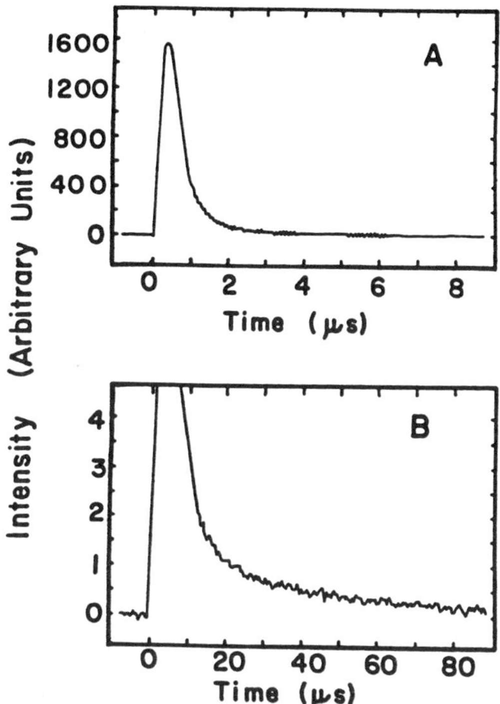

Figure 4. Time course of 1270-nm emission from L1210 leukemia cells incubated with polyporphyrin and then exposed to light from a pulsed laser. Conditions were 510-nm excitation wavelength, 15-mJ laser pulse energy, 300-ns laser pulse width, 2×10^7 cells/ml, phosphate-buffered saline made with deuterium oxide solvent, 37°C, average of nine experiments. Note the differences in time scale and sensitivity scale between **A** and **B**. Reprinted with permission of Academic Press from Baker A, Kanofsky JR (1991): Direct observation of singlet oxygen phosphorescence at 1270 nm from L1210 leukemia cells exposed to polyporphyrin and light. *Arch Biochem Biophys* 286:70–75.

molecules, but react rapidly with oxygen (Truscott et al., 1988; Firey et al., 1988). This fact is responsible for the relatively common photochemical generation of singlet oxygen.

Acknowledgments. This work was supported by the Department of Veterans Affairs and by the Potts Estate. I thank Anita Osis for help with preparation of the manuscript.

References

Baker A, Kanofsky JR (1992): Quenching of singlet oxygen by biomolecules from L1210 leukemia cells. *Photochem Photobiol* 55:523–528.

Baker A, Kanofsky JR (1991): Direct observation of singlet oxygen phosphorescence at 1270 nm from L1210 leukemia cells exposed to polyporphyrin and light. *Arch Biochem Biophys* 286:70–75.

Bonnett R, Lambert C, Land EJ, Scourides PA, Sinclair RS, Truscott TG (1983): The triplet and radical species of haematoporphyrin and some of its derivatives. *Photochem Photobiol* 38:1–8.

Dahl TA (1991): Direct exposure of mammalian cells to pure gas-phase singlet oxygen ($^1\Delta$ O_2). *Photochem Photobiol* 53(Suppl):119S.

Dahl TA, Midden WR, Hartman PE (1987): Pure singlet oxygen cytotoxicity for bacteria. *Photochem Photobiol* 46:345–352.

Egorov SY, Kamalov VF, Koroteev NI, Krasnovskii AA Jr., Toleutaeu BN, Zinukou SU (1989): Rise and decay kinetics of photosensitized singlet oxygen luminescence in water. Measurements with nanosecond time-correlated single photon counting technique. *Chem Phys Lett* 163:421–424.

Firey PA, Jones TW, Jori G, Rodgers MAJ (1988): Photoexcitation of zinc phthalocyanine in mouse myeloma cells: the observation of triplet states, but not of singlet oxygen. *Photochem Photobiol* 48:357–360.

Firey PA, Rodgers MAJ (1987): Photoproperties of a silicon naphthalocyanine: a potential photosensitizer for photodynamic therapy. *Photochem Photobiol* 45:535–538.

Foote CS (1968): Mechanisms of photosensitized oxidation. *Science* 162:963–970.

Foote CS (1981): Photooxidation of biological model compounds. In: *Oxygen and Oxy-Radicals in Chemistry and Biology*, Rodgers MAJ, Powers EL, eds., New York: Academic Press.

Foote CS (1987): Type I and type II mechanisms of photodynamic action. In: *Light-Activated Pesticides (ACS Symposium Series 339)*, Heitz JR, Downum KR, eds., Washington, DC: American Chemical Society.

Kanofsky JR (1983): Singlet oxygen production by lactoperoxidase. Evidence from 1270 nm chemiluminescence. *J Biol Chem* 258:5991–5993.

Kanofsky JR (1984a): Singlet oxygen production by chloroperoxidase-hydrogen peroxide-halide systems. *J Biol Chem* 259:5596–5600.

Kanofsky JR (1984b): Singlet oxygen production by lactoperoxidase: halide dependence and quantitation of yield. *J Photochem* 25:105–113.

Kanofsky JR (1988): Singlet oxygen production from the peroxidase-catalyzed oxidation of indole-3-acetic acid. *J Biol Chem* 263:14171–14175.

Kanofsky JR (1989a): Bromine derivatives of amino acids as intermediates in the peroxidase-catalyzed formation of singlet oxygen. *Arch Biochem Biophys* 274:229–234.

Kanofsky JR (1989b): Singlet oxygen production by biological systems. *Chem Biol Interact* 70:1–28.

Kanofsky JR (1990): Peroxidase-catalyzed generation of singlet oxygen and of free radicals. In: *Peroxidases; Chemistry and Biology*, Everse J, Everse K, Grisham MB, eds., Boca Raton: CRC Press, Inc.

Kanofsky JR (1991): Quenching of singlet oxygen by human red cell ghosts. *Photochem Photobiol* 53:93–99.

Kanofsky JR, Axelrod B (1986): Singlet oxygen production by soybean lipoxygenase isozymes. *J Biol Chem* 261:1099–1104.

Kanofsky JR, Hoogland H, Wever R, Weiss SJ (1988): Singlet oxygen production by human eosinophils. *J Biol Chem* 263:9692–9696.

Kanofsky JR, Sima P (1991): Singlet oxygen production from the reactions of ozone with biological molecules. *J Biol Chem* 266:9039–9042.

Kanofsky JR, Wright J, Miles-Richardson GE, Tauber AI (1984): Biochemical requirements for singlet oxygen production by purified human myeloperoxidase. *J Clin Invest* 74:1489–1495.

Keir WF, Land EJ, MacLennan AH, McGarvey DJ, Truscott TG (1987): Pulsed radiation studies of photodynamic sensitizers: the nature of DHE. *Photochem Photobiol* 46:587–589.

Khan AU (1984): Myeloperoxidase singlet molecular oxygen generation detected by direct infrared electronic emission. *Biochem Biophys Res Commun* 122:668–675.

Khan AU, Gebauer P, Hager LP (1983): Chloroperoxidase generation of singlet Δ molecular oxygen observed directly by spectroscopy in the 1- to 1.6 μm region. *Proc Natl Acad Sci USA* 80:5195–5197.

Land EJ, McDonagh AF, McGarvey DJ, Truscott TG (1988): Photophysical studies of tin (IV)-protoporphyrin: potential phototoxicity of a chemotherapeutic agent proposed for the prevention of neonatal jaundice. *Proc Natl Acad Sci USA* 85:5249–5253.

Moan J, Berg K (1991): The photodegradation of porphyrins in cells can be used to estimate the lifetime of singlet oxygen. *Photochem Photobiol* 53:549–553.

Moan J, Pettersen EO, Christensen T (1979): The mechanism of photodynamic inactivation of human cells in vitro in the presence of haematoporphyrin. *Br J Cancer* 39:398–407.

Ogilby PR, Foote CS (1982): Chemistry of singlet oxygen. 36. Singlet molecular oxygen ($^1\Delta_g$) luminescence in solution following pulsed laser excitation. Solvent deuterium isotope effects on the lifetime of singlet oxygen. *J Am Chem Soc* 104:2069–2070.

Parker JG (1987): Optical monitoring of singlet oxygen generation during photodynamic treatment of tumors. *IEEE Circuits and Devices Magazine* 10–21.

Parker JG (1990): Factors of importance in the generation of singlet oxygen during photo-dynamic treatment of tumors. *Society Photo Optical Instrumentation Engineers (Photodynamic Therapy: Mechanisms II)* 1203:19–31.

Parker JG, Stanbro WD (1984a): Optical determination of the rates of formation and decay of O_2 ($^1\Delta_g$) in H_2O, D_2O and other solvents. *J Photochem* 25:545–547.

Parker JG, Stanbro WD (1984b): Dependence of photosensitized singlet oxygen production on porphyrin structure and solvent. In: *Porphyrin Localization and Treatment of Tumors (Progress in Clinical & Biological Research Ser: Vol 170)*, Doiron DR, Gomer CJ, eds., New York: Alan R. Liss.

Patterson MS, Madsen SJ, Wilson BC (1990): Experimental tests of the feasibility of singlet oxygen luminescence monitoring *in vivo* during photodynamic therapy. *J Photochem Photobiol B:Biol* 5:69–84.

Pryor WA, Giamalva DH, Church DF (1984): Kinetics of ozonation. 2. Amino acids and model compounds in water and comparisons to rates in nonpolar solvents. *J Am Chem Soc* 106:7094–7100.

Rodgers MAJ (1988): On the problems involved in detecting luminescence from singlet oxygen in biological systems. *J Photochem Photobiol B:Biol* 1:371–378.

Rougee M, Benasson RV (1986): Determination of the decay rate constants of singlet oxygen ($^1\Delta_g$) deactivation in the presence of biomolecules. *C R Acad Sci Paris* 302, Ser. II:1223–1226.

Straight RC, Spikes JD (1985): Photosensitized oxidation of biomolecules. In: *Singlet O_2, Vol. 4*, Frimer AA, ed. Boca Rotan, FL: CRC Press.

Truscott TG, McLean AJ, Phillips AMR, Foulds WS (1988): Detection of haematoporphyrin derivative and haematoporphyrin excited states in cell environments. *Cancer Lett* 41:31–35.

Wefers H, Schulte-Frohlinde D, Sies H (1987): Loss of transforming activity of plasmid DNA (pBR322) in E. coli caused by singlet molecular oxygen. *Fed Eur Biochem Soc Lett* 211:49–52.

Weiss SJ, Test ST, Eckmann CM, Roos D, Regiani S (1986): Brominating oxidants generated by human eosinophils. *Science* 234:200–203.

Chapter 5

In Vivo Detection of Oxygen Free Radical Species

Robert A. Floyd

Over the last several years an increasing amount of research has implicated the importance of oxygen free radicals in the etiology of several diseases and pathological conditions. Proof of the importance of oxygen free radicals and the oxidative damage they initiate depends on methods that will unequivocally establish not only the presence of oxygen free radicals but a clear association of their formation with the induced oxidative damage and a clear-cut relationship with the induction or progression of the disease or pathological condition. This has been a difficult research area where newer methods have begun to yield results. This brief chapter has three aims: define the problems facing the experimentalist in this area, present basic concepts underlying newer methods available, and summarize recent results obtained on a model system. The results clearly add credence to the notion that oxygen free radicals and oxidative damage are etiological agents in the development of tissue injury caused by oxidative insults.

Oxygen and Oxidative Stress

Life first developed in the absence of oxygen. However, oxygen became omnipresent soon after the appearance of photosynthetic activity about 3 to 4 billion years ago and thus biological systems had to adapt to its presence. Fortunately, oxygen does not react with (i.e., directly oxidize) most biological molecules. This is because molecular oxygen is in a triplet spin state in its ground energy state. In contrast, most biological molecules are in a singlet state in their lowest energy state

Oxygen Free Radicals in Tissue Damage
Merrill Tarr and Fred Samson, Editors

and thus do not directly react with oxygen (Floyd, 1990a). Life adapted to the presence of oxygen by using it as an acceptor for reductants and evolved the process of respiration; in so doing it thus became possible to capture more energy (ATP) than was previously possible. Thus, the two extreme reduction states of oxygen (i.e., molecular oxygen and water) are common in biological systems, but the "semireduced" species of oxygen are also present, in very low levels, and they are responsible for the oxidative damage that occurs. Thus, superoxide, hydrogen peroxide, and hydroxyl free radicals are formed in biological systems. In addition, lipid peroxidation products as well as other species are formed during oxidative damage, which also presents a danger to biological systems. Thus, aerobic systems experience an oxidative damage potential, P_o, applied at all times.

Nature has evolved systems to protect itself from the applied oxidative damage potential. The protective system that has evolved as an antioxidant defense capacity, A_c, consists of several enzymes and antioxidant molecules. The protective enzymes superoxide dismutase (SOD), catalase (CAT), and glutathione peroxidase (GSHPx), in combination with the antioxidant molecules vitamin E, ascorbate, glutathione, and uric acid, represent the antioxidant defense capacity of the system. The oxidative damage potential is opposed by the antioxidant defense capacity. These processes are in dynamic equilibrium with each other but occur such that they favor the oxidative damage potential, as represented in Fig. 1. That is, the balance between the two is such that a small amount of oxidative damage occurs constantly and thus a small amount of oxygen free radical species is able to bypass the antioxidant defense capacity and present a danger to the system at all times. In addition, certain conditions imposed on the system cause a large increase in the amount of oxygen free radicals produced, thus causing an enhanced amount of oxidative damage.

Oxygen Free Radicals In Vivo: The Problem

It is important to realize that the amount of oxygen free radicals present at any specific time in tissue is extremely small. It is possible to make an informed estimate of the amount of oxygen free radicals that may be present in the brain. It has been noted that brain uses about 20% of the total oxygen demand of the body, consuming about 50 ml/min for a weight of about 1.4 kg (Frisell, 1982). If we assume that oxygen has a half-life in brain of 1 s, then the concentration of oxygen

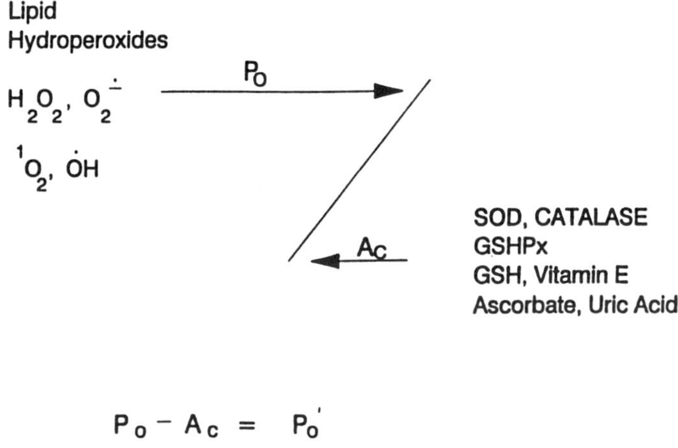

Figure 1 Diagrammatic representation that the oxidative damage potential, P_o, is experienced by aerobic organisms at all times and that this is opposed by the antioxidant defense capacity, A_c, of the system with the ultimate result that a small amount of oxidant species, P_o', is being produced (i.e., not quenched) at all times. Reprinted with permission of Little Brown and Company, from *Annals of Neurology* (1992), 32:S22–S27.

in the tissue would be about 26 μM. If one assumes that only one out of every thousand oxygen molecules consumed by the brain becomes a reactive semireduced oxygen molecule and that its half-life is 1 s, then the instantaneous concentration would be 25 μM.

Certainly a half-life of 1 s is much longer than would be expected for most oxygen free radicals; that is, hydroxyl free radicals react at diffusion limited rates so their half-life would be a few nanoseconds at most, and therefore one could safely state that 10^{-11}M would probably be an overestimate of the instantaneous amount of oxygen free radicals present at any specific time. This low level of oxygen free radicals represents a major analytical problem, for it is impossible to directly detect 10^{-11}M oxygen free radicals. Not only is this amount extremely low, but oxygen free radicals have nondistinctive magnetic and spectroscopic properties. Therefore, it is necessary to use indirect methods to ascertain their presence.

The reactions governing the amount of oxygen free radicals present in tissue are summarized by the following equations, where P_o' refers to reactive oxygen species and B and B_{ox} refer to a biological molecule and an oxidized biological molecule respectively:

$$O_2 + \text{tissue} \xrightarrow{k_o} P'_o \tag{1}$$

$$P'_o + B \xrightarrow{k_b} B_{ox} \tag{2}$$

If $k_b > k_o$, then the amount of oxygen free radicals present will be much lower than their rate of formation would predict. The difference between P_o and A_c predicts a flux of oxygen free radicals:

$$P_o - A_c = P'_o \tag{3}$$

Thus, P'_o refers to the flux of oxygen free radicals in biological systems.

General Approaches to Detect Oxygen Free Radicals and Assess Oxidative Damage

The fact that the amount of oxygen free radicals present at any specific time is very low has made it necessary to use methods other than direct detection. Two general approaches that have been used include (a) use of exogenous traps that react with oxygen free radicals to produce stable and unique products, and (b) measurement of the unique products produced when oxygen free radicals react with biological molecules.

Conditions that are necessary when exogenous traps are used include the requirement that the trap reacts rapidly with the oxygen free radicals produced. Ideally, the trap must react and compete with biological molecules:

$$P'_0 + \text{Trap} \xrightarrow{k_T} \text{Trap}_{ox} \tag{4}$$

If the Trap_{ox} is stable and if $k_T > k_b > k_o$, then it is possible to have a buildup of Trap_{ox} such that it can be measured some time after the trap has been added to the biological system. It is necessary to be able to measure Trap_{ox} very sensitively. It is also necessary that the trap or Trap_{ox} be nontoxic and/or does not alter biological function in the concentrations needed to assess the free radical formation rate.

Figure 2 shows two exogenous traps that we have used and the reactions they undergo when trapping free radicals in biological systems. Salicylate reacts at diffusion limited rates with hydroxyl free radicals and is relatively nontoxic in the amounts used (i.e., 50 mg/kg, administered

I.P. 20 min before an oxidative insult). The hydroxylation products 2, 3- and 2, 5-dihydroxy benzoic acid (DHBA) as well as salicylate can be quantitated using HPLC with an electrochemical detector and a fluorescence detector, respectively (Floyd et al., 1986a, 1990). As little as 2.0 pmol of DHBA can be measured with an electrochemical detector system (Floyd et al., 1984). DHBA is formed at a level of about 0.1% of the amount of salicylate present.

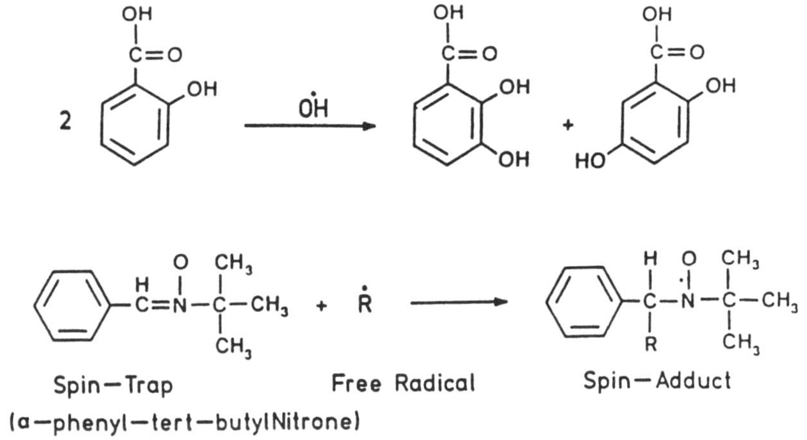

Figure 2 Equations demonstrating the two traps we have most commonly used to assess free radical production in biological systems. Salicylate reacts with hydroxyl free radicals at a diffusion limited rate to yield the 2, 3- and 2, 5-dihydroxy benzoic acid products. The spin-trap, α-phenyl-*tert*-butyl nitrone (PBN), reacts with free radical $\dot{R}$ to yield a relatively stable spin-adduct, which is a nitroxide. Reprinted with permission of Little Brown and Company, from *Annals of Neurology* (1992), 32:S22–S27.

PBN reacts with certain free radicals to yield products, termed spin-adducts, that are relatively stable nitroxides in most cases, depending on the particular free radicals that the spin trap has reacted with. Thus, with time there is usually a build-up of the amount of spin-adduct(s) such that it is possible using electron paramagnetic resonance (EPR) to ascertain the amount present as well as ideally identify the particular free radicals that have been trapped. In some cases the spin-adduct(s) can be assayed within the biological matrix, but in most cases they must be extracted and then characterized in an organic solvent extract.

Oxidative damage to proteins occurs during the normal oxidative stress experienced constantly by aerobic systems (Stadtman, 1990). They are damaged much more during an increased oxidative stress condition, such as for instance during an ischemia/reperfusion insult (IRI). These notions have been reviewed previously (Stadtman, 1990). In general, oxidative damage to proteins involves the combined action of the trace metals Fe and/or Cu and hydrogen peroxide (Stadtman, 1990). The metal binding site on the protein will in general dictate the specific amino acids that may be damaged by the oxidatively damaging species released by the reaction of hydrogen peroxide with Fe or Cu. An excellent example is the enzyme glutamine synthetase, whose activity is sensitive to oxidative damage (Stadtman, 1990). Oxidative damage to glutamine synthetase (GS) causes oxidation of histidine-269, which is located near the enzymes active center, which contains Fe.

In general, the amino acids arginine, histidine, proline, and lysine are the most sensitive to oxidative damage (Stadtman, 1990). Oxidative damage causes oxidation of specific amino acids to yield an increased carbonyl content of the protein. The increased carbonyl content can be analyzed by several different methods (Levine et al., 1990).

Oxidized proteins are degraded to individual amino acids by a unique class of proteases termed "neutral proteases," discovered in 1985 by Rivett (1985). These proteases appear to be the same as the so-called macro-oxy protease characterized by Davies and colleagues (Pacifici et al., 1989). Figure 3 shows the general relationship of protein oxidation resulting in an increased carbonyl content and the degradation of oxidized proteins by neutral proteases.

Nucleic acids are damaged by oxygen free radicals. Nucleic acid strand breakage is the most frequent lesion caused by oxygen free radical species. Superoxide (Brawn and Fridovich, 1981) and hydroxyl free radicals (Schneider et al., 1988, 1989) have been shown to cause plasmid DNA strand breakage. In addition to strand breakage, chemical modifications of the nucleic acid bases also may occur. The base modification products most studied in association with oxidative damage to DNA and RNA include 5-hydroxymethyl uracil, thymine glycol, and 8-hydroxyguanine. Oxidation of guanine in DNA and RNA to form 8-hydroxyguanine has been an intense area of research recently. The presence of this modified base in DNA has been shown to increase under several conditions that also lead to tumor development (Floyd, 1990b). Hydroxyl free radicals (Floyd et al., 1988) as well as singlet oxygen

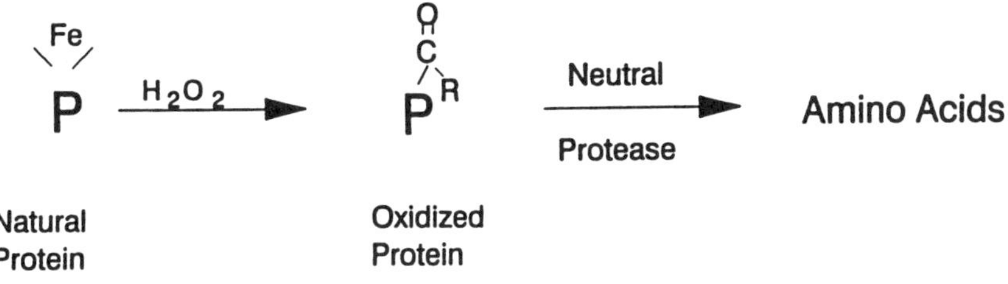

Figure 3 Representation of the oxidation of proteins by the action of Fe and H_2O_2 to yield oxidized proteins which have an increased level of carbonyl groups present. Neutral protease(s) degrades oxidized proteins. Reprinted with permission of Little Brown and Company, from *Annals of Neurology* (1992), 32:S22–S27.

(Floyd et al., 1989) mediate formation of 8-hydroxyguanine in DNA.

Analysis of the presence of 8-hydroxyguanine in DNA and RNA has been used as an indicator of oxidative damage to these macromolecules. This oxidized base has been studied intensely recently, because it can now be measured very sensitively. Thus, the development of the use of HPLC with electrochemical detection (HPLC-ED) to analyze sensitively for the nucleoside (Floyd et al., 1990) has been a great advantage in this area. It is possible using HPLC-ED to detect 20 fmol of 8-hydroxy-2′-deoxyguanosine (8-OHdG) with a signal to noise of 3 (Floyd et al., 1986b). This means that with 10 μg of DNA it is possible to determine as little as one 8-OHdG/10^5dG. This amount of 8-OHdG seems to be a normal lower limit value observed in most biological systems (Floyd, 1990c). Some of the problems that one must address using this methodology has been presented recently (Floyd et al., 1990).

Results Obtained in Brain Ischemia/Reperfusion Model

We have applied several different methodologies of assessing oxidative damage to the Mongolian gerbil brain undergoing ischemia/reperfusion-mediated injury (Cao et al., 1988, Oliver et al., 1990). The virtues of this animal model is that ligation of the two common carotids causes cessation of blood flow into the forebrain but in contrast it does not alter blood flow into the cerebellum or brain stem, thus allowing comparisons between brain regions to assess region specificity of the damage brought on by an ischemia/reperfusion insult (IRI). Figures 4 and 5 provide summaries of the original data obtained. The following methodologies, (a) salicylate trapping of hydroxyl free radicals, (b) spin-trapping of free radicals using PBN, (c) assessing protein oxidative damage by measuring protein carbonyl content, and (d) measuring the loss of glutamine synthetase activity, have all been used in the gerbil brain model (Cao et al., 1988, Oliver et al., 1990).

Figure 4 shows the summary of results on specific brain regions of gerbils given an ischemia (10 min)/reperfusion (60 min) insult. The data clearly show that brain cortex of IRI-treated gerbils underwent oxidative damage as assessed by salicylate hydroxylation, protein oxidation and loss of GS activity; but in contrast, brain stem, which is not perfused by the common carotids, did not show oxidative damage. Salicylate hydroxylation as well as protein oxidation nearly doubled in cortex of IRI-treated animals as compared to sham-operated controls, whereas IRI treatment caused no change in brain stem. Protein oxidation essentially doubled in cortex but no change was noted in the brain stem of IRI-lesioned animals when they were compared to sham-operated controls. The activity of glutamine synthetase (GS) fell about one third in IRI-lesioned animals when compared to controls, but no change was noted in the brain stem of IRI-lesioned gerbils.

In the course of these investigations, we have made a surprising discovery, namely, that the spin-trapping agent, PBN, offers protection from the oxidative damage that occurs during an IRI-induced lesion to gerbil brain. Data illustrating this point are presented in summary form in Fig. 5. Thus, PBN administration (300 mg/kg, I.P.) 20 min before an IRI prevented the lethality caused by a long ischemia period. Ischemia was more lethal to old gerbils than younger gerbils, but PBN protected both young and old animals. PBN also diminished the IRI-induced increase in protein oxidation and the decrease in GS activty in brain cortex.

Summary and Future Possibilities

It is clear that as ever more sensitive methodologies are brought to bear on biomedical problems, a clear-cut role for oxygen free radicals and oxidative damage is becoming evident. The surprising discovery that the

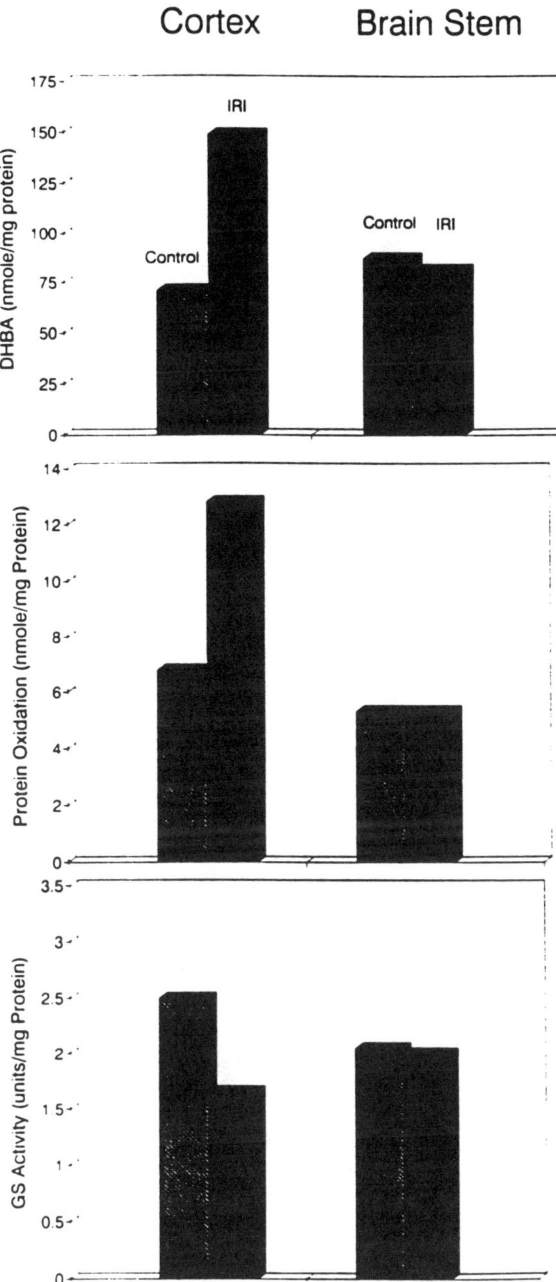

Figure 4 A summary presentation of research results (Cao et al., 1988, Oliver et al., 1990) that demonstrates that brain cortex is oxidatively damaged during earlier (Floyd and Carney, 1992) and is reproduced here with permission of the publisher. Reprinted with permission of Little Brown and Company, from *Annals of Neurology* (1992), 32:S22–S27.

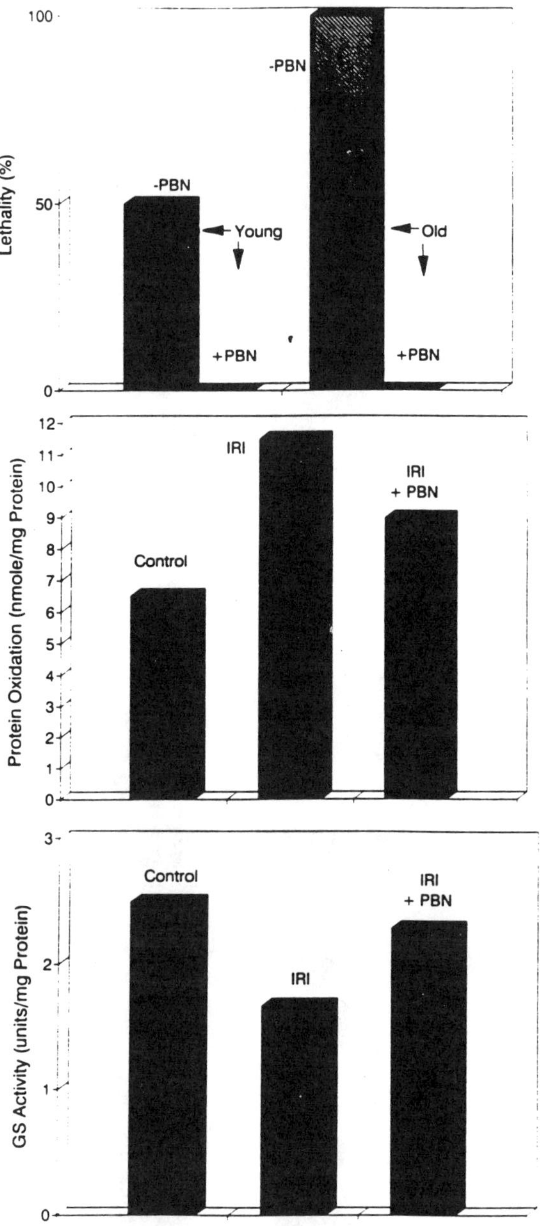

Figure 5 A summary presentation of research results (Cao et al., 1988, Oliver et al., 1990) that demonstrates that PBN protects the brain cortex from the oxidatively damaging events that occur during an IRI. This summary was presented earlier (Floyd and Carney, 1992). Reprinted with permission of Little Brown and Company, from *Annals of Neurology* (1992), 32:S22–S27.

spin-trapping agent, PBN, protects from oxidative damage in brain may open many new possibilities for therapy. It is clear that this area of research will provide many new challenges and opportunities ahead.

Acknowledgments. This research was supported in part by NIH grants NS 23307 and AG 09690. The excellent collaborative efforts of Drs. J. M. Carney, E. R. Stadtman, C. N. Oliver, and P. Stark-Reed made the research possible.

References

Brawn K, Fridovich I (1981): DNA Strand Scission by Enzymically Generated Oxygen Radicals. *Arch Biochem Biophys* 206(2):414–419.

Cao W, Carney JM, Duchon A, Floyd RA, Chevion M (1988): Oxygen Free Radical Involvement in Ischemia and Reperfusion Injury to Brain. *Neurosci Lett* 88:233–238.

Floyd RA (1990a): Role of oxygen free radicals in carcinogenesis and brain ischemia. *FASEB J* 4:2587–2597.

Floyd RA (1990b): The role of 8-hydroxyguanine in carcinogenesis. *Carcinogenesis* 11:1447–1450.

Floyd RA (1990c): The development of the sensitive analysis for 8-hydroxy-2′-deoxyguanosine FRRC HIGHLIGHT: this note highlights a seminal paper published in Free Rad Res Comms 1:163–172 (1986). *Free Rad Res Commun* 8:139–141.

Floyd RA, Carney JM (1992): Free radical damage to protein and DNA; mechanisms involved and relevant observations on brain undergoing oxidative stress. *Ann Neurol* 32:522–527.

Floyd RA, Henderson R, Watson JJ, Wong PK (1986a): Use of salicylate with high pressure liquid chromatography and electrochemical detection (LCED) as a sensitive measure of hydroxyl free radicals in adriamycin treated rats. *J Free Rad Biol Med* 2:13–18.

Floyd RA, Watson JJ, Harris J, West M, Wong PK (1986b): Formation of 8-hydroxydeoxyguanosine, hydroxyl free radical adduct of DNA in granulocytes exposed to the tumor promoter, tetradeconylphobolacetate. *Biochem Biophys Res Commun* 137(2):841–846.

Floyd RA, Watson JJ, Wong PK (1984): Sensitive assay of hydroxyl free radical formation utilizing high pressure liquid chromatography with electrochemical detection of phenol and salicylate hydroxylation productions. *J Biochem Biophys Meth* 10:221–235.

Floyd RA, West MS, Eneff KL, Hogsett WE, Tingey DT (1988): Hydroxyl free radical mediated formation of 8-hydroxyguanine in isolated DNA. *Arch Biochem Biophys* 262(1):266–272.

Floyd RA, West MS, Eneff KL, Schneider JE (1989): Methylene blue plus light mediates 8-hydroxyguanine formation in DNA. *ABB* 273(1):106–111.

Floyd RA, West MS, Eneff KL, Schneider JE, Wong PK, Tingey DT, Hogsett WE (1990): Conditions influencing yield and analysis of 8-hydroxy-2′-deoxyguanosine in oxidatively damaged DNA. *Anal Biochem* 188:155–158.

Frisell WR (1982): Human Biochemistry. New York: MacMillan Publishing Co.

Levine RL, Garland D, Oliver CN, Amici A, Climent I, Lenz A-G, Ahn B-W, Shaltiel S, Stadtman ER (1990): Determination of carboxyl content in oxidatively modified proteins. *Methods Enzymol* 186:464–478.

Oliver CN, Starke-Reed PE, Stadtman ER, Liu GJ, Carney JM, Floyd RA (1990): Oxidative damage to brain proteins, loss of glutamine synthetase activity, and production of free radicals during ischemia/reperfusion-induced injury to gerbil brain. *Proc Natl Acad Sci USA* 87:5144–5147.

Pacifici RE, Salo DC, Davies KJA (1989): Macroxyproteinase (M. O. P): A 670 kDa proteinase complex that degrades oxidatively denatured proteins in red blood cells. *Free Rad Biol Med* 7:521–536.

Rivett AJ (1985): Preferential degradation of the oxidatively modified form of glutamine synthetase by intracellular mammalian proteases. *J Biol Chem* 260:300–305.

Schneider JE, Browning MM, Floyd RA (1988): Ascorbate iron mediation of hydroxyl free radical damage to pBR322 plasmid DNA. *Free Rad Biol Med* 5:287–295.

Schneider JE, Browning MM, Zhu X, Eneff K, Floyd RA (1989): Characterization of hydroxyl free radical mediated damage to plasmid pBR322 DNA. *Mutat Res* 214:23–31.

Stadtman ER (1990): Metal ion catalyzed oxidation of proteins: biochemical mechanism and biological consequences. *Free Rad Biol Med* 9:315–325.

Chapter 6

Reactive Oxidant Species in Rat Brain Extracellular Fluid

Matthew E. Layton and Thomas L. Pazdernik

Reactive oxidant species (ROS) include oxygen free radicals, hydrogen peroxide, lipid peroxides and hydroperoxides, singlet oxygen, oxygen redox-cycling molecules such as quinones and hydroquinones, hypochlorite, and peroxynitrite (Ames et al., 1989; Ames et al., 1981; Aruoma et al., 1989; Beckman et al., 1990; McCord, 1968; Pryor, 1986; Puppo et al., 1990). ROS contribute to damage in a variety of tissues during pathophysiological states, including insults to the central nervous system that result in neuropathology (Kontos, 1989; Siesjo et al., 1989a). Proving ROS participation in tissue damage and occurrence *in vivo* is difficult due to the characteristic short half-lives of these reactive molecules. Methods for ROS detection *in vivo* include: (a) administering free radical spin trapping agents to animals followed by electron spin resonance spectroscopy of tissue homogenates (Carney and Floyd, 1991; Imaizumi et al., 1986; Lai et al., 1986; Oliver et al., 1990), (b) salicylate administration to animals, which can form adducts with ROS, followed by high performance liquid chromatography with electrochemical detection (Cao et al., 1988; Floyd et al., 1984), (c) nitroblue tetrazolium and cytochrome *c* reduction by electron transfer reactions with ROS in biological systems with subsequent spectrophotometric quantification (Armstead et al., 1989; Kennedy et al., 1989; Kontos et al., 1985; Kontos and Povlishock, 1986; McCord, 1968), and (d) ROS-initiated chemiluminescence, or light production by excited molecules, which can be assayed in various animal models (Boveris et al., 1980; Flecha et al., 1991). A widely used assay to detect byproducts of ROS formation is the thiobarbituric acid-reactive substance test for conjugated dienes and malondialdehyde (Janero and

Oxygen Free Radicals in Tissue Damage
Merrill Tarr and Fred Samson, Editors

Burghardt, 1989; Pellmar et al., 1989). Experiments designed to inhibit the generation or propagation of ROS by using specific enzymes and/or antioxidants also measure ROS indirectly (Anderson et al., 1985; Aruoma et al., 1991; Buzadzic et al., 1990; Frei et al., 1988; Kahl et al., 1987; Lemke et al., 1990).

ROS are considered as a class, since the entire range of reactive species mentioned may be generated during tissue damage, especially when coupled with the inflammatory process. ROS generation in tissue results in cascade reactions. A free radical molecule can abstract hydrogen atoms from membrane components, which can form hydroxyl, peroxyl, alkoxyl, and alkyl radicals capable of initiating the disruption of lipid membrane structure (Pryor, 1986). Lipid-rich regions may be oxidized by ROS, subsequently demonstrating cross-linked sulfhydryl groups, also found in proteins and other macromolecules, and instability of the lipid bilayer from the liberation of free fatty acids (Asano et al., 1989; Babbs and Steiner, 1990). DNA damage mediated by ROS includes strand cross-linking, strand scission and may indirectly involve the inactivation of proteins responsible for strand repair (Oliver et al., 1990; Starke-Reed and Oliver, 1988, 1989).

The effect of ROS on tissues is inversely related to the antioxidant protective or defensive capacity of the tissue; that is, when antioxidants are overwhelmed or depleted, ROS are no longer neutralized and toxicity ensues (Frei et al., 1988, 1990; Niki, 1991). Endogenous antioxidants include intramembranous or hydrophobic molecules (e.g., α-tocopherol, β-carotene), aqueous antioxidants in extracellular fluid, cytosolic fluid, blood and bile (ascorbic acid, uric acid, bilirubin, hypotaurine and the histidine-like molecules carnosine and homocarnosine), and enzymes found in primarily intracellular locations (glutathione peroxidases and transferases, superoxide dismutase, and catalase) (Ames et al., 1981; Aruoma et al., 1988; Frei et al., 1990; Kohen et al., 1988; McCord and Fridovich, 1969). Antioxidant molecules exert their neutralizing effects in a cooperative manner. Hydrophobic molecules such as a α-tocopherol function primarily as antioxidants in membranes where their solubility is the greatest. Ascorbic acid molecules in the aqueous environment next to the membrane can donate electrons to molecules of α-tocopherol oxidized by ROS to regenerate the reduced antioxidant form (Niki, 1991). The existence of ROS *in vivo* at "basal" levels is debatable, considering the endogenous antioxidants available for homeostatically neutralizing and dismantling ROS (Barja et al., 1990).

Biochemical and pharmacological tools such as antioxidants help with the characterization of specific molecules within the diverse ROS classification. Exogenous antioxidant molecules such as dimethylthiourea, dimethyl sulfoxide (DMSO), and Trolox-C are frequently employed to inhibit ROS activity in assays with tissue or protein damage as an end-point. If antioxidants provide protective effects and result in less damage, then ROS activity may be inferred (Pellmar et al., 1989; Rao et al., 1988; Wasil et al., 1987). Similar indirect means of measuring ROS include inhibition of enzymes known to generate ROS (e.g., xanthine oxidase and cyclooxygenase) by allopurinol (Kanemitsu et al., 1989), aspirin and/or indomethacin (Armstead et al., 1989; Joshita et al., 1989). Transition metal chelators [e.g., desferrioxamine, ethylenediamine tetraacetic acid (EDTA)] (Patt et al., 1990), free radical spin traps [e.g., 5,5-dimethyl-l-pyrroline N-oxide (DMPO), alpha-phenyl-*tert*-butyl nitrone (PBN) (Carney and Floyd, 1991; Carney et al., 1991; Imaizumi et al., 1986; Lai et al., 1986; Oliver et al., 1990; Pritsos et al., 1985; Tominaga et al., 1987), and free radical scavengers (e.g., *N*-acetyl cysteine, oxypurinol, pyrroloquinoline quinone) (Aruoma et al., 1989; Joshita et al., 1989; Hamagishi et al., 1990; Mizoi et al., 1986; Moorhouse et al., 1987) are also agents that inhibit the damaging effects of ROS in biological assays. Usually the result of adding ROS scavengers or inhibitors is decreased oxidative damage to tissues, membranes, and proteins when compared to untreated controls.

The actual nature of ROS in brain extracellular fluid is unknown. Frequently ROS assays involve the disruption of the tissue of interest by homogenization or freezing before the quantification of the ROS-derived analytes. Damage to the tissue during sample handling can complicate the analysis, since experimental models for ROS detection are primarily oriented toward tissue injury. Examples include ischemia-reperfusion injury (Asano et al., 1985; Cao et al., 1988; Floyd, 1990; Kanemitsu et al., 1989; Nakano et al., 1990; Patt et al., 1988; Rothman and Olney, 1986; Uyama et al., 1990), trauma to cerebral cortex and spinal cord (Anderson et al., 1985; Braughler and Hall, 1989; Demediuk et al., 1985a, b; Wei et al., 1980; Hall and Braughler, 1989; Kontos and Wei, 1986), aging (Ames, 1989a; Ames and Saul, 1986; Benzi et al., 1988, 1989, 1990; Bourre, 1988; Carney et al., 1991; Fraga et al., 1990; Stadtman et al., 1988; Starke-Reed and Oliver, 1988, 1989), Parkinson's disease (Halliwell, 1989; Riederer et al., 1989; Olanow, 1990), Alzheimer's disease (Backon, 1991), subarachnoid hemorrhage (Takeuchi et al., 1991), hypoglycemia (Siesjo and Bengtsson, 1989), radiation (Halliwell and Aruoma, 1991; Lai et al., 1986; Pellmar et al., 1988; Riesz et al., 1990), and seizures (Armstead et al., 1989; Hiramatsu et al., 1986; Siesjo et al., 1989b).

Oxidative stress may be induced in central nervous system (CNS) tissue by applying substrates and their corresponding enzymes known to produce ROS during metabolism (e.g., arachidonic acid with prostaglandin synthase or xanthine with xanthine oxidase) (Kontos, 1987; Patt et al., 1988; Wei et al., 1986). Substances such as hemoglobin (Prat and Turrens, 1990; Sadrzadeh et al., 1987, 1988), myoglobin (Aruoma et al., 1991), transition metals like iron and copper (Gutteridge and Halliwell, 1989; Miller and Aust, 1989; Patt et al., 1990; Willmore et al., 1986), and direct-acting oxidants (e.g., chloramine-T, N-chlorosuccinimide) generate ROS when added to CNS tissue (Pellmar and Neel, 1989; Knobeloch et al., 1990).

Specific assays to detect superoxide radicals ($O2^-$) include superoxide dismutase (SOD)-inhibitable ferricytochrome *c* reduction (McCord, 1968), SOD-inhibitable reduction and precipitation of nitroblue tetrazolium (NBT) (Kontos and Povlishock, 1986; Kontos and Wei, 1986), and diacyldeuteroheme-substituted horse radish peroxidase adduct formation (Makino et al., 1986). Hydroxyl radicals (OH·) are indirectly detected *in vivo* by the formation of a specific hydroxylated salicylate adduct, dihydroxybenzoic acid (DHBA) (Cao et al., 1988; Floyd et al., 1984; Halliwell et al., 1991) and hydroxylated DNA nucleotides such as 8-hydroxydeoxyguanosine (Ames, 1989b; Ames and Saul, 1986; Fraga et al., 1990; Halliwell and Aruoma, 1991; Park et al., 1989). Additional methods for OH· detection include deoxyribose oxidation to form thiobarbituric acid–reactive products such as malondialdehyde (Gutteridge and Halliwell, 1988; Halliwell et al., 1987; Halliwell et al., 1988) and DMSO oxidation to formaldehyde (Beckman et al., 1990). Peroxyl radicals (ROO·) are detected indirectly by epoxidation of polycyclic aromatic hydrocarbons (Pruess et al., 1989) and by spin trapping with electron spin resonance (ESR) spectroscopy (Kohen et al., 1988; Sugata et al., 1989).

Non–free radical ROS such as lipid hydroperoxides (ROOH; e.g., 15-hydroperoxyeicosatetraenoic acid) are generated in biological systems during arachidonic acid metabolism by prostaglandin synthase and lipoxygenase, enzymes that form prostaglandins and leukotrienes, respectively (Leffler et al., 1987). Arachidonic acid is liberated from membranes during oxidative stress, which provides increased amounts of substrate for ROS production (Asano et al., 1985; Demediuk and Faden, 1988; Demediuk et al., 1985; Koide et al., 1985; Siesjo et al., 1989). Eicosanoids and prostanoids can be detected and identified directly by radiochromatography, gas chromatography, and liquid chromatography with fluorescence detection (Anderson et al., 1985; Asano et al., 1989; Faden et al., 1988; Kontos, 1987; Saunders et al., 1987). Other assays for hydroperoxides utilize chemiluminescence, spontaneous or induced, from organs in situ (Boveris et al., 1980; Imaizumi et al., 1986) or homogenates (Prat and Turrens, 1990). Hydroperoxides can interact with transition metal-containing substances such as hematin to produce free radicals (Reed et al., 1988) and subsequently disrupt critical cellular organelle functions by lipid peroxidation (Demediuk and Faden, 1988; Demediuk et al., 1989; Nakano et al., 1990; Rehncrona et al., 1980; Siesjo et al., 1989b; Tanaka et al., 1987). Hydrogen peroxide (H_2O_2) may be indirectly quantified by H_2O_2-dependent 3-aminotriazole inhibition of catalase (Barja et al., 1990; Benzi et al., 1990; Patt et al., 1988). When H_2O_2 is present as substrate, 3-aminotriazole forms an irreversible inhibitory complex with catalase. The degree of catalase inhibition in tissue homogenates implies a corresponding quantity of H_2O_2 necessary to form the inhibitory complex. Catalase-inhibitable isoluminol/microperoxidase-amplified chemiluminescence may also be used to measure H_2O_2 in samples (Yamamoto and Ames, 1987). Lipid hydroperoxides and quinones may also be detected by this method (Frei et al., 1988; Wieland et al., 1989).

To detect ROS in rat brain extracellular fluid, we analyzed intracerebral microdialysis samples for ROS by adapting an isoluminol/microperoxidase chemiluminescence assay (Yamamoto and Ames, 1987). Intracerebral microdialysis perfusion yields a sample ready for analysis of low molecular weight constituents from CNS extracellular fluid in awake, freely moving animals (Amberg and Lindefors, 1989; Benveniste, 1990). We analyzed intracerebral microdialysis samples collected from nonanesthetized rats before ("basal perfusion") and during kainic acid (KA)-induced seizures. Systemic administration of KA to rats produces intense seizures that cause extensive, necrotic neuropathology in the piriform cortex and hippocampus (Chastain et al., 1989; Lehmann et al., 1985; Wade et al., 1987). If indeed ROS represent an element of the final common pathway in brain damage, neuropathology only associated with KA-induced seizures might involve ROS generation. By using microdialysis in combination with chemiluminescence, we have developed an assay to detect changes in ROS in brain extracellular fluid during seizures. The assay evolved with improvements in techniques and instrumentation as observations were made through three distinct sets of experiments. Initial microdialysis probes were constructed with materials later found to generate ROS, which necessitated the substitution of probes made from inert materials. In addition, the first two instruments used for chemiluminescence quantification were found to have disadvantages after experimental data were collected, which led to the acquisition of an instrument designed specifically for the task. The results obtained using intracerebral microdialysis with chemiluminescence have provided evidence that ROS are generated in rat brain extracellular fluid during seizures and may contribute to seizure-associated brain damage.

Methods

Animals. Adult male Wistar rats weighing 250 to 350 g were obtained from Sasco (Omaha, NE). Food and water were provided ad libitum, and a 12-hr light/dark cycle was maintained. Experiments were conducted in accordance with guidelines of the National Research Council DHEW publication #(NIH) 80–23 (1980). The animals were anesthetized with pentobarbital (80 mg/kg i.p.) before stereotaxic insertion of microdialysis probes into the right piriform cortex (measurements: A= -1.8; L=-5.7; V= -9.0) (Paxinos and Watson, 1986). Dental acrylic secured the apparatus to the skull. Animals were awake and freely moving when basal perfusion was initiated 24 h after probe insertion and kainic acid (16 mg/kg i.p.) was administered to induce seizures.

Microdialysis probes. Two styles of microdialysis probes, loop and concentric tubes, were prepared from Dow 50 cellulose membranes (MW cutoff = 5000; O.D. = 250 μm). The membranes for loop probes were glued with epoxy into two 2-cm sections of 23-gauge stainless steel tubing, leaving 6 mm of exposed surface area for dialysis. Concentric tube probes consisted of fused silica hollow tubing (Polymicro Technology; I.D. = 75 μm; O.D. = 145 μm) inside teflon tubing and used the same membranes as described above with a length of 4 mm exposed as dialysis surface area.

Probe perfusion. The perfusion medium for both *in vitro* and *in vivo* microdialysis experiments was modified Kreb's Ringer Bicarbonate (KRB;

pH 7.4; NaCl 144.5 mM; KCl 3 mM; $CaCl_2$ 1.2 mM; KH_2PO_4 0.4 mM; $MgSO_4$ 1.2 mM; $NaHCO_3$ 2.5 mM). A Hamilton Microliter 1000 Series Gastight syringe with a teflon-tipped dowel rod to accommodate a standard syringe fitting and a Harvard syringe infusion pump provided constant flow delivery of perfusion medium through the microdialysis probes at a flow rate of 2 μl/min. For *in vitro* experiments, microdialysis probes were placed in a vessel containing 25 ml of KRB perfusion medium. Samples were collected at 20-min intervals.

Chemiluminescence assay. An isoluminol/microperoxidase cocktail (100 mM sodium borate, 1 mM isoluminol, 0.01 mM microperoxidase in 70% water and 30% methanol at pH 10) (Frei, 1988) was added to hydrogen peroxide (H_2O_2) standard solutions or microdialysis samples. The chemiluminescence produced was quantified by conversion of the number of photons that struck photomultiplier tubes to integrated electrical signals (Roswell and White, 1978). Kreb's Ringer Bicarbonate (KRB) and KRB perfused through microdialysis probes, both *in vitro* and *in vivo*, were assayed for chemiluminescence in the presence and absence of catalase (50 U/ml). Catalase is an enzyme that specifically reduces H_2O_2 to water and oxygen. KRB media values and KRB perfusate values refer to the chemiluminescence produced by KRB media before and after *in vitro* perfusion through the microdialysis probes. An equal volume of KRB perfusion medium, which was the catalase vehicle, was added to samples not treated with catalase.

Instrumentation. Three instruments were used to quantify chemiluminescence initiated by hydrogen peroxide (H_2O_2) standard solutions and by intracerebral microdialysates from rat piriform cortex during basal and seizure phase perfusion: (a) a Packard Tricarb C2425 liquid scintillation counter in the out-of-coincidence mode, (b) a Chronolog Model 500 platelet aggregometer in the luminescence mode, and (c) an MGM Opticomp I luminometer. Concentration-dependent H_2O_2-initiated chemiluminescence is shown in Fig. 1 for each set of experiments.

Liquid scintillation counter. A liquid scintillation counter is able to quantify photons produced by isoluminol/microperoxidase chemiluminescence when used in the out-of-coincidence mode (Pekoe, 1985). In-coincidence refers to recording counts only when two photomultiplier tubes are struck by particles simultaneously, and is used to quantify beta radiation. By

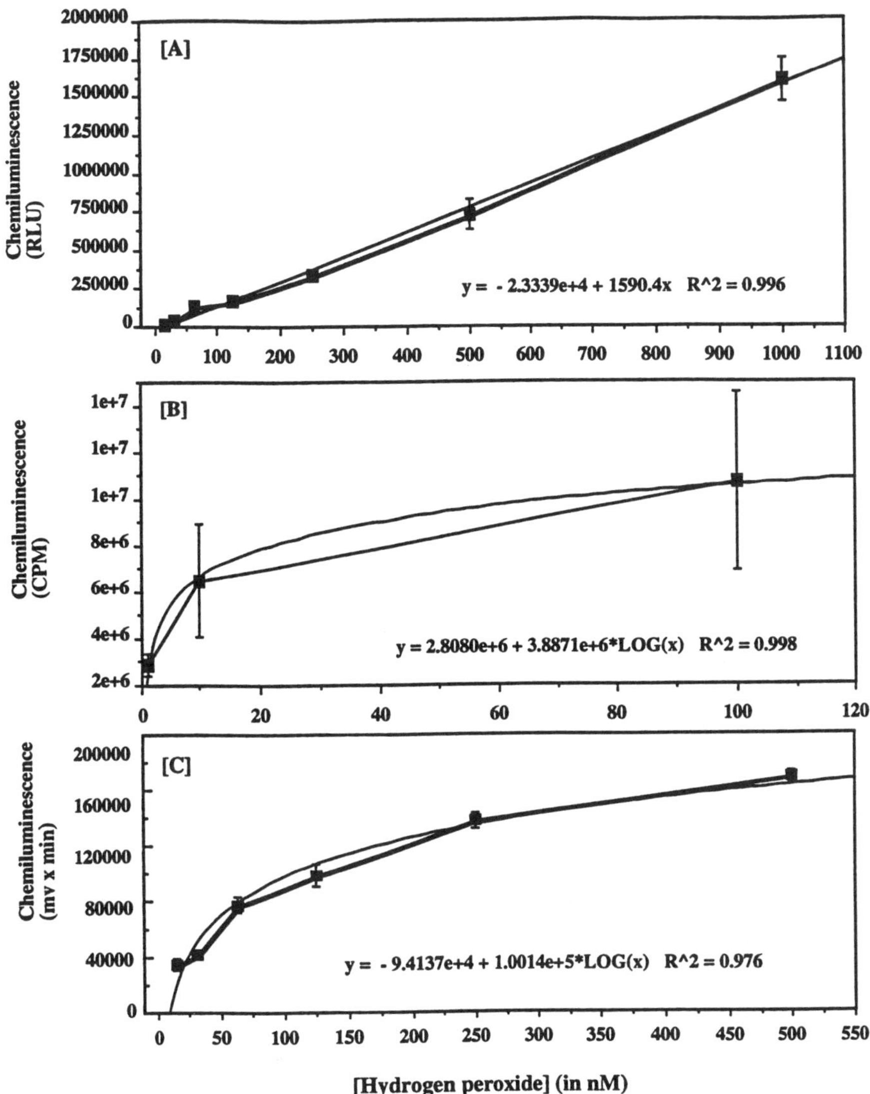

Figure 1 Hydrogen peroxide concentration-dependent isoluminol/microperoxidase chemiluminescence is linear ($R^2 = 0.99$) when measured with an MGM Opticomp I luminometer [**A**], but logarithmic with a Tri-Carb liquid scintillation counter in the out-of coincidence mode [**B**] and a Chronolog platelet aggregometer [**C**] in the luminescence mode. Values represent mean ± SEM; n= 9, 3, and 6, respectively. RLU, relative light units; CPM, counts per minute.

placing the counter in the out-of-coincidence mode, single photons can be counted as they strike either photomultiplier tube. H_2O_2 standards or microdialysis samples were mixed with the isoluminol/microperoxidase cocktail, and counted for nine intervals of 6 s each with a 10-s delay between intervals. Chemiluminescence was expressed as counts per minute (CPM).

Platelet aggregometer. A platelet aggregometer in the luminescence mode was also used to quantify photons produced during isoluminol/microperoxidase chemiluminescence. The isoluminol cocktail was placed in a stirred, constant temperature cuvette in the sample compartment of the platelet aggregometer. H_2O_2 standards or microdialysis samples were injected directly into the cuvette and the chemiluminescence produced was converted to a peak on a Shimadzu C-R3A Chromatopac integrating recorder. The light produced was quantified for 4 min after addition of both standards and samples. Chemiluminescence was expressed as peak area in millivolts $\times$ minutes (mV $\times$ min).

Luminometer. A third instrument to quantify chemiluminescence was a luminometer. Designed specifically for bioluminescence and chemiluminescence assays, the luminometer has a photomultiplier tube capable of quantifying intense light production in a variety of assays. H_2O_2 standards or microdialysis samples are placed in a test tube, inserted into the sample compartment of the luminometer, and the isoluminol/microperoxidase cocktail was automatically injected into the test tube from a reservoir when the lid was closed. Chemiluminescence quantification began immediately using an integrated kinetic photon counting protocol of 60 4-s counting intervals with no delays between intervals, a total counting period of 4-min. Chemiluminescence was expressed as relative light units (RLU).

Results

Concentration-dependent chemiluminescence produced by H_2O_2 standards was linear when measured with a luminometer (A), but logarithmic when measured with a liquid scintillation counter (B) or a platelet aggregometer (C), a shown in Fig. 1. The H_2O_2 standard curve measured in a luminometer was linear from 15 nM to 1 μM (r^2 = 0.996). The hyperbolic standard curves in (B) and (C) can be described by logarithmic equations with high correlations, $r^2 = 0.998$ and 0.976, respectively. The KRB media produced 39,038 $\pm$ 3746 RLU in this chemiluminescence assay when measured with a luminometer in the absence of catalase and 19,638 ± 2203 RLU in the presence of catalase (mean ± SEM; n=16). The difference in the RLU between KRB media analyzed in the presence and absence of catalase indicated that the perfusion media itself contained approximately 30 nM H_2O_2.

The chemiluminescence of microdialysates collected during *in vitro* perfusion of the apparatus system with KRB depended on the materials used for probe construction, as depicted in Fig. 2. When assayed with a luminometer, perfusates from steel and tygon probes generated more

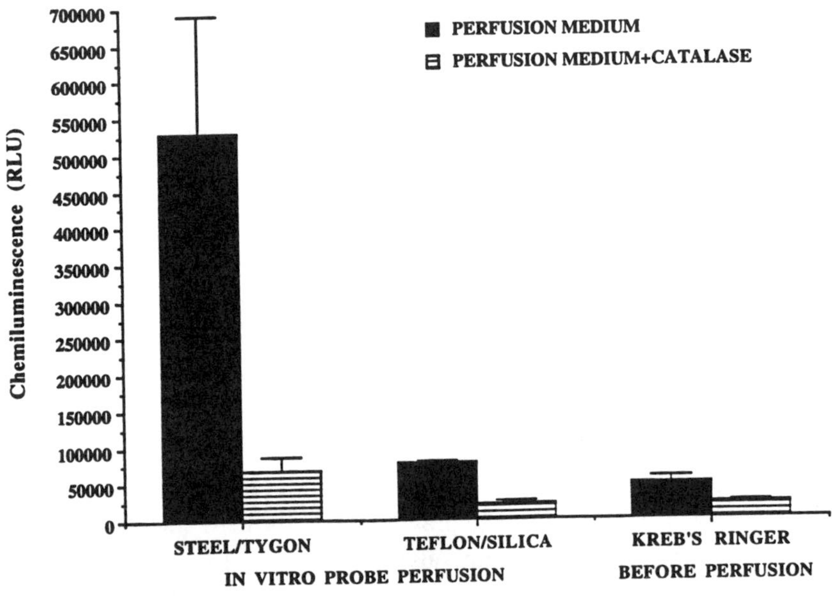

Figure 2 *In vitro* microdialysate chemiluminescence is substantially higher in perfusates from probes constructed with steel and tygon tubing than from probes made of teflon and fused silica tubing. Microdialysates were treated with KRB perfusion medium (■) or KRB containing catalase (□; 50 U/ml) before analysis with an MGM Opticomp I luminometer. The difference in chemiluminescence in the presence and absence of catalase corresponds to hydrogen peroxide-dependent chemiluminescence. Values represent mean ± SEM; n=3, 8, and 16 for steel/tygon probes, teflon/fused silica probes, and KRB medium alone before perfusion, respectively.

chemiluminescence compared to perfusates from teflon and fused silica probes, both in the presence and absence of catalase. Probes constructed from tygon, stainless steel tubing, and epoxy yielded *in vitro* chemiluminescence values of 529,168 ± 160,285 RLU in the absence of catalase and 66,636 ± 20,348 RLU in the presence of catalase (mean ± SEM; n=3). Using probes constructed from teflon, fused silica, and contact adhesive, *in vitro* perfusate chemiluminescence values were approximately 700% less in the absence of catalase (73,229 ± 11,616 RLU) and 300% less in the presence of catalase (18,205 ± 2251 RLU; mean ± SEM; n=10). H_2O_2 perfusate concentrations after *in vitro* perfusion of the microdialysis apparatus were calculated to be approximately 300 nM for steel/tygon probes and 50 nM for teflon/fused silica probes.

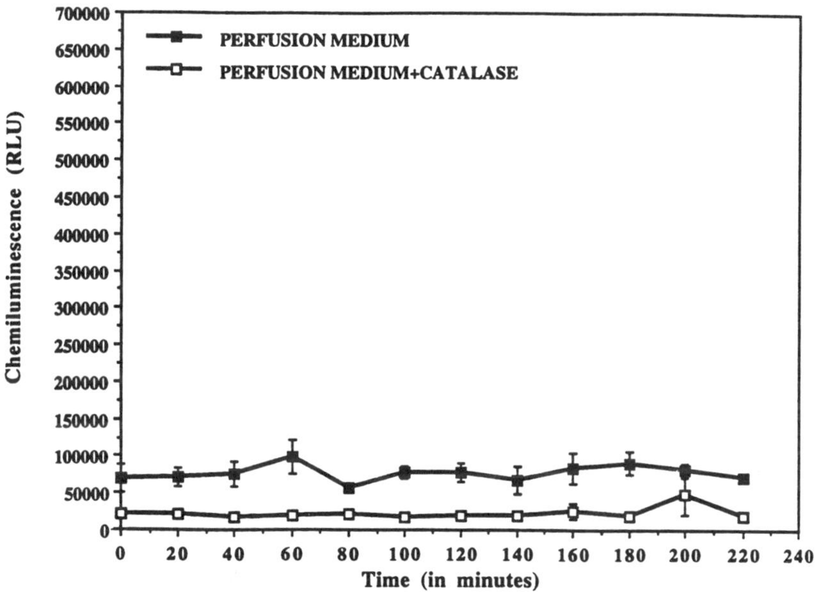

Figure 3 Chemiluminescence of microdialysates during *in vitro* perfusion of teflon/fused silica probes with KRB is relatively stable over time. Microdialysates were collected at 20-min intervals and treated with KRB perfusion medium (■) or KRB containing catalase (□; 50 U/ml) before analysis in an MGM Opticomp I luminometer. The difference in chemiluminescence in the presence and absence of catalase corresponds to hydrogen peroxide-dependent chemiluminescence. Values represent mean ± SEM; n=3.

Microdialysate chemiluminescence assayed with a luminometer during *in vitro* perfusion of teflon/fused silica probes with KRB was relatively stable over a period of several hours, as shown in Fig. 3. This was true for both total chemiluminescence and the catalase-resistant component of chemiluminescence. Although the presence of substantial H_2O_2

in samples was evident, the results show the relative temporal stability of the assay both in the presence and absence of catalase.

The chemiluminescence initiated by intracerebral microdialysates collected with steel/tygon probes and quantified in a liquid scintillation counter increased after systemic administration of KA to rats, as shown in Fig. 4. Average microdialysate chemiluminescence values from perfusates collected during basal perfusion (before KA) were 7.6 $\pm 1.4 \times 10^{-6}$ CPM in the absence of catalase, which increased during KA-induced seizures to $11.3 \pm 1.1 \times 10^{-6}$ CPM (mean ± SEM; n=4).

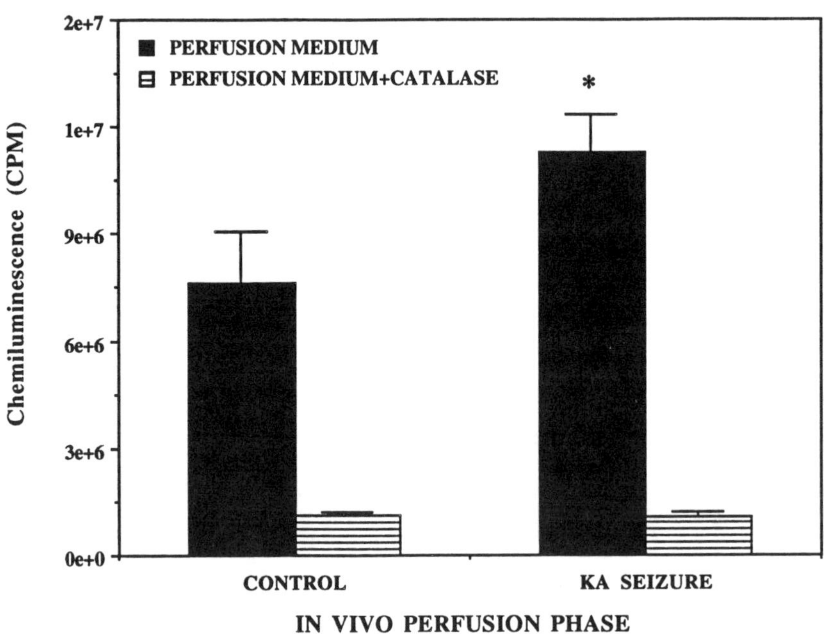

Figure 4 Intracerebral microdialysates from a rat piriform cortex showed significantly increased chemiluminescence during seizures induced by systemic administration of kainic acid (KA 16 mg/kg i.p.). Microdialysates were obtained with steel/tygon microdialysis probes and assayed in a Tri-Carb liquid scintillation counter in the out-of-coincidence mode. Samples were treated with KRB perfusion medium (■) or KRB with catalase (□; 50 U/ml). KA-induced seizures caused an increase in intracerebral microdialysate catalase-sensitive chemiluminescence, i.e., an increase in hydrogen peroxide. Values represent mean ± SEM; n=4.
*p=.01, paired *t*-test vs. controls.

Microdialysate chemiluminescence values of samples treated with catalase (i.e., the catalase-resistant component) were the same before and after KA administration ($1.1 \pm 0.1 \times 10^{-6}$ CPM). By this method a significant increase in brain microdialysate chemiluminescence was demonstrated during KA-induced seizures ($p = 0.01$, paired t-test), which was catalase-sensitive (i.e., an increase in H_2O_2).

The intracerebral microdialysate chemiluminescence from the experiment depicted in Fig. 5 was quantified with a platelet aggregometer

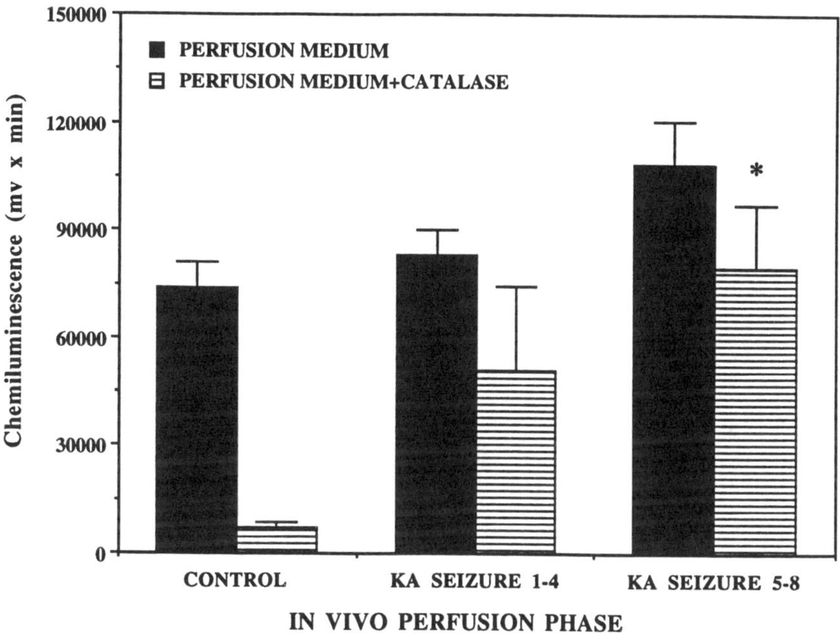

Figure 5 Chemiluminescence of intracerebral microdialysates from rat piriform cortex increased during kainic acid (KA)-induced seizures. Samples were obtained using teflon/fused silica microdialysis probes and were analyzed in a Chronolog platelet aggregometer in the luminescence mode. Samples were treated with KRB perfusion medium (■) or KRB containing catalase (□ 50 U/ml). Systemic administration of KA (16 mg/kg i.p.) increased intracerebral microdialysate chemiluminescence significantly in catalase-treated samples ($p = .007$) and apparently in perfusion medium–treated samples ($p=.09$). Values represent mean ± SEM (n=9) for pooled samples collected during basal perfusion and after KA administration during 15 minute collection intervals 1–4 and 5–8.
*p= .007, paired t-test vs. catalase-treated controls.

in the luminescence mode. Samples were collected with teflon/fused silica probes during basal perfusion and KA-induced seizures. Pooled sample values during basal perfusion were 73,991 ± 6790 mV × min in the absence of catalase (n=9) and 6952 ± 1661 mV × min in the presence of catalase (n=7). Chemiluminescence in microdialysates increased during seizures to an average of 83,032 ± 7254 mV × min in the absence of catalase in pooled samples 1 to 4 post-KA administration, and to 108,168 ± 12,099 mV × min in pooled samples 5 to 8 (n=9). Microdialysate chemiluminescence in the presence of catalase increased during seizures to 50,922 ± 23,390 and 79,734 ± 17,653 mV × min over post-KA sample periods 1 to 4 and 5 to 8, respectively (mean ± SEM; n=7). Thus, KA-induced seizures significantly increased intracerebral microdialysate chemiluminescence in the presence of catalase; that is, a catalase-resistant component (p = .007, paired *t*-test) and apparently increased total chemiluminescence of intracerebral microdialysates in the absence of catalase (p = .09).

As illustrated in Fig. 6, intracerebral microdialysates obtained with teflon/fused silica probes and assayed with a luminometer showed increased chemiluminescence during KA-induced seizures in both the presence and absence catalase. The chemiluminescence of intracerebral microdialysates obtained during basal perfusion 24 hr after probe insertion was 44,118 ± 2294 RLU in the absence of catalase just before KA injection and increased to 66,546 ± 6108 RLU during seizures 80 min after injection (n=7). Injection of the KA vehicle had no effect on intracerebral microdialysate chemiluminescence. For example, microdialysate chemiluminescence was 49,135 ± 1412 RLU in the absence of catalase just before vehicle injection in control animals and 42,586 ± 5802 80 min after injection (mean ± SEM; n=2).

Intracerebral microdialysate chemiluminescence in the presence of catalase, that is, catalase-resistant chemiluminescence, was 20,741 ± 1859 RLU during basal perfusion just before KA injection. Catalase-resistant microdialysate chemiluminescence increased to 54,420 ± 10,427 RLU during seizures 80 min after KA injection (n=7). Catalase-resistant microdialysate chemiluminescence in control animals did not change significantly from values just before vehicle injection (20,400 ± 2450 RLU) to values obtained 80 min after injection of vehicle (19,721 ± 1739; n=2). Thus, intracerebral microdialysate chemiluminescence increased significantly over time in both the presence (p = .0001, ANOVA with repeated measures) and absence (p = .0002) of catalase during KA-induced seizures and did not increase over time in vehicle-injected control animals.

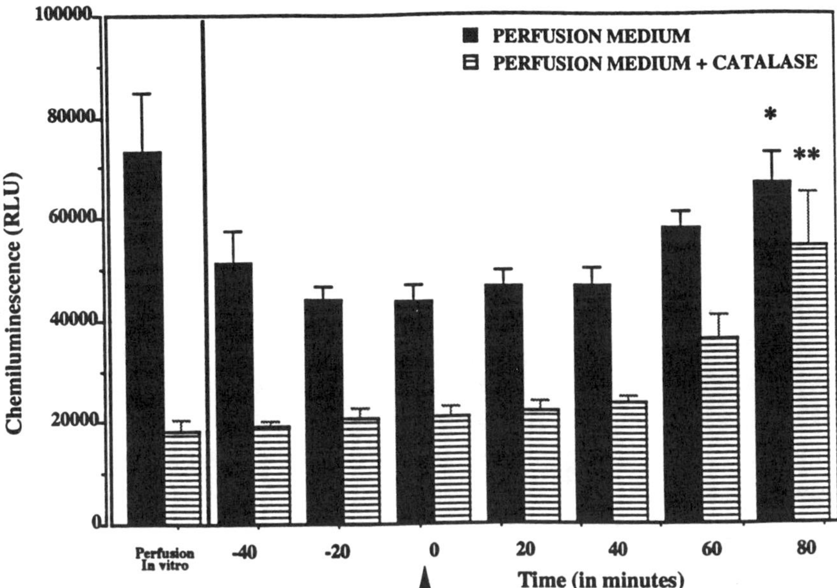

Figure 6 Intracerebral microdialysates from rat piriform cortex showed increased chemiluminescence during KA induced seizures. Chemiluminescence was predominantly catalase-sensitive during *in vitro* perfusion and basal perfusion *in vivo*, but the catalase-resistant component increased dramatically during intense seizure activity. Microdialysates were obtained with teflon/fused silica probes perfused with KRB at a flow rate of 2 μl/min. Samples were treated with KRB perfusion medium (■) or KRB containing catalase (□; 50 U/ml) and analyzed in an MGM Opticomp I luminometer. KA was administered at time 0 (▲; 16 mg/kg i.p.). Values represent mean $\pm$ SEM; n=7
* p=.0002, **p=.0001, ANOVA with repeated measures.

Discussion

In our experiments, reactive oxidant species were detected in intracerebral microdialysates from rat piriform cortex analyzed with an isoluminol/microperoxidase chemiluminescence assay. The goals of these studies were to adapt the isoluminol/microperoxidase assay for plasma analysis described by Ames' group (Frei et al., 1988, 1990; Yamamoto et al., 1987) to detect ROS in brain microdialysis samples. ROS such as hydroperoxides are cleaved by microperoxidase to form hydroxyl and alkoxyl radicals, which initiate the release of photons and energy from the excited state of isoluminol molecules. ROS were found to increase in intracerebral microdialysis samples of rat brain extracellular fluid during KA-induced seizures. The ROS that increased during seizures were primarily catalase-resistant (i.e., ROS other than H_2O_2). Reactive oxidative molecules generated in brain extracellular fluid during seizures may contribute to the conspicuous brain damage in the rat piriform cortex associated with KA seizures.

The first task was to find appropriate instrumentation to demonstrate linear standard curves using H_2O_2 to initiate concentration-dependent chemiluminescence. H_2O_2 standard-initiated chemiluminescence was tested under a variety of conditions as described in the methods section. Light produced by H_2O_2 in this reaction consists of an intense early burst of light emission followed by a rapid decline phase and a prolonged, sloping decay phase. Difficulty arose in using a liquid scintillation counter to quantify the initial burst of chemiluminescence because there was a delay between the addition of H_2O_2 standards or microdialysis samples to the cocktail and the onset of counting. The initial burst of light was not quantified and counting began on the steeply falling portion of the hyperbolic chemiluminescence curve. The advantage of using a platelet aggregometer over a liquid scintillation counter was the accessibility of the sample compartment. Injecting H_2O_2 standards or microdialysis samples directly into the cuvette containing the cocktail in the sample compartment decreased the delay between mixing and light quantification to 1 to 2 s. In this manner, the initial burst of chemiluminescence was captured.

Further instrument adaptation for chemiluminescence detection was necessary since H_2O_2 standard-initiated chemiluminescence curves were hyperbolic when measured with a liquid scintillation counter and a platelet aggregometer, but linear when measured with a luminometer. The light produced at the higher concentrations of H_2O_2 may have saturated the photomultiplier tubes in the first two instruments. The luminometer was designed specifically to measure a wide light intensity range in luminescence assays. Our methods were improved by adapting the assay to more appropriate sampling techniques and instruments for chemiluminescence quantification as observations were made during data collection.

The difference in chemiluminescence produced in the presence and absence of catalase, an enzyme that inactivates H_2O_2, corresponds to the H_2O_2 concentration in the microdialysis sample. The perfusion medium itself (KRB) contained a concentration of about 30 nM H_2O_2. When this medium was passed through the perfusion apparatus and microdialysis probes that were routinely used in our laboratory, extensive amounts of H_2O_2 (up to 300 nM) were generated (see Fig. 2). We identified stainless steel, tygon tubing, and epoxy adhesive as the major contributors to the generation of H_2O_2. Therefore, the currently used perfusion apparatus and microdialysis probes are constructed out of relatively inert materials including glass, fused silica, teflon, and contact adhesive. With this newly designed system, the amount of H_2O_2 in perfusates after *in vitro* perfusion was reduced to about 50 nM. Moreover, the amount of H_2O_2 remained constant over several hours of *in vitro* perfusion (see Fig. 3) or when the samples sat on the benchtop for several minutes after collection and before analysis. When the perfusion media was passed through probes implanted in the rat piriform cortex, the amount of H_2O_2 in the perfusate was reduced from the amount routinely found in samples after *in vitro* perfusion from approximately 50 nM to 30 nM. This is most likely because of the low molecular weight antioxidants (e.g., ascorbic acid, uric acid) in brain extracellular fluid that we have detected in intracerebral microdialysates (unpublished observations). When present in high concentrations, ascorbic acid is known to reduce H_2O_2 (Frei, 1988). We have demonstrated increases in ascorbic acid in rat piriform cortex during KA-induced seizures of approximately 1500% (data not shown). This phenomenon may explain the associated decrease in intracerebral microdialysate H_2O_2 concentrations from 30 nM before KA-induced seizures to 18 nM during seizures. A surprising observation was that *in vivo* microdialysates collected and allowed to sit on the benchtop for several minutes demonstrated increased catalase-sensitive (i.e., H_2O_2) chemiluminescence over samples assayed immediately (data not shown). These findings point to the dynamic nature of the brain extracellular fluid. Although these studies are in their infancy, it appears that the chemistry of the brain extracellular fluid changes so that during seizures, antioxidants such as ascorbic acid increase to compensate for increased oxidative stress, as reflected by increased intracerebral microdialysate chemiluminescence. Thus, during insults leading to neuropathology (i.e., KA-induced seizures) a dynamic change occurs in the brain extracellular fluid chemistry. The nature of these changes and the biochemical constituents involved are under current investigation.

The catalase-resistant component of microdialysate chemiluminescence (i.e., initiated by ROS other than H_2O_2) was approximately equal during *in vitro* perfusion of the microdialysis apparatus and basal perfusion before seizures *in vivo*. Catalase-resistant chemiluminescence increased dramatically in intracerebral microdialysates during seizures (see Fig. 6) and may have represented the generation of oxidative species such as lipid hydroperoxides or redox-active quinones in brain extracellular fluid. These molecules have been found in plasma using isoluminol/microperoxidase chemiluminescence analysis (Frei et al., 1988) and may also be present in brain extracellular fluid. During neurological insults, arachidonic acid has been shown to be metabolized to hydroperoxides during leukotriene and prostaglandin synthesis (Asano et al., 1985). These oxidative species have relatively long half-lives compared to other ROS and would be more likely to be stable enough in microdialysates to be detected.

The increased catalase-sensitive (i.e., H_2O_2) chemiluminescence during KA seizures observed in microdialysates obtained with steel and tygon probes (see Fig. 4) is difficult to interpret in light of the large amounts of H_2O_2 found to be generated during *in vitro* perfusion of these probes. When seizures were induced under these conditions, the steel present in the probes may have catalyzed the formation of H_2O_2 in the perfusion medium from brain extracellular fluid constituents. The effects of the materials used in designing experiments to detect ROS are important when physiological fluids are in contact with commonly used laboratory equipment. For example, nitroxide radicals have been detected in saline from plastic syringes by spin traps and electron spin resonance spectroscopy (Buettner et al., 1991). In our experiments, many commonly used laboratory materials, including stainless steel, plastics, polyethylene pipette tips and tubing, Silastic and tygon tubing, and epoxy adhesive generated substantial chemiluminescence when perfused with KRB media. Careful precautions must be taken during attempts to measure ROS, especially *in vivo*. Samples for ROS detection must be handled extremely carefully and results obtained must be interpreted cautiously.

Finally, as is obvious from these studies, the detection of ROS directly *in vivo* is difficult. Biochemical constituents must be monitored in an environment as unperturbed as possible if one wishes to assess accurately the status of reactive species in a dynamic biological system. The importance of ROS in vivo is only beginning to be understood. ROS may represent a final common pathway leading to the disruption of cellular homeostasis in important and diverse disease states. The benefits of detecting and measuring ROS *in vivo* will be apparent as our understanding of ROS in pathophysiological processes is increased, thus leading to effective intervention strategies.

Acknowledgments. The authors wish to thank the Ralph L. Smith Mental Retardation Research Center, Kansas University Medical Center, Fred E. Samson and Stanley R. Nelson for enlightening discussion, Robert Cross and James Rengel for their invaluable technical assistance, and John Wood for assistance and equipment used during the development of the assays described herein. This work was supported in part by DAMD-17-90-C-0041 and ES07079.

References

Amberg G, Lindefors N (1989): Intracerebral microdialysis: II. Mathematical studies of diffusion kinetics. *J Pharmacol Meth* 22(3):157–183.

Ames B, Cathcart R, Schwiers E, Hochstein P (1981): Uric acid provides an antioxidant defense in humans against oxidant- and radical-caused aging and cancer. A hypothesis. *Proc Natl Acad Sci USA* 78(11):6858–6862.

Ames BN (1989a): Endogenous DNA damage as related to cancer and aging. *Mutant Res* 214(1):41–46.

Ames BN (1989b): Endogenous oxidative DNA damage, aging, and cancer. *Free Radic Res Commun* 7(3–6):121–128.

Anderson DK, Saunders RD, Demediuk P, Dugan LL, Braughler JM, Hall ED, Means ED, Horrocks LA (1985): Lipid hydrolysis and peroxidation in injured spinal cord: partial protection with methylprednisolone or vitamin E and selenium. *Cent Nerv Syst Trauma* 2(4):257–267.

Armstead WM, Mirro R, Leffler CW, Busija DW (1989): Cerebral superoxide anion generation during seizures in newborn pigs. *J Cereb Blood Flow Metab* 9(2):175–179.

Aruoma OI, Smith C, Cecchini R, Evans PJ, Halliwell B (1991): Free radical scavenging and inhibition of lipid peroxidation by beta-blockers and by agents that interfere with calcium metabolism. A physiologically-significant process? *Biochem Pharmacol* 42(4):735–743.

Aruoma OI, Akanmu D, Cecchini R, Halliwell B (1991): Evaluation of the ability of the angiotensin-converting enzyme inhibitor captopril to scavenge reactive oxygen species. *Chem Biol Interact* 77(3):303–314.

Aruoma OI, Halliwell B, Hoey BM, Butler J (1989): The antioxidant action of N-acetylcysteine: its reaction with hydrogen peroxide, hydroxyl radical, superoxide, and hypochlorous acid. *Free Rad Biol Med* 6(6):593–597.

Aruoma OI, Halliwell B, Hoey BM, Butler J (1988): The antioxidant action of taurine, hypotaurine and their metabolic precursors. *Biochem J* 256(1):251–255.

Asano T, Gotoh O, Koide T, Takakura K (1985): Ischemic brain edema following occlusion of the middle cerebral artery in the rat. II: Alteration of the eicosanoid synthesis profile of brain microvessels. *Stroke* 16(1):110–113.

Asano T, Koide T, Gotoh O, Joshita H, Hanamura T, Shigeno T, Takakura K (1989): The role of free radicals and eicosanoids in the pathogenic mechanism underlying brain edema. *Mol Chem Neuropathol* 10(2):101–133.

Babbs CF, Steiner MG (1990): Stimulation of free radical reactions in biology and medicine: a new two-compartment kinetic model of intracellular lipid peroxidation. *Free Rad Biol Med* 8(5):471–485.

Backon J (1991): Dementia in cancer patients undergoing chemotherapy: implication of free radical injury and relevance to Alzheimer disease. *Med Hypotheses* 35(2):146–147.

Barja dQG, Perez CR, Lopez TM (1990): Changes in cerebral antioxidant enzymes, peroxidation, and the glutathione system of frogs after aging and catalase inhibition. *J Neurosci Res* 26(3):370–376.

Beckman JS, Beckman TW, Chen J, Marshall PA, Freeman BA (1990): Apparent hydroxyl radical production by peroxynitrite: implications for endothelial injury from nitric oxide and superoxide. *Proc Natl Acad Sci USA* 87(4):1620–1624.

Beneviste H, Hüttemeier P (1990): Microdialysis—theory and application. *Prog Neurobiol* 35:195–215.

Benzi G, Marzatico F, Pastoris O, Villa RF (1990): Influence of oxidative stress on the age-linked alterations of the cerebral glutathione system. *J Neurosci Res* 26(1):120–128.

Benzi G, Pastoris O, Marzatico F, Villa RF (1988): Influence of aging and drug treatment on the cerebral glutathione system. *Neurobiol Aging* 9(4):371–375.

Benzi G, Pastoris O, Marzatico F, Villa RF (1989): Age-related effect induced by oxidative stress on the cerebral glutathione system. *Neurochem Res* 14(5):473–481.

Benzi G, Pastoris O, Villa RF (1988): Changes induced by aging and drug treatment on cerebral enzymatic antioxidant systems. *Neurochem Res* 13(5):467–478.

Bourre JM (1988): Free radicals, polyunsaturated fatty acids, cell death, brain aging. *C R Soc Biol Paris* 182(1):5–36.

Boveris A, Cadenas E, Reiter R, Filipowski M, Nakase Y, Chance B (1980): Organ chemiluminescence: noninvasive assay for oxidative radical reactions. *Proc Natl Acad Sci USA* 77(1):347–351.

Braughler JM, Hall ED (1989): Central nervous system trauma and stroke. I. Biochemical considerations for oxygen radical formation and lipid peroxidation. *Free Rad Biol Med* 6(3):289–301.

Buettner G, Scott B, Kerber R, Mugge A (1991): Free radicals from plastic syringes. *Free Rad Biol Med* 11:69–70.

Buzadzic B, Spasic M, Saicic Z, Radojicic SR, Halliwell B, Petrovic VM (1990): Antioxidant defenses in the ground squirrel Citellus citellus. 1. A comparison with the rat. *Free Rad Biol Med* 9(5):401–406.

Cao W, Carney JM, Duchon A, Floyd RA, Chevion M (1988): Oxygen free radical involvement in ischemia and reperfusion injury to brain. *Neurosci Lett* 88(2):233–238.

Carney JM, Floyd RA (1991): Protection against oxidative damage to CNS by alpha-phenyl-tert-butyl nitrone (PBN) and other spin-trapping agents: a novel series of nonlipid free radical scavengers. *J Mol Neurosci* 3(1):47–57.

Carney JM, Starke-Reed P, Oliver CN, Landum RW, Cheng MS, Wu JF, Floyd RA (1991): Reversal of age-related increase in brain protein oxidation, decrease in enzyme activity, and loss in temporal and spatial memory by chronic administration of the spin-trapping compound N-tert-butyl-alpha-phenylnitrone. *Proc Natl Acad Sci USA* 88(9):3633–3636.

Chastain J, Samson F, Nelson SR, Pazdernik TL (1989): Kainic acid-induced seizures: changes in brain extracellular ions as assessed by intracranial microdialysis. *Life Sci* 45(9):811–817.

Demediuk P, Saunders RD, Anderson DK, Means ED, Horrocks LA (1985a): Membrane lipid changes in laminectomized and traumatized cat spinal cord. *Proc Natl Acad Sci USA* 82(20):7071–7075.

Demediuk P, Saunders RD, Clendenon NR, Means ED, Anderson DK, Horrocks LA (1985b): Changes in lipid metabolism in traumatized spinal cord. *Prog Brain Res* 63(211):211–226.

Demediuk P, Faden AI (1988): Traumatic spinal cord injury in rats causes increases in tissue thromboxane but not peptidoleukotrienes. *J Neurosci Res* 20(1):115–121.

Demediuk P, Daly MP, Faden AI (1989): Changes in free fatty acids, phospholipids, and cholesterol following impact injury to the rat spinal cord. *J Neurosci Res* 23(1):95–106.

Faden AI, Lemke M, Demediuk P (1988): Effects of BW755C, a mixed cyclo-oxygenase-lipoxygenase inhibitor, following traumatic spinal cord injury in rats. *Brain Res* 463(1):63–68.

Flecha B, Llesuy S, Boveris A (1991): Hydroperoxide-initiated chemiluminescence: an assay for oxidative stress in biopsies of heart, liver, and muscle. *Free Rad Biol Med* 10:93–100.

Floyd RA (1990): Role of oxygen free radicals in carcinogenesis and brain ischemia. *FASEB J* 4(9):2587–2597.

Floyd RA, Watson JJ, Wong PK (1984): Sensitive assay of hydroxyl free radical formation utilizing high pressure liquid chromatography with electrochemical detection of phenol and salicylate hydroxylation products. *J Biochem Biophys Meth* 10(3-4):221–235.

Fraga CG, Shigenaga NK, Park JW, Degan P, Ames BN (1990): Oxidative damage to DNA during aging: 8-hydroxy-2′-deoxyguanosine in rat organ DNA and urine. *Proc Natl Acad Sci USA* 87(12):4533-4537.

Frei B, Stocker R, Ames BN (1988): Antioxidant defenses and lipid peroxidation in human blood plasma. *Proc Natl Acad Sci USA* 85(24):9748–9752.

Frei B, Stocker R, England L, Ames BN (1990): Ascorbate: the most effective antioxidant in human blood plasma. *Adv Exp Med Biol* 264(155):155–163.

Frei B, Yamamoto Y, Niclas D, Ames BN (1988): Evaluation of an isoluminol chemiluminescence assay for the detection of hydroperoxides in human blood plasma. *Anal Biochem* 175(1):120–130.

Gutteridge JM, Halliwell B (1988): The deoxyribose assay: an assay both for 'free' hydroxyl radical and for site-specific hydroxyl radical production [letter]. *Biochem J* 253(3):932–933.

Gutteridge JM, Halliwell B (1989): Iron toxicity and oxygen radicals. *Baillieres Clin Haematol* 2(2):195–256.

Hall ED, Braughler JM (1989): Central nervous system trauma and stroke. II. Physiological and pharmacological evidence for involvement of oxygen radicals and lipid peroxidation. *Free Rad Biol Med* 6(3):303–313.

Halliwell B (1989): Oxidants and the central nervous system: some fundamental questions. Is oxidant damage relevant to Parkinson's disease, Alzheimer's disease, traumatic injury or stroke? *Acta Neurol Scand Suppl* 126(23):23–33.

Halliwell B, Aruoma OI (1991): DNA damage by oxygen-derived species. Its mechanism and measurement in mammalian systems. *FEBS Lett* 281(1-2): 9–19.

Halliwell B, Grootveld M, Gutteridge JM (1988): Methods for the measurement of hydroxyl radicals in biomedical systems: deoxyribose degradation and aromatic hydroxylation. *Meth Biochem Anal* 33(59):59–90.

Halliwell B, Gutteridge JM, Aruoma OI (1987): The deoxyribose method: a simple "test-tube" assay for determination of rate constants for reactions for hydroxyl radicals. *Anal Biochem* 165(1):215–219.

Halliwell B, Kaur H, Ingelman SM (1991): Hydroxylation of salicylate as an assay for hydroxyl radicals: a cautionary note [letter]. *Free Rad Biol Med* 10(6):439–441.

Hamagishi Y, Murata S, Kamei H, Oki T, Adachi O, Ameyama M (1990): New biological properties of pyrroloquinoline quinone and its related compounds: inhibition of chemiluminescence, lipid peroxidation and rat paw edema. *J Pharmacol Exp Ther* 255(3):980–985.

Hiramatsu M, Edamatsu R, Kohno M, Mori A (1986): The possible involvement of free radicals in seizure mechanism. *Jpn J Psychiatry Neurol* 40(3):349–352.

Imaizumi S, Tominaga T, Uenohara H, Yoshimoto T, Suzuki J, Fujita Y (1986): Initiation and propagation of lipid peroxidation in cerebral infarction models. Experimental studies. *Neurol Res* 8(4):214–220.

Janero DR, Burghardt B (1989): Thiobarbituric acid-reactive malondialdehyde formation during superoxide-dependent, iron-catalyzed lipid peroxidation: influence of peroxidation conditions. *Lipids* 24(2):125–131.

Joshita H, Asano T, Hanamura T, Takakura K (1989): Effect of indomethacin and a free radical scavenger on cerebral blood flow and edema after cerebral artery occlusion in cats. *Stroke* 20(6):788–794.

Kahl R, Weimann A, Weinke S, Hildebrandt AG (1987): Detection of oxygen activation and determination of the activity of antioxidants towards reactive oxygen species by use of the chemiluminigenic probes luminol and lucigenin. *Arch Toxicol* 60(1-3):158–162.

Kanemitsu H, Tamura A, Kirino T, Oka H, Sano K, Iwamoto T, Yoshiura M, Iriyama K (1989): Allopurinol inhibits uric acid accumulation in the rat brain following focal cerebral ischemia. *Brain Res* 499(2):367–370.

Kennedy T, Rao N, Hopkins C, Pennington L, Tolley E, Hoidal J (1989): Role of reactive oxygen species in reperfusion injury of the rabbit lung. *J Clin Invest* 83:1326–1335.

Knobeloch LM, Blondin GA, Lyford SB, Harkin JM (1990): A rapid bioassay for chemicals that induce pro-oxidant states. *J Appl Toxicol* 10(1):1–5.

Kohen R, Yamamoto Y, Cundy KC, Ames BN (1988): Antioxidant activity of carnosine, homocarnosine, and anserine present in muscle and brain. *Proc Natl Acad Sci USA* 85(9):3175–3179.

Koide T, Gotoh O, Asano T, Takakura K (1985): Alterations of the eicosanoid synthetic capacity of rat brain microvessels following ischemia: relevance to ischemic brain edema. *J Neurochem* 44(1):85–93.

Kontos HA (1987): Oxygen radicals from arachidonate metabolism in abnormal vascular responses. *Am Rev Respir Dis* 136(2):474–477.

Kontos HA (1989): Oxygen radicals in CNS damage. *Chem Biol Interact* 72(3):229–255.

Kontos HA, Wei EP, Ellis EF, Jenkins LW, Povlishock JT, Rowe GT, Hess ML (1985): Appearance of superoxide anion radical in cerebral extracellular space during increased prostaglandin synthesis in cats. *Circ Res* 57(1):142–151.

Kontos HA, Povlishock JT (1986): Oxygen radicals in brain injury. *Cent Nerv Syst Trauma* 3(4):257–263.

Kontos HA, Wei EP (1986): Superoxide production in experimental brain injury. *J Neurosurg* 64(5):803–807.

Lai EK, Crossley C, Sridhar R, Misra HP, Janzen EG, McCay PB (1986): *in vivo* spin trapping of free radicals generated in brain, spleen, and liver during gamma radiation of mice. *Arch Biochem Biophys* 244(1):156–160.

Leffler CW, Busija DW, Armstead WM, Mirro R (1987): Prostanoids and pial arteriolar diameter in hypotensive newborn pigs. *Am J Physiol* 252:H687–H691.

Lehmann A, Hagberg H, Jacobson I, Hamberger A (1985): Effects of status epilepticus on extracellular amino acids in the hippocampus. *Brain Res* 359(1-2):147–151.

Lemke M, Frei B, Ames BN, Faden AI (1990): Decreases in tissue levels of ubiquinol-9 and -10, ascorbate and alpha-tocopherol following spinal cord impact trauma in rats. *Neurosci Lett* 108(1-2):201–220.

Makino R, Tanaka T, Iizuka T, Ishimura Y, Kanegasaki S (1986): Stoichiometric conversion of oxygen to superoxide anion during the respiratory burst in neutrophils. Direct evidence by a new method for measurement of superoxide anion with diacetyldeuteroheme-substituted horseradish peroxidase. *J Biol Chem* 261(25):11444–11447.

McCord J, Fridovich I (1969): Superoxide dismutase: an enzymic function for erythrocuprein. *J Biol Chem* 244(22):6049.

McCord J, Fridovich I (1968): The reduction of cytochrome c by milk xanthine oxidase. *J Biol Chem* 213(21):5753–5760.

Miller DM, Aust SD (1989): Studies of ascorbate-dependent, iron-catalyzed lipid peroxidation. *Arch Biochem Biophys* 271(1):113–119.

Mizoi K, Suzuki J, Imaizumi S, Yoshimoto T (1986): Development of new cerebral protective agents: the free radical scavengers. *Neurol Res* 8(2):75–80.

Moorhouse PC, Grootveld M, Halliwell B, Quinlan JG, Gutteridge JM (1987): Allopurinol and oxypurinol are hydroxyl radical scavengers. *FEBS Lett* 213(1):23–28.

Nakano S, Kogure K, Abe K, Yae T (1990): Ischemia-induced alterations in lipid metabolism of the gerbil cerebral cortex: I. Changes in free fatty acid liberation. *J Neurochem* 54(6):1911–1916.

Niki E (1991): Action of ascorbic acid as a scavenger of active and stable oxygen radicals. *Am J Clin Nutr* 54:1119S–1124S.

Olanow CW (1990): Oxidation reactions in Parkinson's disease. *Neurology* 40(10):S32–S37.

Oliver CN, Starke-Reed P, Stadtman ER, Liu GJ, Carney JM, Floyd RA (1990): Oxidative damage to brain proteins, loss of glutamine synthetase activity, and production of free radicals during ischemia/reperfusion-induced injury to gerbil brain. *Proc Natl Acad Sci USA* 87(13):5144-5147.

Park JW, Cundy KC, Ames BN (1989): Detection of DNA adducts by high-performance liquid chromatography with electrochemical detection. *Carcinogenesis* 10(5):827–832.

Patt A, Harken AH, Burton LK, Rodell TC, Piermattei D, Schorr WJ, Parker NB, Berger EM, Horesh IR, Terada LS, Stuart L, Cheronis JC, Repine JE (1988): Xanthine oxidase-derived hydrogen peroxide contributes to ischemia reperfusion-induced edema in gerbil brains. *J Clin Invest* 81(5):1556–1562.

Patt A, Horesh IR, Berger EM, Harken AH, Repine JE (1990): Iron depletion or chelation reduces ischemia/reperfusion-induced edema in gerbil brains. *J Pediatr Surg* 25(2):224–227.

Paxinos G, Watson C (eds.) (1986): *The Rat Brain in Stereotaxic Coordinates*, 2nd ed. North Ryde, NSW, Australia: Academic Press.

Pekoe GM (1985): Adaptation of the liquid scintillation counter to measure rapid burst chemiluminescence reactions. In: *Bioluminescence and Chemiluminescence: Instruments and Applications*, vol 2, Van Dyke K, ed. Boca Raton: CRC Press, p. 147–158.

Pellmar TC, Neel KL (1989): Oxidative damage in the guinea pig hippocampal slice. *Free Rad Biol Med* 6(5):467–472.

Pellmar TC, Neel KL, Lee KH (1989): Free radicals mediate peroxidative damage in guinea pig hippocampus *in vitro* . *J Neurosci Res* 24(3):437–444.

Pellmar TC, Tolliver JM, Neel KL (1988): Radiation-induced impairment of neuronal excitability. *Fundam Appl Toxicol* 11(4):577–578.

Prat AG, Turrens JF (1990): Ascorbate- and hemoglobin-dependent brain chemiluminescence. *Free Rad Biol Med* 8(4):319–325.

Pritsos CA, Constantinides PP, Tritton TR, Heimbrook DC, Sartorelli AC (1985): Use of high-performance liquid chromatography to detect hydroxyl and superoxide radicals generated from mitomycin C. *Anal Biochem* 150(2):294–299.

Pruess SD, Nimesheim A, Marnett LJ (1989): Peroxyl radical- and cytochrome P-450-dependent metabolic activation of (+)-7,8-dihydroxy-7,8-dihydrobenzo-(a)pyrene in mouse skin *in vitro* and *in vivo*. *Cancer Res* 49(7):1732–1737.

Pryor WA (1986): Oxy-radicals and related species: their formation, lifetimes, and reactions. *Annu Rev Physiol* 48:657–667.

Puppo A, Cecchini R, Aruoma OI, Bolli R, Halliwell B (1990): Scavenging of hypochlorous acid and of myoglobin-derived oxidants by the cardioprotective agent mercaptopropionylglycine. *Free Rad Res Commun* 10(6):371–381.

Rao PS, Rujikarn N, Luber JJ (1988): High-performance liquid chromatographic method for the direct quantitation of oxy radicals in myocardium and blood by means of 1,3-dimethylthiourea and dimethyl sulfoxide. *J Chromatogr* 459(269):269–273.

Reed GA, Layton ME, Ryan MJ (1988): Metabolic activation of cyclopenteno[c,d]pyrene by peroxyl radicals. *Carcinogenesis* 9(12):2291–2295.

Rehncrona S, Smith D, Akesson B, Westerberg E, Siesjo B (1980): Peroxidative changes in brain cortical fatty acids and phospholipids, as characterized during Fe^{+2-} and ascorbic acid-stimulated lipid peroxidation *in vitro*. *J Neurochem* 34(6):1630–1638.

Riederer P, Sofic E, Rausch W, Schmidt B, Reynolds G, Jellinger K, Youdim M (1989): Transition metals ferritin, glutathione, and ascorbic acid in Parkinsonian brains. *J Neurochem* 52:515–520.

Riesz P, Kondo T, Krishna CM (1990): Free radical formation by ultrasound in aqueous solutions. A spin trapping study. *Free Rad Res Commun* 10 (1-2):27–35.

Roswell D, White EH (1978): The chemiluminescence of luminol and related hydrazides. *Methods Enzymol* 57:409–455.

Rothman SM, Olney JW (1986): Glutamate and the pathophysiology of hypoxic–ischemic brain damage. *Ann Neurol* 19(2):105–111.

Sadrzadeh S, Anderson D, Panter S, Hallaway P, Eaton J (1987): Hemoglobin potentiates central nervous system damage. *J Clin Invest* 79:662–664.

Sadrzadeh SM, Eaton JW (1988): Hemoglobin-mediated oxidant damage to the central nervous system requires endogenous ascorbate. *J Clin Invest* 82(5):1510–1515.

Saunders RD, Dugan LL, Demediuk P, Means ED, Horrocks LA, Anderson DK (1987): Effects of methylprednisolone and the combination of alpha-tocopherol and selenium on arachidonic acid metabolism and lipid peroxidation in traumatized spinal cord tissue. *J Neurochem* 49(1):24–31.

Siesjo BK, Agardh CD, Bengtsson F (1989a): Free radicals and brain damage. *Cerebrovasc Brain Metab Rev* 1(3):165–211.

Siesjo BK, Agardh CD, Bengtsson F, Smith ML (1989b): Arachidonic acid metabolism in seizures. *Ann NY Acad Sci* 559(323):323–339.

Siesjo BK, Bengtsson F (1989c): Calcium fluxes, calcium antagonists, and calcium-related pathology in brain ischemia, hypoglycemia, and spreading depression: a unifying hypothesis. *J Cereb Blood Flow Metab* 9(2):127–140.

Stadtman ER, Oliver CN, Levine RL, Fucci L, Rivett AJ (1988): Implication of protein oxidation in protein turnover, aging, and oxygen toxicity. *Basic Life Sci* 49(331):331–339.

Starke-Reed P, Oliver C (1988): Oxidative modification of enzymes during aging and acute oxidative stress. *Basic Life Sci* 49(537):537–540.

Starke-Reed P, Oliver C (1989): Protein oxidation and proteolysis during aging and oxidative stress. *Arch Biochem Biophys* 275(2):559–567.

Sugata R, Iwahashi H, Ishii T, Kido R (1989): Separation of polyunsaturated fatty acid radicals by high-performance liquid chromatography with electron spin resonance and ultraviolet detection. *J Chromatogr* 487(1):9–16.

Takeuchi H, Handa Y, Kobayashi H, Kawano H, Hayashi M (1991): Impairment of cerebral autoregulation during the development of chronic cerebral vasospasm after subarachnoid hemorrhage in primates. *Neurosurgery* 28:41–48.

Tanaka T, Kanegasaki S, Makino R, Iizuka T, Ishimura Y (1987): Saturated and trans-unsaturated fatty acids elicit high levels of superoxide generation in intact and cell-free preparations of neutrophils. *Biochem Biophys Res Commun* 144(2):606–612.

Tominaga T, Imaizumi S, Yoshimoto T, Suzuki J, Fujita Y (1987): Application of spin-trapping study to rat ischemic brain homogenate incubated with NADPH and Fe-EDTA. *Brain Res* 402(2):370–372.

Uyama O, Shiratsuki N, Matsuyama T, Nakanishi T, Matsumoto Y, Yamada T, Narita M, Sugita M (1990): Protective effects of superoxide dismutase on acute reperfusion injury of gerbil brain. *Free Rad Biol Med* 8(3):265–268.

Wade JV, Samson FE, Nelson SR, Pazdernik TL (1987): Changes in extracellular amino acids during soman- and kainic acid-induced seizures. *J Neurochem* 49(2):645–650.

Wasil M, Halliwell B, Grootveld M, Moorhouse CP, Hutchison DC, Baum H (1987): The specificity of thiourea, dimethylthiourea and dimethyl sulphoxide as scavengers of hydroxyl radicals. Their protection of alpha 1-antiproteinase against inactivation by hypochlorous acid. *Biochem J* 243(3):867–870.

Wei EP, Dietrich WD, Povlishock JT, Navari RM, Kontos HA (1980): Functional, morphological, and metabolic abnormalities of the cerebral microcirculation after concussive injury in cats. *Circ Res* 46:37–47.

Wei EP, Ellison MD, Kontos HA, Povlishock JT (1986): O_2 radicals in arachi-donate-induced increased blood-brain barrier permeability to proteins. *Am J Physiol* 251(4, Pt.2):H693–699.

Wieland E, Niedmann D. Diedrich F, Seidel D, Kather H (1989): Luminescence in the study of lipid metabolism. *J Biolumin Chemilumin* 4(1): 436–445.

Willmore L, Triggs W, Gray J (1986): The role of iron-induced hippocampal peroxidation in acute epileptogenesis. *Brain Res* 382:422–426.

Yamamoto Y, Ames BN (1987): Detection of lipid hydroperoxides and hydrogen peroxide at picomole levels by an HPLC and isoluminol chemiluminescence assay. *Free Rad Biol Med* 3(5):359–361.

Chapter 7

Oxygen Radicals Mediate Ischemia–Reperfusion–Induced Leukocyte–Endothelial Cell Adhesive Interactions

Norman R. Harris, Barbara J. Zimmerman
and D. Neil Granger

A decade ago, we proposed that reactive oxygen metabolites play an important role in mediating the microvascular dysfunction associated with reperfusion of the ischemic small intestine (Granger et al., 1981). In subsequent studies, we attempted to determine the source and identity of the reactive oxygen metabolites that contribute to reperfusion injury in the intestine. The results of our studies and published work from other laboratories suggested that xanthine oxidase is a major source of the superoxide and hydrogen peroxide produced in the postischemic intestine. It was initially held that the superoxide, hydrogen peroxide, and hydroxyl radicals (formed via the iron-catalyzed Haber-Weiss reaction) produced by xanthine oxidase were directly responsible for the microvascular dysfunction induced by ischemia-reperfusion; however, recent studies from our laboratory indicate that xanthine oxidase–derived oxidants may play a more important role in initiating the formation of substances that attract, activate, and promote the adherence of granulocytes to microvascular endothelium. The adherent and activated granulocytes are now considered to be primary mediators of tissue injury, with xanthine oxidase–derived oxidants contributing to the initiation of the inflammatory response (Fig. 1), (Granger, 1988). In this chapter, we summarize some of the key observations that led to our current view regarding the

Oxygen Free Radicals in Tissue Damage
Merrill Tarr and Fred Samson, Editors

importance of oxygen radicals in modulating the leukocyte-endothelial cell adhesive interactions elicited by ischemia-reperfusion.

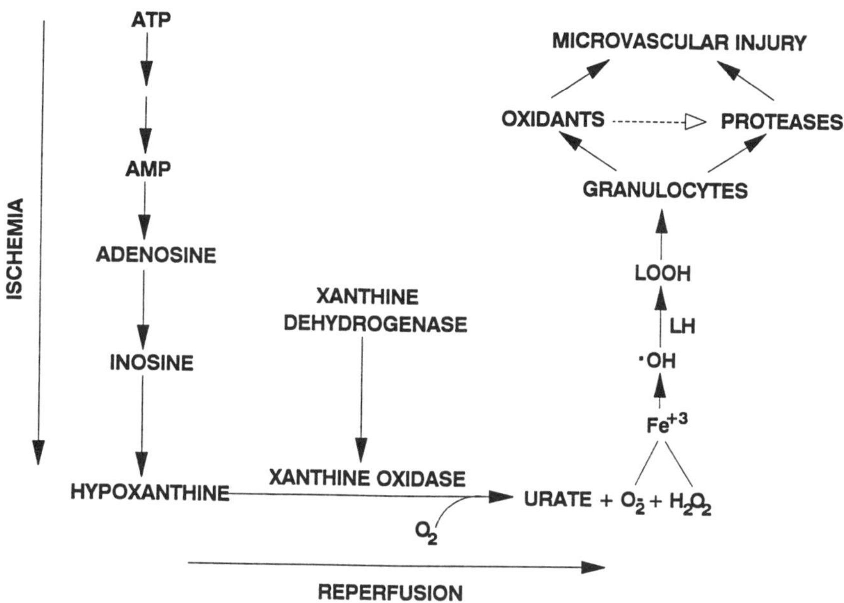

Figure 1 Mechanism proposed to explain the link between ischemia/reperfusion–induced oxygen radical production, recruitment of granulocytes, and microvascular injury. Reprinted with permission of The American Physiological Society from Granger DN (1988): Role of xanthine oxidase and granulocytes in ischemia-reperfusion injury. *Am J Physiol* 255:H1269-H1275

Ischemia-Reperfusion Promotes Granulocyte Recruitment in Intestine

Our first attempt to examine the possible role of granulocytes in ischemia/reperfusion injury focused on whether granulocytes accumulate in intestinal mucosa exposed to ischemia and reperfusion (Grisham et al., 1986). To achieve this objective, we measured tissue-associated myeloperoxidase (MPO) activity in mucosal biopsies obtained from feline intestine exposed to 3 hrs of ischemia (blood flow reduced to 20% of baseline) and 1 hr of reperfusion. The data derived from these experiments indicate that significant granulocyte accumulation occurs during the ischemic period, with a more dramatic increase observed after reperfusion.

Since previous studies have demonstrated that feline eosinophils are devoid of peroxidase activity, it would appear that the ischemia/reperfusion (I/R)-induced increase in MPO activity largely reflects an accumulation of neutrophils, which make up 90% to 95% of circulating granulocytes in the cat. Based on the tissue MPO measurements and the number of circulating neutrophils per unit of MPO activity (10^6 PMNs per unit of MPO activity), we estimate that there are approximately 40 million neutrophils per gram of postischemic intestinal mucosa.

To determine whether xanthine oxidase and superoxide play a role in I/R-induced neutrophil infiltration, cats were pretreated with either superoxide dismutase (SOD) or allopurinol (Grisham et al., 1986). Both agents, when administered at concentrations that were previously shown

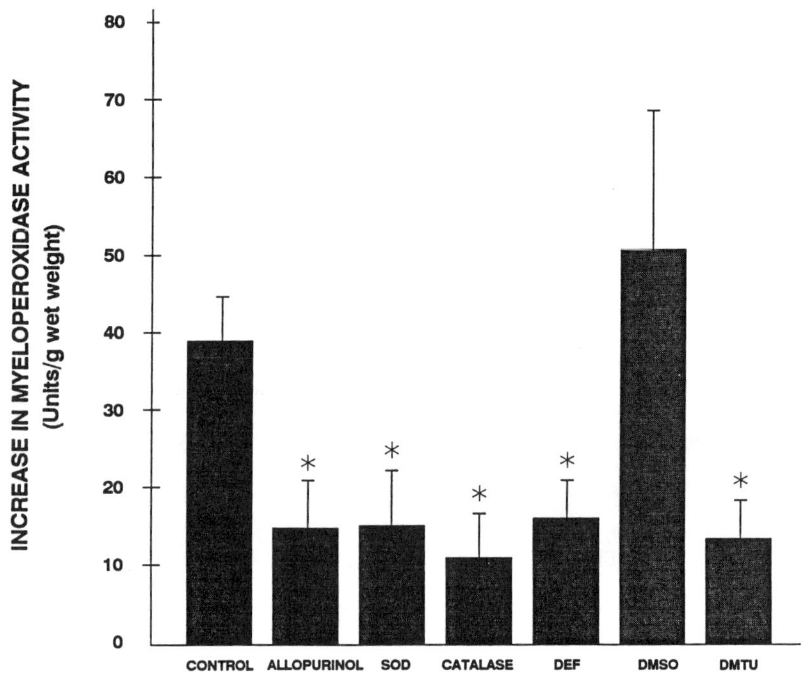

Figure 2 Effects of ischemia-reperfusion on tissue-associated myeloperoxidase activity in feline intestine. SOD, superoxide dismutase; DEF, desferrioxamine; DMSO, dimethylsulfoxide; DMTU, dimethylthiourea. Reprinted with permission of The American Physiological Society from Zimmerman BJ et al. (1990): Role of oxidants in ischemia/reperfusion-induced granulocyte infiltration. *Am J Physiol* 258:G185-G190.

to attenuate I/R injury in feline intestine (Parks and Granger, 1983; Parks et al., 1982), largely abolished the rise in tissue MPO normally observed in the postischemic intestine (see Fig. 2). These observations were recently extended to include an analysis of the effects of pretreatment with either catalase, desferrioxamine, dimethylsulfoxide (DMSO), and dimethylthiourea (DMTU) on I/R-induced neutrophil infiltration (Zimmerman et al., 1990). All of these agents, with the exception of DMSO, reduced the reperfusion-induced rise in tissue MPO in a manner similar to SOD and allopurinol (Fig. 2). Thus, the results obtained from the two studies (Grisham et al., 1986; Zimmerman et al., 1990) suggest that xanthine oxidase–derived oxygen radicals contribute significantly to the neutrophil recruitment elicited by I/R.

Leukocyte Adhesion is a Prerequisite for Reperfusion-Induced Microvascular Dysfunction

An important question that arises from the observation that I/R elicits neutrophil infiltration into the gut mucosa is whether the recruited neutrophils are a cause or merely a consequence of I/R injury in the small bowel. Two approaches were used to address whether neutrophils mediate the microvascular dysfunction associated with reperfusion of the ischemic bowel; that is, neutrophil depletion with polyclonal antiserum and prevention of neutrophil adherence with a monoclonal antibody (MoAb 60.3) directed against the common β-subunit (CD18) of the leukocyte adhesion glycoprotein, CD11/CD18 (Hernandez et al., 1987). Intestinal microvascular permeability to plasma proteins was measured in control preparations, in preparations subjected to 1 hr ischemia and 1 hr reperfusion, and in two experimental groups that received either anti-neutrophil serum or MoAb 60.3 (Fig. 3). The data indicate that either depletion of circulating leukocytes or inhibition of PMN adherence to postcapillary venules significantly attenuates the increased microvascular permeability induced by I/R. The observation that neutrophil depletion and prevention of neutrophil adherence are equally effective in attenuating the microvascular injury suggests that neutrophil adherence to microvascular endothelium is an important and rate-limiting step in neutrophil-mediated, I/R-induced microvascular injury.

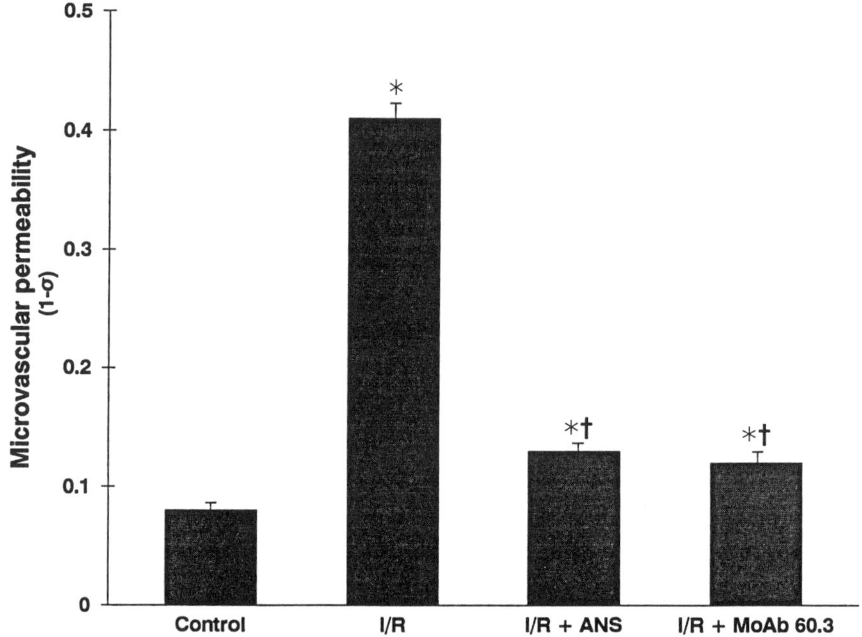

Figure 3 Role of neutrophils in ischemia-reperfusion–induced increase in microvascular permeability (σ = osmotic reflection coefficient). Monoclonal antibody 60.3 immunoneutralizes the common β-subunit (CD18) of the leukocyte adhesion glycoprotein CD11/CD18. Data from Hernandez et al., 1987.

Ischemia-Reperfusion Initiates Leukocyte Adherence and Emigration in Postcapillary Venules

Based on the aforementioned observations that antibodies which prevent leukocyte adherence also prevent I/R-induced microvascular injury, the importance of defining the influence of ischemia and reperfusion on leukocyte–endothelial cell adhesive interactions became apparent. To achieve this objective, the technique of intravital microscopy was used to monitor leukocyte adherence and emigration in the cat mesenteric microcirculation. Straight unbranched segments of postcapillary venules ranging from 25 to 45 μm in diameter were observed using a video camera mounted on a microscope. This allowed the image to be displayed on a monitor and recorded on videotape for subsequent analysis. Playback of the videotape allowed for measurement of several leukocyte parameters including leukocyte rolling velocity, leukocyte adherence (stationary for $\geq$30 s), and the extravasation (emigration) of leukocytes into the perivascular interstitium. These intravital microscopic techniques were used to examine directly the influence of ischemia (1 hr) and reperfusion (1 hr) on leukocyte adherence and emigration in mesenteric postcapillary venules (Granger et al., 1989). We observed (see Fig. 4) that 1 hr of ischemia resulted in significant increases in leukocyte adherence (4.4-fold) and emigration (3.4-fold). After reperfusion, further increments in leukocyte adherence (7-fold) and emigration (8-fold) were observed.

Superoxide Contributes to Ischemia–Reperfusion–Induced Leukocyte Adherence

In 1982, Del Maestro et al. (1982) demonstrated an effect of reactive oxygen metabolites on leukocyte–endothelial cell adhesion. In their study, superoxide and hydrogen peroxide were generated by superfusion of hamster cheek pouch with hypoxanthine (HX) and xanthine oxidase (XO). The enzymatically derived oxidants attenuated leukocyte rolling velocity by 50% and caused a concomitant increase in the number of adherent leukocytes in postcapillary venules. They also demonstrated that the HX-XO–induced alterations in leukocyte rolling velocity and leukocyte adhesion were completely prevented by SOD. However, neither catalase nor L-methionine were able to attenuate the changes induced by HX-XO. These observations provided the first evidence that superoxide may be an important modulator of leukocyte–endothelial cell adhesive interactions during acute and/or chronic inflammation.

In view of the large body of evidence implicating xanthine oxidase–derived superoxide in I/R-induced tissue injury, it was hypothesized that the reactive oxygen metabolites generated by xanthine oxidase may mediate the leukocyte–endothelial cell interactions initiated by ischemia-reperfusion. To address this possibility, we examined the leukocyte adherence and emigration responses to I/R in animals pretreated with either allopurinol or SOD (Fig. 4). Allopurinol treatment did not alter the responses to ischemia per se; however, it largely prevented the further increment in leukocyte adherence and emigration associated with reperfusion. SOD, on the other hand, attenuated the leukocyte adhesion responses elicited by both ischemia and reperfusion. The reductions in reperfusion-induced leukocyte emigration in SOD and allopurinol-treated animals suggests that these treatments interfere either with the ability of

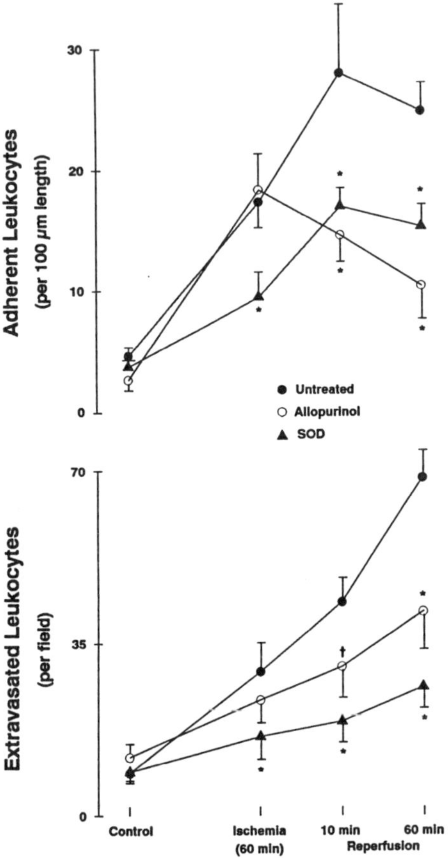

Figure 4 Effects of ischemia and reperfusion on leukocyte adhesion and emigration in postcapillary venules. SOD, superoxide dismutase. Reprinted with permission of The American Physiological Society from Granger DN et al. (1989): Leukocyte adherence to venular endothelium during ischemia-reperfusion. *Am J Physiol* 257:G683-G688.

leukocytes to adhere to endothelium or with their ability to move across the endothelial cell barrier, or both. However, we observed that the linear relationships between leukocyte emigration and adherence in allopurinol- and SOD-treated animals were not different from the relationships derived from untreated animals (Granger et al., 1989). This observation argues against the possibility that allopurinol and SOD interfere directly with the leukocyte's ability to emigrate from the vasculature. Also, the observation that, in contrast to allopurinol, SOD decreased adhesion and extravasation during both ischemia and reperfusion indicates that superoxide is produced by a source that is not inhibitable by allopurinol, possibly neutrophils.

Endothelial Cells are Required for Superoxide-Mediated Leukocyte Adherence

In the aforementioned studies on I/R-induced, superoxide-mediated leukocyte adherence, SOD was administered before the induction of ischemia. To delineate whether superoxide simply initiates I/R-induced leukocyte adherence or whether it contributes to the sustained leukocyte adhesion observed long after reperfusion, another series of adhesion experiments were performed in which SOD was administered 1 hr after reperfusion (Suzuki et al., 1989). We observed that human recombinant CuZn-SOD reduced reperfusion-induced leukocyte adherence by 35% to 45% within 10 min after intravenous administration (Fig. 5). Human recombinant SOD (hSOD) also increases the ratio of leukocyte rolling velocity-to-erythrocyte velocity, indicating that the weaker adhesive interactions that mediate leukocyte rolling are also modulated by superoxide. Peroxide-inactivated hSOD did not affect either leukocyte adherence or rolling. Monoclonal antibody IB_4, which is directed against the common β-subunit (CD18) of the leukocyte adhesion glycoprotein CD11/CD18, reduced reperfusion-induced leukocyte adherence by 75%, yet it did not alter leukocyte rolling velocity.

Additional studies measured neutrophil adhesion to cultured microvascular endothelium exposed to anoxia (simulating ischemia) and reoxygenation (simulating reperfusion). The endothelial cells were exposed to an anoxic gas mixture and the pH adjusted to 6.5 to simulate ischemic conditions. After 30 min of anoxia, followed by 60 min of reoxygenation, the endothelial cell monolayers were exposed to either feline neutrophils alone or neutrophils in the presence of either hSOD (4 and 8 mg/kg), peroxide-inactivated hSOD, or MoAb IB_4. The results obtained in the cell culture model of ischemia-reperfusion were similar to those obtained in our *in vivo* experiments; that is, hSOD decreased neutrophil adhesion by 19% (low dose) and 27% (high dose), whereas inactive hSOD was ineffective. MoAb IB_4 had a more profound effect (38% reduction) on anoxia-reoxygenation-induced neutrophil adhesion. Neutrophil adherence to a biologically inactive surface (plastic) was unaffected by hSOD.

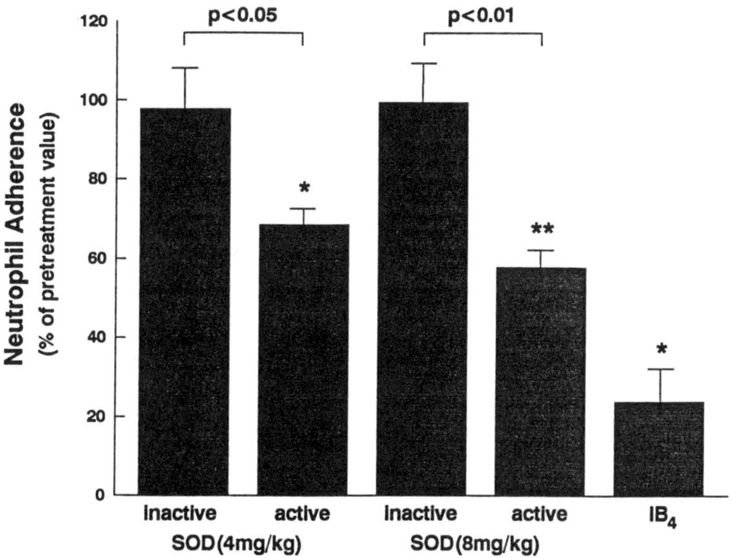

Figure 5 Effects of superoxide dismutase (SOD), peroxide-inactivated SOD, and an anti-CD18 monoclonal antibody (IB_4) on reperfusion-induced neutrophil adherence in mesenteric venules. Reprinted with permission of The American Physiological Society from Suzuki M, et al. (1989): Superoxide mediates reperfusion-induced leukocyte–endothelial cell interactions. *Am J Physiol* 257:H1740-H1745.

Several conclusions were drawn from these results. Since only active hSOD decreased leukocyte adhesion to endothelial cells (*in vivo* and *in vitro*), the catalytic activity of hSOD appears to be responsible for its antiadherence properties. The observations that active hSOD resulted in an increased leukocyte rolling velocity and a reduction in leukocyte adherence suggest that hSOD reduces the interactions between leukocytes and endothelium. However, MoAb IB_4, which is directed against the leukocyte adhesion glycoprotein CD11/CD18, also decreased leukocyte adherence, yet it did not affect leukocyte rolling velocity. These results suggest that leukocyte rolling is dependent on O_2^- and independent of CD11/CD18, whereas I/R-induced leukocyte adherence is influenced by both O_2^- and CD11/CD18. Finally, since hSOD does not affect neutrophil adhesion to plastic, yet IB_4 virtually eliminates it, the presence of endothelial cells seems to be required for O_2^--mediated neutrophil adhesion.

Xanthine Oxidase–Derived Hydrogen Peroxide Contributes to Reperfusion-Induced Leukocyte Adherence in Postcapillary Venules

The observation that SOD reduces leukocyte adherence in mesenteric venules raises the interesting possibility that other agents that have been used either to scavenge or prevent the production of oxygen radicals in postischemic tissues may also exert an influence on the adhesive interactions between leukocytes and venular endothelium. To address this possibility, we determined whether catalase, desferrioxamine, and oxypurinol alter leukocyte adhesion to microvascular endothelium when administered at concentrations that have been shown to protect against reperfusion injury (Suzuki et al., 1991). We observed that catalase and oxypurinol significantly attenuate leukocyte adherence when administered 1 hr after reperfusion, whereas neither inactive catalase nor desferrioxamine altered leukocyte adhesion. These observations indicate that at least part of the protective effects previously observed with oxypurinol and catalase in models of I/R injury may be attributable to the antiadhesive actions of these drugs, while the beneficial effects of desferrioxamine cannot be ascribed to an influence on leukocyte adhesion. These results also indicate that xanthine oxidase and H_2O_2 play an important role in mediating the leukocyte–endothelial cell adhesive interactions elicited by I/R.

Although CuZn-SOD is efficient in catalyzing the dismutation of superoxide, the enzyme is known to exhibit low level, nonspecific peroxidase activity, a characteristic not shared by Mn-SOD (Hodgson and Fridovich, 1975). Thus, the similarity between the antiadhesive actions of CuZn-SOD and that of a catalase raised the concern that CuZn-SOD attenuates leukocyte adherence as a result of its limited ability to decompose H_2O_2 rather than its ability to dismutate superoxide. To address this issue, we examined the influence of Mn-SOD on reperfusion-induced leukocyte adherence in a manner similar to that previously employed for CuZn-SOD (Suzuki et al., 1991). Our observation that Mn-SOD reduces leukocyte adhesion comparable to CuZn-SOD suggests that the antiadhesive actions of CuZn-SOD can be attributed to its ability to scavenge superoxide.

Inactivation of Nitric Oxide May Explain Superoxide-Mediated Leukocyte Adherence

There are several possible explanations for the ability of SOD to reduce leukocyte adhesion in postcapillary venules. SOD may interfere with either superoxide-induced (a) formation of inflammatory mediators, (b) expression of leukocyte and/or endothelial cell adhesion molecules, or (c) inactivation of an endothelial cell-derived antiadhesive substance. We have recently addressed the possibility that nitric oxide is an endogenous molecule that accounts for the antiadhesive actions of SOD (Kubes et al., 1991). Nitric oxide (NO) is normally produced by microvascular endothelium and is rapidly inactivated by superoxide. Thus, if NO normally attenuates leukocyte adhesion to vascular endothelium, then conditions associated with enhanced formation of superoxide (e.g., ischemia/reperfusion) should lead to increased leukocyte adherence. To test this hypothesis, we examined the influence of inhibitors of NO production of leukocyte adherence in normal postcapillary venules. N^G-nitro-L-arginine methylester (L-NAME) and N^G-monomethyl-L-arginine (L-NMMA), analogues of L-arginine that inhibit NO production, were superfused onto the surface of cat mesentery. Both inhibitors increased leukocyte adherence more than 15-fold (Fig. 6). Leukocyte emigration was also increased and venular wall shear was reduced by nearly half. The CD18-specific antibody MoAb IB_4 abolished the leukocyte adherence induced by both L-NAME and L-NMMA. Incubation of isolated feline neutrophils with L-NMMA, but not L-NAME, led up to upregulation of CD11/CD18 as assessed by flow cytometry. Only a minor component of the increased leukocyte adherence induced by NO inhibitors could be attributed to the reduction in venular shear rate. The L-NAME–induced leukocyte adherence was inhibited by L-arginine but not by D-arginine. *In vitro* tests showed that L-NMMA, but not L-NAME, causes about 10% up-regulation of CD18. Therefore, the increase in adherence caused by L-NMMA was likely due to both NO inhibition and CD18 up-regulation. Another factor that may have contributed to the increase in adhesion after both L-NMMA and L-NAME treatment was the decrease in shear rate: venular diameter remained unchanged, but arteriolar constriction caused about a 50% reduction in red blood cell centerline velocity. Therefore, additional tests were performed in which partial arterial occlusion was used to decrease venular shear rate. The results indicated that reduced shear rates could not explain the increase in leukocyte adherence observed during L-NAME treatment. Overall, these results indicate that NO may be an important endogenous inhibitor of leukocyte adherence in postcapillary venules and that inactivation of NO by superoxide may contribute to the increased leukocyte adherence observed in conditions associated with enhanced production of superoxide by endothelial cells and/or neutrophils.

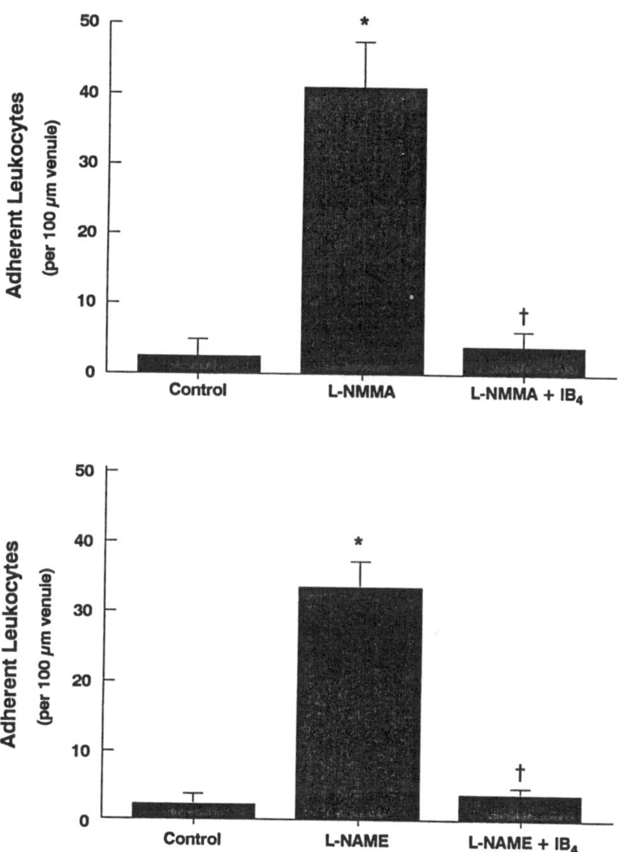

Figure 6 Effect of nitric oxide synthesis inhibitors (L-NAME and L-NMMA) on leukocyte adherence in postcapillary venules. IB_4, a CD18-specific monoclonal antibody. Reprinted with permission of the authors from Kubes P et al. (1991): Nitric oxide: an endogenous modulator of leukocyte adhesion. *Proc Natl Acad Sci USA* 88:4651-4655.

Hydrogen Peroxide and Monochloramine Promote Leukocyte Adherence in Postcapillary Venules Via CD18 Activation

The antiadhesive action of catalase in mesenteric venules exposed to I/R raises the interesting possibility that neutrophil-derived hydrogen peroxide may promote leukocyte adherence either directly or indirectly via the formation of MPO derived oxidants such as hypochlorous acid (HOCl) and monochloramine (NH_2Cl). It has been estimated that most of the hydrogen peroxide produced by activated neutrophils is consumed by MPO, which is released into extracellular fluid from azurophilic granules. MPO catalyzes the oxidation of chloride by hydrogen peroxide to yield HOCl, which in turn rapidly reacts with primary amines to yield N-chloramines such as monochloramine and taurine monochloramine. HOCl and N-chloramines are highly reactive oxidants that may be more or less cytoxic than hydrogen peroxide, depending on the lipophilicity of the compound. To address the possible role of neutrophil-derived oxidants in modulating leukocyte adhesion, we determined whether hydrogen peroxide (H_2O_2), hypochlorous acid (HOCl), and monochloramine (NH_2Cl), at concentrations produced by activated neutrophils, promote leukocyte adherence to microvascular endothelium in postcapillary venules (Suzuki et al., 1991). The results derived from these experiments indicate that H_2O_2 and NH_2Cl, but not HOCl, promote leukocyte adhesion to venular endothelium. Incubation of isolated cat neutrophils with either NH_2Cl or H_2O_2 resulted in up-regulation of CD11/CD18, as assessed by flow cytometry. Although the leukocyte adhesion induced by both H_2O_2 and NH_2Cl was associated with a reduction in venular wall shear rate, corresponding decrements in shear rate induced by partial occlusion of the mesenteric artery did not lead to similar levels of leukocyte adherence. The leukocyte adherence induced by either H_2O_2 or NH_2Cl was largely prevented by the CD18-specific monoclonal antibody MoAb IB_4 (Fig. 7), which indicates that both oxidants promote leukocyte adherence via up-regulation of CD11/CD18. Thus, it appears that H_2O_2 and NH_2Cl promote leukocyte adhesion at physiologically relevant concentrations and that the adhesion results from up-regulation and/or activation of CD11/CD18.

Epilogue

With the advent of sophisticated technology for monitoring leukocyte adhesion to endothelial cells in microvessels has come an appreciation for the role of reactive oxygen metabolites in modulating leukocyte–endothelial cell adhesion. It is now apparent that both superoxide and hydrogen peroxide have the potential for amplifying the inflammatory response via their ability to promote the formation of stronger adhesive bonds between circulating neutrophils and microvascular endothelium. Although the cellular and molecular basis for these proadhesive interactions remain undefined, it appears reasonable to propose that novel therapeutic strategies for the management of acute and chronic inflammatory conditions could be devised based on the observation that oxygen free radical scavengers reduce leukocyte adherence and emigration.

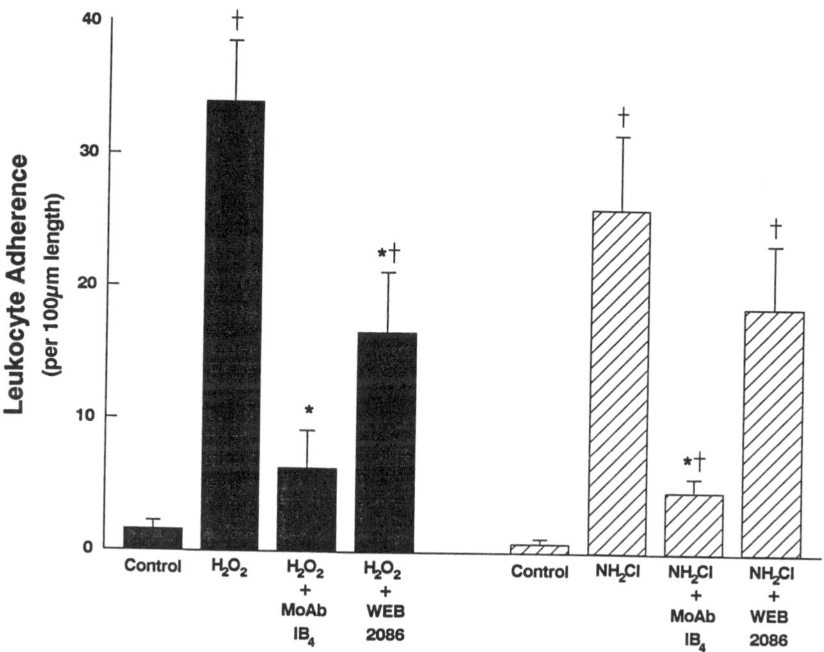

Figure 7 Effects of H_2O_2 and NH_2Cl on leukocyte adhesion to mesenteric venules. Also shown are the effects of treatment with either MoAb IB_4 or WEB 2086, a PAF receptor antagonist. Reprinted with permission of the Academic Press from Suzuki M, et al. (1991): Neutrophil-derived oxidants promote leukocyte adherence in postcapillary venules. *Microvasc Res* 42:125-138.

Acknowledgment. Supported by a grant from the National Institutes of Health (DK33594)

References

Del Maestro RF, Planker M, Arfors KE (1982): Evidence for the participation of superoxide anion radical in altering the adhesive interaction between granulocytes and endothelium, in vivo. *Int J Microcirc Clin Exp* 1:105–120.

Granger DN (1988): Role of xanthine oxidase and granulocytes in ischemia-reperfusion injury. *Am J Physiol* 255:H1269–H1275.

Granger DN, Benoit JN, Suzuki M, Grisham MB (1989): Leukocyte adherence to venular endothelium during ischemia-reperfusion. *Am J Physiol* 257:G683–G688.

Granger DN, Rutili G, McCord JM (1981): Superoxide radicals in feline intestinal ischemia. *Gastroenterology* 81:22–29.

Grisham MB, Hernandez LA, Granger DN (1986): Xanthine oxidase and neutrophil infiltration in intestinal ischemia. *Am J Physiol* 251:G567–G574.

Hernandez LA, Grisham MB, Twohig B, Arfors KE, Harlan JM, Granger DN (1987): Role of neutrophils in ischemia-reperfusion-induced microvascular injury. *Am J Physiol* 253:H699–H703.

Hodgson EK, Fridovich I (1975): The interaction of bovine erythrocyte superoxide dismutase with hydrogen peroxide: chemiluminescence and peroxidation. *Biochemistry* 14:5299–5302.

Kubes P, Suzuki M, Granger DN (1991): Nitric oxide: an endogenous modulator of leukocyte adhesion. *Proc Natl Acad Sci USA* 88:4651–4655.

Parks DA, Bulkley GB, Granger DN, Hamilton SR, McCord JM (1982): Ischemic injury in the cat small intestine: role of superoxide radicals. *Gastroenterology* 82:9–15.

Parks DA, Granger DN (1983): Ischemia-induced vascular changes: role of xanthine oxidase and hydroxyl radicals. *Am J Physiol* 245:G285–G289.

Suzuki M, Asako H, Kubes P, Jennings S, Grisham MB, Granger DN (1991): Neutrophil-derived oxidants promote leukocyte adherence in postcapillary venules. *Microvasc Res* 42:125–138.

Suzuki M, Grisham MB, Granger DN (1991): Leukocyte-endothelial cell adhesive interactions: Role of xanthine oxidase-derived oxidants. *J Leuk Biol* 50:488–494.

Suzuki M, Inauen W, Kvietys PR, Grisham MB, Meininger C, Schelling ME, Granger HJ, Granger DN (1989): Superoxide mediates reperfusion-induced leukocyte-endothelial cell interactions. *Am J Physiol* 257:H1740–H1745.

Zimmerman BJ, Grisham MB, Granger DN (1990): Role of oxidants in ischemia/reperfusion-induced granulocyte infiltration. *Am J Physiol* 258:G185–G190.

Chapter 8

The Respiratory Burst and Drug Metabolism: Implications for Idiosyncratic Drug Reactions and Antiinflammatory Effects

Jack P. Uetrecht

The Respiratory Burst

A major function of neutrophils and monocytes is the destruction of pathogenic organisms (Klebanoff and Clark, 1978). The activation of neutrophils by bacteria or other stimuli leads to a large increase in the uptake in oxygen, referred to as the respiratory burst, and much of this oxygen is converted to superoxide by NADPH oxidase. The superoxide is converted to hydrogen peroxide, both spontaneously and through the action of superoxide dismutase (SOD). Activation of neutrophils also leads to the release of myeloperoxidase (MPO) from granules, and this is converted by hydrogen peroxide to an oxidized form of the enzyme referred to as compound I. Compound I is a strong oxidant and can presumably oxidize many substrates; however, *in vivo* the major substrate appears to be chloride ion, which is oxidized to hypochlorous acid. It appears that most of the hydrogen peroxide generated by neutrophils is converted to hypochlorous acid (Weiss, 1989). Hypochlorous acid has potent antimicrobial activity, and it also appears to play a role in the modulation of other active agents released by neutrophils such as collagenase. Monocytes are capable of a respiratory burst and also release MOP when activated; however, their activity is somewhat less than that of neutrophils. In general, monocytes appear to lose MOP when they become tissue macrophages (van Furth, 1980).

Oxygen Free Radicals in Tissue Damage
Merrill Tarr and Fred Samson, Editors

The oxidants that are produced by activated neutrophils and monocytes can not only kill microorganisms but they also can cause tissue damage and inflammation, as discussed by other speakers at this conference. In addition, we and others have shown that these oxidants can oxidize drugs (Uetrecht, 1990). The oxidation of drugs by activated neutrophils often leads to chemically reactive metabolites. This process appears to mediate some of the antiinflammatory effects of drugs, and it may also be responsible for some idiosyncratic drug reactions.

Idiosyncratic Drug Reactions

Idiosyncratic drug reactions represent a major medical problem (Mathews, 1984). They are often life-threatening and their idiosyncratic nature makes them essentially impossible to prevent. Little is known about the mechanism of such reactions; however, they usually have characteristics that suggest involvement of the immune system (Park et al., 1987; Parker, 1982; Pohl et al., 1988; Uetrecht, 1990). These characteristics include a delay between initiation of therapy and the development of toxicity of more than a week, but the reaction is usually immediate if a patient is reexposed to the drug. The reactions are often said to be independent of dose. Although this is not strictly true, their dose dependency is often not apparent because the therapeutic dose may already be at the top of the dose-response curve for the idiosyncratic reaction. In addition, most patients will not have an idiosyncratic reaction at any dose of the drug and this makes it even more difficult to detect a dose dependency in patients who are at risk of an idiosyncratic reaction. Most animals also do not have idiosyncratic reactions to drugs that cause idiosyncratic reactions in humans. The idiosyncratic nature and almost complete lack of animal models makes the study of idiosyncratic drug reactions very difficult.

There is a large body of evidence to suggest that many types of adverse drug reactions are due to reactive metabolites of drugs (Nelson, 1982; Nelson and Pearson, 1990; Park and Kitteringham,1990). The major site of drug metabolism is the liver. Although many idiosyncratic drug reactions involve the liver, even more involve the skin, and involvement of the bone marrow, kidneys, and several other organs is also common (Mathews, 1984; Parker, 1982). By their nature, reactive metabolites usually have short biological half-lifes. If they were only formed in the liver one might guess that toxicity would be confined to the liver, such as halothane-induced hepatic necrosis. Halothane-induced hepatic necrosis appears to involve the formation of trifluoroacetyl chloride and reaction of this metabolite with lysine groups on hepatic protein (Kenna et al., 1988). In some patients this altered protein leads to immune-mediated damage of liver cells. It is likely that metabolism of halothane occurs almost exclusively in the liver, and this toxicity is also confined to the liver. In idiosyncratic reactions that involve other organs it is likely that the reactive metabolite either has a longer half-life or is formed in other organs. Since many idiosyncratic reactions appear to involve the immune system, formation of reactive metabolites by cells of the immune system, such as monocytes, may be important.

Metabolism of Drugs to Reactive Metbolites by Activated Neutrophils and Monocytes

The metabolism of drugs by activated leukocytes or the myeloperoxidase system appears to be determined by the presence of a functional group that is readily oxidized by compound I or hypochlorous acid rather than by specific binding of the drug to myeloperoxidase (Uetrecht, 1990). An example of such a functional group is an arylamine. An example of drug that has an arylamine functional group is procainamide. Procainamide is oxidized, first to a hydroxylamine, then to a nitroso derivative, and finally to a nitro derivative by the combination of myeloperoxidase and hydrogen peroxide or by activated neutrophils or monocytes (Uetrecht et al., 1988a). If chloride ion is added to the myeloperoxidase/hydrogen peroxide system, the major product is N-chloroprocainamide (Uetrecht and Zahid, 1991). Even though there is a chloride present in the incubations of procainamide with activated cells, the N-chloro-metabolite is not observed, presumably because it reacts rapidly with the cells. In support of this hypothesis, synthetic N-chloroprocainamide disappears rapidly when added to a suspension of neutrophils. These reactions are summarized in Fig. 1.

Other aromatic amines metabolized in a similar manner include dapsone (Uetrecht et al., 1988b), sulfamethoxazole (Cribb et al., 1990), and a metabolite of chloramphenicol in which the nitro group has been reduced to an amine (unpublished observations). These pathways appear to be the general pathways of the metabolism of primary arylamines by the myeloperoxidase system or activated neutrophils, and it is likely that other arylamines would be metabolized in a similar manner. The nitroso and chloramine metabolites are chemically reactive and could be involved

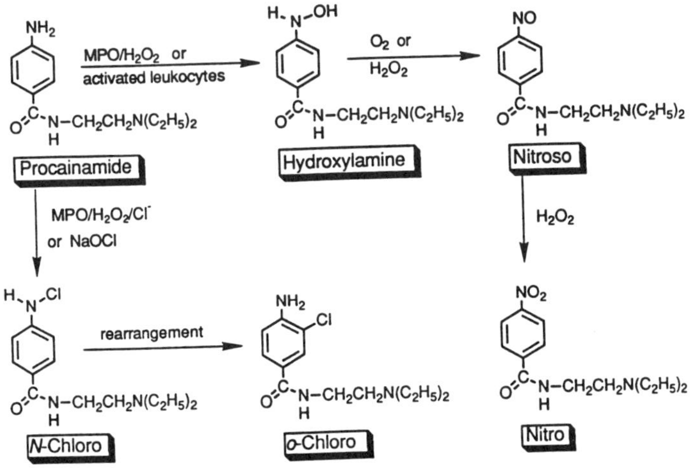

Procainamide → (MPO/H_2O_2 or activated leukocytes) → Hydroxylamine → (O_2 or H_2O_2) → Nitroso → (H_2O_2) → Nitro

Procainamide → ($MPO/H_2O_2/Cl^-$ or NaOCl) → *N*-Chloro → (rearrangement) → *o*-Chloro

Figure 1 The metabolic pathways of procainamide mediated by activated neutrophils or the myeloperoxidase system.

in the adverse reactions associated with these drugs.

Other nitrogen-containing drugs that have been shown to be metabolized by activated neutrophils include vesnarinone, clozapine, and phenytoin. Vesnarinone is a tertiary arylamine and a new drug used to treat congestive heart failure. It also appears to be oxidized by activated neutrophils or the myeloperoxidase system and the product(s) covalently bind to the cells (Uetrecht et al., 1989). Clozapine is a new antipsychotic agent with a diazepine ring that is metabolized by activated neutrophils. It appears that a free radical is produced, leading to a glutathione adduct, and other reactive metabolites may be formed (Fischer et al., 1991). Phenytoin is chlorinated by the myeloperoxidase system to N,N′-dichlorophenytoin, and this appears to be responsible for covalent binding of the drug to activated neutrophils (Uetrecht and Zahid, 1988).

In addition to oxidation of nitrogen, oxidation of sulfur-containing drugs also occurs. Specifically, propylthiouracil is oxidized to a reactive sulfonic acid by activated neutrophils, and there are also several intermediates that may also be chemically reactive (Waldhauser and Uetrecht, 1991). Even carbon–carbon double bonds can be metabolized by activated neutrophils. The metabolism of carbamazepine involves chlorination of the 10,11 double bond to give a carbonium ion intermediate, which undergoes rearrangement that leads to contraction of the azepine ring and eventual loss of a ring carbon (submitted for publication). Thus, although the metabolism of drugs activated neutrophils is limited, a relatively wide range of drugs can be oxidized by this system.

Possible Implications of Drug Metabolism by Leukocytes

The types of functional groups that are metabolized by activated neutrophils and monocytes are also associated with a relatively large incidence of idiosyncratic drug reactions. In particular, arylamines are usually associated with a high incidence of agranulocytosis and, with the exception of dapsone, they are associated with a high incidence of drug-induced lupus (Uetrecht, 1988, 1989b, 1990). Other drugs, such as sulfasalazine, chloramphenicol, and practolol, which are metabolized to primary aryl amines, are probably also metabolized by activated neutrophils. Drugs that contain thiono or thiol groups such as propylthiouracil and penicillamine are also associated with a high incidence of such adverse reactions. These drugs are also associated with a relatively high incidence of more generalized idiosyncratic drug reactions (Uetrecht, 1990).

It appears that metabolism of drugs by activated leukocytes often involves reactive intermediates, and since reactive intermediates are believed to be involved in the mechanism of many idiosyncratic reactions, this metabolism may be responsible for some idiosyncratic reactions, especially those that involve neutrophils such as agranulocytosis. Drug-induced lupus is an autoimmune syndrome, and since macrophages are essential for processing and presenting antigen to T cells in the induction of an immune reaction (Unanue and Allen, 1987), metabolism of drugs to reactive metabolites by activated monocytes may be involved in the mechanism of drug-induced lupus (Uetrecht, 1990). At present we have no direct evidence for such hypotheses. They are based soley on the circumstantial evidence such as the observation that functional groups that are easily oxidized by this system are associated with a high incidence of certain adverse reactions. Furthermore, it makes more sense to investigate reactive metabolite formation by the target organ than by the liver.

One important aspect of idiosyncratic reactions is the determination of the risk factors involved since at present these reactions are totally unpredictable. If oxidation of drugs to reactive metabolites by neutrophils or monocytes is involved in the mechanism of some of these reactions, since activation of the cells is necessary for metabolism, an infection or other inflammatory condition could be a risk factor. In the case of vesnarinone, it appears as if the use of influenza vaccine during treatment greatly increased the risk of agranulocytosis. Furthermore, we found that *in vitro* influenza vaccine activated neutrophils and led to covalent binding of the drug to the cells (Uetrecht et al., 1989). It also appears that the risk of drug-induced agranulocytosis may be higher in the winter months, but this is not documented.

It appears that metabolism of drugs by neutrophils can also be responsible for some of the antiinflammatory effects of drugs. For example, drugs such as dapsone have antiinflammatory effects in diseases, such as dermatitis herpetiformis, in which the inflammation is due to neutrophils (Uetrecht, 1989a). The mechanism of this antiinflammatory effect appears to be due to inhibition of myeloperoxidase (Stendahl et al., 1978). Kettle and Winterbourn have demonstrated that many antiinflammatory drugs reduce compound I of myeloperoxidase to compound II, which is relatively inactive, and this is probably one mechanism by which some drugs exert their antiinflammatory effects (Kettle and Winterbourn, 1991). The reactive metabolites of drugs generated by myeloperoxidase may also inhibit the enzyme by irreversibly binding to it. Other antiinflammatory drugs that are metabolized by the myeloperoxidase system include phenylbutazone, diclofenac, 5-aminosalicylate and tenoxicam (Dull et al., 1987; Ichihara et al., 1985, 1986; Zuurbier et al., 1990).

Summary

In addition to the other manifestations of the respiratory burst, it now appears that it can result in the metabolism of a wide range of drugs. The major cells involved are neutrophils and monocytes. The exact extent of this metabolism *in vivo* is not clear because the studies have been done *in vitro*. Although it is unlikely that such metabolism makes a significant contribution to the overall clearance of drugs, the reactive metabolites that are formed could be responsible for many idiosyncratic drug reactions, especially reactions such as agranulocytosis where the target organ is the bone marrow. Metabolism of drugs by neutrophils also appears to mediate some of the antiinflammatory effects of drugs.

Acknowledgment. This work was supported by grants from the Medical Research Council of Canada (MA-9336 and MA-10036).

References

Cribb AE, Miller M, Tesoro A, Spielberg SP (1990): Peroxidase-dependent oxidation of sulfonamides by monocytes and neutrophils from humans and dogs. *Mol Pharmacol* 38:744–751.

Dull BJ, Salata K, Van Langenhove A, Goldman P (1987): 5-Aminosalicylate: oxidation by activated leukocytes and protection of cultured cells from oxidative damage. *Biochem Pharmacol* 36:2467–2472.

Fischer V, Haar JA, Greiner L, Loyd RV, Mason RP (1991): Possible role of free radical formation in clozapine (Clozaril^R)-induced agranulocytosis. *Mol Pharmacol.* 40:846–853.

Ichihara S, Tomisawa H, Fukazawa H, Tateishi M (1985): Involvement of leukocyte peroxidases in the metabolism of tenoxicam. *Biochem Pharmacol* 34: 1337–1338.

Ichihara S, Tomisawa H, Fukazawa H, Tateishi M, Joly R, Heintz R (1986): Involvement of leukocytes in the oxygenation and chlorination reaction of phenylbutazone. *Biochem Pharmacol* 35:3935–3939.

Kenna JG, Satoh H, Christ DD, Pohl LR (1988): Metabolic basis for a drug hypersensitivity: antobodies in sera from patients with halothane hepatitis recognize liver neoantigens that contain the trifluoroacetyl group derived from halothane. *J Pharmacol Exp Ther* 245:1103–1109.

Kettle AJ, Winterbourn CC (1991): Mechanism of inhibition of myeloperoxidase by anti-inflammatory drugs. *Biochem Pharmacol* 41:1485–1492.

Klebanoff SJ, Clark RA (1978): *The Neutrophil: Function and Clinical Disorders.* Amsterdam: Elsevier/North-Holland Inc.

Mathews KP (1984): Clinical spectrum of allergic and pseudoallergic drug reactions, *J Allergy Clin Immunol* 74:556–558.

Nelson SD (1982): Metabolic activation and drug toxicity. *J Med Chem* 25:753–765.

Nelson SD, Pearson PG (1990): Covalent and noncovalent interactions in acute lethal cell injury caused by chemicals. *Annu Rev Pharmacol Toxicol* 30:169–195.

Park BK, Coleman JW, Kitteringham NR (1987): Drug disposition and drug hypersensitivity. *Biochem Pharmacol* 36:581–590.

Park BK, Kitteringham NR (1990): Drug-protein conjugation and its immunological consequences. *Drug Metab Rev* 22:87–144.

Parker CW (1982): Allergic reactions in man. *Pharmacol Rev* 34:85–104.

Pohl LR, Satoh H, Christ DD, Kenna JG (1988): The immunologic and metabolic basis of drug hypersensitivities. *Annu Rev Pharmacol* 28:367–387.

Stendahl O, Molin L, Dahlgren C (1978): The inhibition of polymorphonuclear leukocyte cytoxicity by dapsone: a possible mechanism in the treatment of dermatitis herpetiformis. *J Clin Invest* 62:214–220.

Uetrecht J (1989a): Dapsone and sulfapyridine. In: *Clinics in Dermatology*, Shear NH, ed., Philadelphia: Lippincott.

Uetrecht J (1989b): Mechanism of hypersensitivity reactions: proposed involvement of reactive metabolites generated by activated leukocytes. *Trends Pharmacol Sci* 10:463–467.

Uetrecht J, Zahid N (1988): N-Chlorination of phenytoin by myeloperoxidase to a reactive metabolite. *Chem Res Toxicol* 1:148–151.

Uetrecht J, Zahid N, Rubin R (1988a): Metabolism of procainamide to a hydroxylamine by human neutrophils and mononuclear leukocytes. *Chem Res Toxicol*1:74–78.

Uetrecht J, Zahid N, Shear NH, Biggar WD (1988b): Metabolism of dapsone to a hydroxylamine by human neutrophils and mononuclear cells. *J Pharmacol Exp Ther* 245:274–279.

Uetrecht JP (1988): Mechanism of drug-induced lupus. *Chem Res Toxicol* 1:133–143.

Uetrecht JP (1990): Drug metabolism by leukocytes, its role in drug-induced lupus and other idiosyncratic drug reactions. *CRC Crit Rev Tox* 20:213–235.

Uetrecht JP, Zahid N (1991): N-Chlorination and oxidation of procainamide by myeloperoxidase:toxicological implications. *Chem Res Toxicol* 4:218–222.

Uetrecht JP, Zahid N, Spielberg SP (1989): Oxidation of OPC-8212 to a reactive intermediate by influenza vaccine-activated neutrophils: possible relationship to agranulocytosis. *Eur J Clin Pharmacol* 36(suppl):A53.

Unanue ER, Allen PM (1987): The basis for the immunoregulatory role of macrophages and other accessory cells. *Science* 236:551–557.

van Furth R (1980): *Mononuclear Phagocytes: Functional Aspects.* The Hague: Martinus Nijhoff.

Waldhauser L, Uetrecht J (1991): Oxidation of propylthiouracil to reactive metabolites by activated neutrophils: implications for agranulocytosis. *Drug Metab Dispos* 19:354–359.

Weiss SJ (1989): Tissue destruction by neutrophils. *N Engl J Med* 320:365–376.

Zuurbier KWM, Bakkenist ARJ, Fokkens RH, Nibbering NMM, Wever R, Muijsers AO (1990): Interaction of myeloperoxidase with diclofenac: inhibition of the chlorinating activity of myeloperoxidase by diclofenac and oxidation of diclofenac to dihydroxyazobenzene by myeloperoxidase. *Biochem Pharmacol* 40:1801–1808.

Chapter 9

Role of Oxygen Radicals in Central Nervous System Trauma

Edward D. Hall

There is now extensive experimental support for the early occurrence and pathophysiological importance of oxygen radical formation and cell membrane lipid peroxidation in the injured nervous system (Braughler and Hall, 1989; Demopoulos et al., 1980; Hall and Braughler, 1986; Kontos and Povlishock, 1986). The radical-initiated peroxidation of neuronal, glial, and vascular cell membranes and myelin is catalyzed by free iron released from hemoglobin, transferrin, and ferritin by either lowered tissue pH or oxygen radicals. If unchecked, lipid peroxidation is a geometrically progressing process that will spread over the surface of the cell membrane causing impairment to phospholipid-dependent enzymes, disruption of ionic gradients and, if severe enough, membrane lysis.

Several criteria for the establishment of the pathophysiological significance of oxygen radical processes have been met. These include (a) the demonstration of increased post-traumatic levels of oxygen radicals and lipid peroxides in brain or spinal cord soon after injury, (b) the spatial and temporal correlation between oxygen radical formation and pathophysiological alterations (e.g., loss of microvascular autoregulation, vasogenic edema, progressive posttraumatic ischemia development, axonal degeneration), (c) the striking similarity between post-traumatic central nervous system (CNS) pathology and that caused by chemical peroxidative insult (e.g., iron micro injection), and (d) the protective efficacy of oxygen radical scavenging agents or compounds that inhibit lipid peroxidation (i.e., lipid antioxidants). The injured CNS provides a fertile environment for the generation of oxygen radical and lipid peroxidation reactions because of a high content of susceptible polyunsaturated fatty acids.

Oxygen Free Radicals in Tissue Damage
Merrill Tarr and Fred Samson, Editors

This chapter reviews some of the evidence for the involvement of free radicals in acute CNS injury and their relationship to specific pathophysiological events. Much of this information has been obtained from investigations demonstrating the protective efficacy of lipid antioxidant agents in experimental models.

Oxygen Radicals in Spinal Cord Injury

Much of the available information concerning the occurrence of oxygen radical-induced lipid peroxidation has been gleaned from studies on experimental acute spinal cord injury. Biochemical indices of early radical reactions in the bluntly injured spinal cord (contusion or compression injuries) include an increase in polyunsaturated fatty acid oxidation products such as malonyldialdehyde (Hall and Braughler, 1982; Kurihara, 1985; Milvy et al., 1973), a decrease in tissue cholesterol and the appearance of cholesterol oxidation products (Anderson et al., 1985; Demopoulos et al., 1982), the radical and lipid hydroperoxide-sensitive activation of guanylate cyclase and consequent increase in cyclic guanosine monophosphate (cGMP) (Hall and Braughler, 1982; Kurihara, 1985), a decrease in spinal tissue antioxidant levels (e.g., alpha-tocopherol, ascorbate) (Pietronigro et al., 1983; Saunders et al., 1987), and the early inhibition of lipid peroxidation-sensitive membrane-bound enzymes such as $Na^{+}+K^{+}$-ATPase (Anderson and Means, 1985; Clendenon et al., 1978).

Post-traumatic ischemia. Strong support for a pathophysiological role of these radical and/or peroxidative reactions has been provided recently by investigations showing that the pharmacological inhibition of lipid peroxidation can attenuate the development of post-traumatic spinal cord hypoperfusion. Post-traumatic hypoperfusion is generally believed to contribute to secondary spinal cord tissue degeneration (Hall and Wolf, 1986; Young, 1985).

The first of these investigations concerns the finding of a close correlation between the ability of a single intravenous (I.V.) dose of the glucocorticoid steroid methylprednisolone to inhibit spinal cord lipid peroxidation (Hall and Braughler, 1981, 1982) and its action to retard post-traumatic hypoperfusion after severe blunt trauma (Hall et al., 1984; Young and Flamm, 1982). In both cases, a 30 mg/kg I.V. dose was required, with lower and higher steroid doses being less effective or ineffective. This indicates that methylprednisolone doses that significantly reduce lipid peroxidation are also associated with a reduction in post-traumatic hypoperfusion. Moreover, the methylprednisolone dose-response curve for inhibition of post-traumatic spinal lipid peroxidation and secondary ischemia is essentially identical to that for the concomitant prevention of early post injury lactic acid accumulation (Braughler and Hall, 1983), a biochemical index of post-traumatic ischemia.

More direct evidence for a role of lipid peroxidation in the development of post-traumatic spinal cord ischemia is obtained from studies that have examined the effect of intensive antioxidant pretreatment on the progressive decline in spinal cord white matter blood flow (SCBF) after contusion injury in cats (Hall and Braughler, 1986; Hall and Wolf, 1986). For instance, cats were pretreated daily with high oral doses of d-alpha tocopherol (1000 iu) and selenium (50 μg) for 5 days before spinal contusion injury. In untreated cats, moderately severe spinal contusion injury resulted in a progressive decrease in SCBF from near normal levels immediately after injury to 53.5% below the preinjury level 4 hrs later. In contrast, vitamin E and selenium pretreatment prevented any post-traumatic decrease in white matter perfusion.

It should also be mentioned that pretreatment of cats with the calcium channel blockers nifedipine and diltiazem or the cyclooxygenase inhibitors meclofenamate and ibuprofen also significantly reduced the post-contusion decrease in SCBF (Hall and Wolf, 1986). Although these results suggest a concomitant role of aberrant calcium fluxes (Stokes et al., 1983; Young et al., 1982) and the generation of vasoactive prostanoids (Anderson et al., 1985; Hsu et al., 1985; Jonsson and Daniel, 1976; Saunders et al., 1987) in the ischemic response to blunt spinal injury, this is not inconsistent with the view that microvascular lipid peroxidation is ultimately responsible for the progressive decrease in spinal cord blood flow. As discussed elsewhere (Braughler and Hall, 1987), calcium can greatly exacerbate nervous tissue lipid peroxidation. Moreover, as noted above, the prostaglandin synthase step during prostanoid formation (Kukreja et al., 1986) provides a potentially important source of superoxide radical that has been purported to be a major contributor to post-traumatic microvascular damage in brain. Figure 1 provides a possible pathogenetic sequence focused on the interrelationships of various factors, including free radical–induced lipid peroxidation in post-traumatic ischemia development after severe spinal injury.

Based on the atypical pharmacological characteristics of methylprednisolone's effects on the injured spinal cord and brain, we postu-

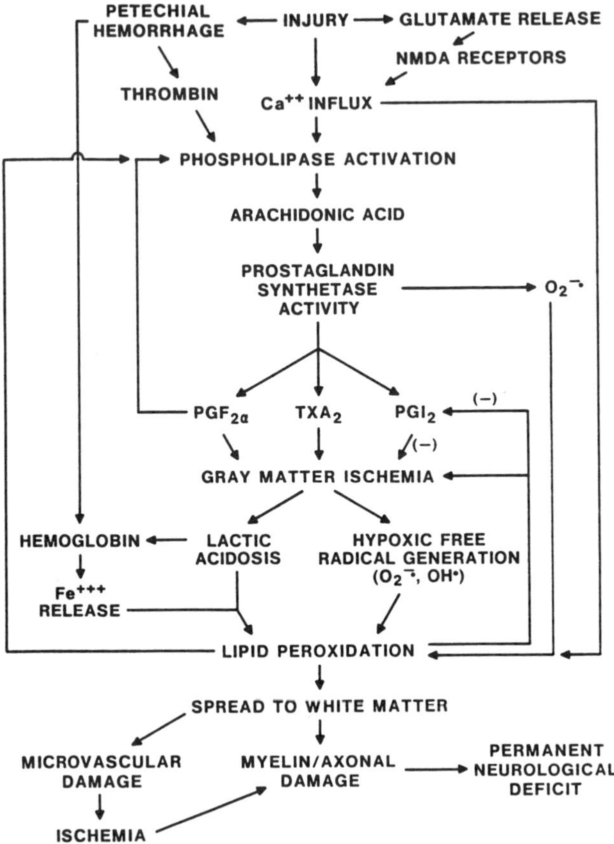

Figure 1 Pathophysiological scheme of acute CNS injury. Reprinted with permission of the *J. Neurosurg* 76:13–22 from Hall ED (1992): Neuroprotective pharmacology of methylprednisolone: a review. *J. Neurosurg.* 76:13–22.

lated that the neuroprotective action (i.e., inhibition of lipid peroxidation and related pathophysiological events) was not glucocorticoid receptor–mediated. Therefore, we further reasoned it ought to be possible to design an analog that would lack the glucocorticoid receptor–based actions and still retain the ability to inhibit peroxidation and protect the injured spinal cord (or brain). This effort lead to the discovery of the 21-aminosteroids ("lazaroids"). It was found that the substitution of a complex amine on nonglucocorticoid (i.e., lacking the 11 beta hydroxyl) steroid nucleus in place of the 21 hydroxyl functionality results in a dramatic enhancement of the lipid antioxidant activity. Many of these 21-aminosteroid compounds effectively inhibit iron-catalyzed lipid peroxidation in rat brain tissue homogenates under assay conditions where the glucocorticoid steroid methylprednisolone is completely ineffective (Braughler et al., 1987; Hall et al., 1988). Of these, U74006F (tirilazad mesylate) has shown excellent activity in experimental models of spinal cord and brain injury.

U74006F has been found to be a potent inhibitor of iron-catalyzed lipid peroxidation *in vitro* in either rat brain homogenates or purified rat brain synaptosomes (Braughler et al., 1987; Hall et al., 1988). U74006F has been further demonstrated to decrease iron-induced damage to cultured cortical neurons (Hall et al., 1991; Monyer et al., 1990). U74006F is also a potent inhibitor of iron-dependent lipid peroxidation in systems that do not contain membranes and are free of iron. By using the free radical generator 2,2′-azobis(2,4-dimethyl valeronitrile) to initiate lipid peroxidation in a homogeneous methanol solution of linoleic acid, the 21-aminosteroids have been found to inhibit lipid peroxidation by scavenging lipid peroxyl radicals (LOO·) and thus block lipid radical reactions in a manner similar to vitamin E (Braughler and Pregenzer, 1989). The LOO· scavenging action of the 21-aminosteroids resides in the amine portion of the molecule and yields an as yet uncharacterized product. During the inhibition of lipid peroxidation, U74006F competes for the same reaction as Vitamin E and slows its degradation (Braughler and Pregenzer, 1989).

In addition to scavenging of lipid peroxyl radicals, U74006F also reacts with hydroxyl radicals generated during *in vitro* Fenton reactions (i.e., $Fe^{++} + H_2O_2 \rightarrow Fe^{+++} + OH^- + \cdot OH$). This reaction is believed to give rise to a hydroxylated metabolite (5 position of the pyrimidine ring). Ironically, this hydroxylated U74006F is probably a more efficient scavenger of lipid peroxyl radicals (i.e., better able to donate an electron from the hydroxyl). Thus, it may be that U74006F can scavenge a hydroxyl radical and in the process becomes a more efficient lipid antioxidant (E. D. Hall, unpublished results). *In vivo* evidence of hydroxyl radical scavenging by U74006F in experimental head injury has been obtained, as detailed later in this chapter.

U74006F also has a stabilizing effect on cell membranes. For example, the iron- or iodoacetate-induced release of free arachidonic acid from pituitary tumor (AtT-20) cell membranes is blocked by U74006F (Braughler et al., 1988). This effect is not due to glucocorticoid activity of the compounds (i.e., phospholipase A2 inhibition), but may be related to its antioxidant actions. In other studies using cultured bovine brain microvascular endothelial cells, U74006F has been shown to localize within the hydrophobic core of cell membranes and cause an increase in lipid ordering (i.e., decreased fluidity) of the phospholipid bilayer (Audus et al., 1991). This action may help to inhibit the propagation of lipid peroxidation by restricting the movement of lipid peroxyl and alkoxyl radicals within the membrane.

The effects of U74006F on the development of post-traumatic spinal cord ischemia have been studied extensively in cat models of compression and contusion-induced spinal injury (Hall, 1988; Hall et al., 1989). In both instances, the compound effectively attenuated the progressive post-traumatic decrease in spinal cord blood flow. The mechanism of action of U74006F in antagonizing post-traumatic ischemia development is believed to involve an inhibition of oxygen radical–mediated microvascular lipid peroxidation. This conclusion is based on the concomitant action of U74006F to attenuate an injury-induced decline in spinal tissue vitamin E at the same doses that reduce post-traumatic ischemia (Hall et al., 1989).

Post-traumatic axonal degeneration. In addition to microvascular damage and consequent development of hypoperfusion in the injured spinal cord, oxygen radical–mediated lipid peroxidation may also be directly involved in the degeneration of spinal axons after injury. Studies have examined this possibility within the context of the anterograde (i.e., Wallerian) degeneration of surgically sectioned cat motor nerve fibers. A model involving the surgical transection of cat soleus motor axons at the greater sciatic foramen, followed 48 hrs later by an assessment of neuromuscular function in the *in vivo* soleus nerve–muscle preparation, has been shown to provide a reproduction of the early consequences of neuronal degeneration (Hall et al., 1977). The 48-hr degenerating motor axons and nerve terminals display subtle, but important, defects in neuromuscular transmission and excitability that can be quantified by a number of functional tests. Studies with this system have shown that intensive pretreatment with glucocorticoids such as triamcinolone (Hall et al., 1977, 1983) or methylprednisolone (Hall and Wolf, 1984) can significantly retard the degenerative process, as evidenced by a significant neuromuscular functional preservation 48 hrs after axon section.

In more recent experiments, an attempt has been made to investigate the possible pharmacological role of lipid peroxidation in the anterograde degeneration process (33). Before nerve section, cats were intensively pretreated for 5 days with daily oral doses of d-alpha tocopherol (200 iu) and selenium (50 g) similar to those employed in the post-traumatic spinal cord hypoperfusion studies (Hall and Wolf, 1986), except that the d-alpha tocopherol dose was lower in the motor axon degeneration experiments. This treatment has been found to retard the anterograde degeneration process significantly. A preservation of neuromuscular function was observed in terms of (a) a greater soleus muscle contractile response to low frequency nerve stimulation, (b) a better maintenance of tetanic contractile tension during high frequency nerve stimulation, and (c) a more rapid recovery from d-tubocurarine–induced neuromuscular block. Although biochemical correlates are still needed, the ability of antioxidant dosing to slow the anterograde degenerative process suggests that peroxidative reactions may be a fundamental mechanism of neuronal degeneration after injury.

In view of the demonstrated ability of at least some glucocorticoids to inhibit neuronal peroxidation directly, an antioxidant mechanism may also underlie the slowing of motor nerve degeneration by intensive glucocorticoid dosing (Hall et al., 1977, 1983; Hall and Wolf, 1984). Relevant to neuronal degeneration in the injured spinal cord, the early and repeated administration of antioxidant doses of methylprednisolone has recently been shown to result in significantly greater long-term structural preservation after blunt cord injury in cats (Braughler et al., 1987). The 21-aminosteroid U74006F has also been demonstrated to affect the rate of anterograde degeneration of motor nerve fibers after experimental injury in the cat soleus motor nerve degeneration model (Hall and Yonkers, 1990).

Post-traumatic conduction failure in surviving axons. *In vitro* studies with frog sciatic nerve (Hall and Telang, 1985) have suggested that lipid peroxidation may also provide an explanation for the failure of surviving spinal axons to recover normal conduction after injury (Blight, 1983). The application of the organic hydroperoxide tertiary butyl hydroperoxide (t-BOOH) to isolated frog nerves leads to a concentration-related block of compound action potential conduction. After removal of low t-BOOH concentrations, conduction recovers fully in terms of compound action potential amplitude, but remains significantly slower than normal. This is reminiscent of the observation that surviving axons from the chronically injured spinal cord are often capable of conduction, but the conduction velocity is slowed in comparison to uninjured spinal axons (Blight, 1983). If the frog nerve is exposed to more intense peroxidation, induced by higher t-BOOH concentrations, followed by washing, action potential conduction partially recovers and then secondarily declines. It is tempting to extrapolate this phenomenon and lipid peroxidation to the early neurophysiological (somatosensory evoked potential) recovery, followed by secondary conduction, loss that takes place within the first few hours after spinal cord contusion (Young and Flamm, 1982).

Effects of antioxidants on post-traumatic neurological recovery. In addition to demonstrating the occurrence of acute post-traumatic oxygen radical reactions and their association with pathophysiological events, the establishment of their importance, relative to other factors, ultimately depends on a demonstration that by interrupting them, chronic neurological recovery can be enhanced. Thus, attempts have been made to show that early treatment of spinal cord injured animals with bonafide antioxidants or agents with antioxidant properties can beneficially affect sensory and motor recovery.

In this vein, a series of experiments was recently conducted to examine the ability of intensive dosing with methylprednisolone begun early (30 min) after moderately severe spinal cord compression injury to improve chronic functional recovery and spinal tissue preservation in cats. As cited earlier, the *in vivo* administration of methylprednisolone has been shown to protect against *in vitro* lipid peroxidation in homogenates of cat spinal cords removed 1 hr after a single large 30 mg/kg I.V. dose (Hall and Braughler, 1981). Probably because of this direct antioxidant effect, the same dose of methylprednisolone had been found to reduce post-traumatic spinal cord lipid peroxidation (Anderson et al., 1985; Demompoulos et al., 1982; Hall and Braughler, 1982). Based on the dose–response and time–action characteristics of methylprednisolone, an intensive methylprednisolone dosing regimen was designed and tested. An initial 30 mg/kg I.V. dose at 30 min post-injury was followed up with a 48-hr maintenance dosing regimen constructed such that effective antioxidant spinal tissue concentrations of the steroid would be maintained over that time period. The cats were then evaluated (blind) on a weekly basis for 4 weeks for walking, running, and stair climbing ability after which histological analysis of the spinal injury site was performed. In comparison with vehicle-treated animals, the methylprednisolone treated cats showed a significantly better recovery by 2 weeks post-injury. A reduction in 4-week post-traumatic spinal tissue loss was also observed, the degree of which was inversely correlated ($r = -0.88$) with the neurological recovery score (Anderson et al., 1985; Braughler et al., 1987). These results show that a dosing regimen centered on the inhibition of post-traumatic lipid peroxidation and related processes (e.g., secondary ischemia) is associated with both enhanced structural preservation and functional recovery.

The results of the second National Acute Spinal Cord Injury Study (NASCIS II) constitutes a major milestone in the search for therapeutic interventions that will interfere significantly with secondary post-traumatic spinal cord degeneration and thereby ameliorate the devastating neurological consequences to improve the 6-month recovery of spinal cord inujury patients, compared to placebo-treated patients, when administered in an intensive 24-hr I.V. antioxidant dosing regimen beginning within 8 hrs after injury. NASCIS II is the first randomized, double-blind, placebo-controlled trial that has unequivocally demonstrated that a pharmacological agent can beneficially modify the course of events after severe central nervous system (CNS) injury. The dose level and dosing regimen (30 mg/kg IV. bolus plus 5.4 mg/kg/hr 23-hr infusion) were derived from careful pharmacological investigations that focused on inhibition of post-traumatic lipid peroxidation as the therapeutic target.

High-dose pretreatment with vitamin E (1000 IU P. O. once daily for 5 days) has been shown to promote the chronic recovery of spinal cord–injured cats (Anderson et al., 1988a). This is consistent with its ability to prevent post-traumatic spinal cord hypoperfusion (Hall and Wolf, 1986) and lipid peroxidation (Anderson et al., 1985) and with the hypothesis that oxygen radical generation and lipid peroxidation are important mediators of post-traumatic spinal cord pathophysiology and degeneration.

Furthermore, the 21-aminosteroid U74006F has been investigated for its ability to promote neurological recovery of cats after a moderately severe compression injury to the lumbar spinal cord (Anderson et al., 1988b). Beginning 30 min after injury, the animals received at 48-hr I.V. regimen of vehicle (sterile water) or U74006F in a random and blinded protocol similar to that employed in earlier studies with methylprednisolone (Braughler et al., 1987). At 4 weeks after injury, vehicle-treated animals uniformly remained paraplegic with a mean recovery score of 2.2 (i.e., 20% of normal). In contrast, cats that received 48-hr doses of U74006F ranging from 1.6 to 160.0 mg/kg showed significantly better recovery, regaining approximately 75% of normal neurological function. The 4-week recovery scores varied from 5.5 to 8.0. Histological examination of the injured spinal cord segment showed a statistically significant correlation (r = –0.57, $p < .001$) between cross-sectional tissue preservation and neurological recovery store, indicating an effect of U74006F to retard post-traumatic tissue degeneration.

Recent studies suggest that U74006F retains its efficacy in promoting post-traumatic recovery after experimental spinal cord injury even when initiation of treatment is delayed to 4 hrs post-injury (Anderson et al., 1991). Thus, this novel nonglucocorticoid 21-aminosteroid antioxidant shows considerable promise as a safer (i.e., lack of glucocorticoid activities) and possibly more effective acute treatment of spinal cord injury than methylprednisolone.

Similarity of peroxidative and mechanical spinal injuries. Additional evidence for a role of oxygen radical formation and lipid peroxidation in post-traumatic spinal cord degeneration has come from studies comparing the spinal neuropathology of mechanical (e.g., compression) injuries with those produced by microinjection of ferrous chloride into the spinal cord (Anderson and Means, 1983). Although circumstantial, the similarities between the neurochemical and neuroanatomical pathology of iron and trauma-induced spinal injury in cats are striking. In both cases, the degenerative changes begin with the central gray matter and spread to include the surrounding white matter. Moreover, as with studies of mechanical trauma, pretreatment with a 30 mg/kg I.V. dose of the glucocorticoid methylprednisolone has been shown to protect significantly against the iron-induced myelopathy (Anderson and Means, 1985). Likewise, co-injection of the hydroxyl radical scavenger mannitol along with the ferrous chloride also exerts a protective effect.

Oxygen Radicals in Head Injury

Post-traumatic microvascular damage. Numerous studies have been carried out using a cat fluid percussion head injury model to investigate the biochemical and physiological mechanisms of post-traumatic brain microvascular damage (Kontos et al., 1980). After moderately severe injury, there is a secondary cerebral (pial) arteriolar dilation together with a loss of reactivity to vasoactive agents or maneuvers including elevations in arterial pCO_2 and a decline in blood pressure. This occurs concomitantly with the appearance of focal endothelial lesions and a reduction in oxygen utilization by the pial vascular wall.

At a biochemical level, one of the earliest effects of this injury is activation of brain phospholipase (PLC) (Wei et al., 1982) and a consequent rise in tissue levels of cyclooxygenase products of arachidonic acid (Ellis et al., 1981). A connection between this and the post-traumatic cerebral microvascular damage is based on the observation that arachidonate or prostaglandin endoperoxide (PGG_2) application to the brain surface causes arteriolar damage similar to that after fluid percussion brain injury (Kontos et al., 1980). However, these investigators believe that the principle mediator of the microvascular damage is superoxide radical generated as a by-product of the prostaglandin synthase reaction during the conversion of arachidonate to the prostaglandins (Kukreja et al., 1986). Specifically, superoxide radical can be detected in the brain extracellular space after injury by the reduction of locally applied nitroblue tetrazolium. Thus, a mechanistic scenario of (a) injury-induced PLC liberation of membrane arachidonate, (b) prostaglandin synthase activation and superoxide generation, and (c) oxygen radical–mediated endothelial damage has been proposed. Pharmacological support for this cascade of events has been provided by studies showing that either cyclooxygenase inhibitors (e.g., indomethacin), which would inhibit arachidonate metabolism and the associated free radical generation, or free radical scavengers (e.g., superoxide dismutase, mannitol), attenuate post-traumatic brain microvascular damage (Kontos and Povlishock, 1986). As indicated in an earlier section of this chapter, post-traumatic spinal cord hypoperfusion can also be attenuated by either cyclooxygenase inhibitors or antioxidants (Hall and Wolf, 1986). Thus, in the case of both blunt head and spinal cord trauma, there is evidence that oxygen radical production during prostanoid formation from injury-liberated arachidonic acid may be largely responsible for lipid peroxidation–induced damage to the microvasculature.

Post-traumatic edema. Related to the post-traumatic microvascular damage is the pathophysiological process of vasogenic brain edema that represents a disruption of blood–brain barrier integrity resulting in sodium and protein accumulation and osmotic fluid expansion of the brain extracellular space. Clinically, this is reflected by an increase in intracranial pressure which, if unchecked, can cause secondary compressive injury to vital brain structures.

The precise mechanism of post-traumatic vasogenic edema is unknown, but Chan et al. (1985) have provided data suggesting an important role for an arachidonic acid–derived oxygen radical–mediated process. In initial *in vitro* experiments, it has been discovered that when rat brain cortical slices are incubated with arachidonic acid, there is a transient burst of superoxide and lipid hydroperoxide formation that correlates with fluid accumulation (i.e., swelling) in the slices. Further *in vivo* studies have shown that arachidonic acid injection into brain can produce vasogenic edema as measured by extravasation of the protein tracking dye Evans' blue (Chan et al., 1985; Hall and Travis, 1988). Consistent with a role of free radical–mediated lipid peroxidation, the 21-aminosteroid lipid antioxidant U74006F has been demonstrated to attenuate arachidonic acid–induced blood–brain barrier disruption (Hall and Travis, 1988; Zuccarello and Anderson, 1989).

Microinjection of ferrous iron (i.e., ferrous chloride) has also been shown to produce focal edema in rat brain, the degree of which is correlated with tissue levels of the lipid peroxidation product malonyldialdehyde. Pretreatment with vitamin E (600 mg/kg I.M. once daily for 5 days) together with selenium (5 ppm in the drinking water) reduced the iron-induced edema and lipid peroxidation (Willmore and Rubin, 1984). Similarly, the 21-aminosteroid U74006F can also reduce the iron-induced blood–brain barrier opening (Zuccarello and Anderson et al., 1989).

More to the point, recent work clearly shows a relationship between post-traumatic free radical formation and induction of vasogenic brain edema. For example, data have been obtained from the mouse head injury model showing that severe concussive injury results in an increase in brain hydroxyl radical levels coincident with an increase in blood–brain barrier permeability (Hall et al., 1992). These studies employed salicylate, which reacts with hydroxyl radicals to form dihydroxybenzoic acid (DHBA). Severe concussive head injury results in an increase in brain levels of salicylate-derived DHBA measured at 30 min post-injury. However, a 3 mg/kg I.V. dose of U74006F leads to a decrease in DHBA formation, implying either an attenuated formation or a chemical scavenging of hydroxyl radicals (Hall et al., 1992). Although the former cannot be ruled out, the latter would seem more likely in view of *in vitro* studies showing that U74006F can indeed react with hydroxyl radical as described above.

Coincident with the reduction in brain hydroxyl radical levels, U74006F (3 mg/kg I.V. within 5 min post-injury) also acts to reduce post-traumatic opening of the blood–brain barrier (i.e., decreased brain uptake of ^{14}C-albumin). This effect of U74006F to close the barrier may be related to the attenuation of hydroxyl radical levels or an antagonism of the effects of free radicals on the barrier endothelium (i.e., decreased membrane lipid peroxidation). Indeed, free radicals are known to increase barrier permeability (Greenwald, 1991). Consistent with this reduction in post-traumatic blood–brain barrier opening which would lead to vasogenic brain edema, U74006F has been shown to attenuate post-traumatic brain edema in a rat model of fluid percussion head injury (McIntosh et al., 1992).

Effects of antioxidants on post-traumatic neurological recovery and survival. There is also evidence that the neuroprotective properties of methylprednisolone, extensively studied in spinal cord injury, are applicable to brain injury. The steroid has been shown to enhance the early recovery of mice subjected to a moderately severe concussive head injury when administered at 5 min post-injury (Hall, 1985). The dose-response curve for this effect is remarkably similar to that discussed above for spinal cord injury. A 30 mg/kg I.V. dose was observed to be optimal whereas lower (15 mg/kg) and higher (60 and 120 mg/kg) were ineffective.

Consistent with the beneficial effect of methylprednisolone in experimental head injury, one controlled clinical trial has shown that a high-dose regimen begun within 6 hrs after severe injury (Glasgow Coma Scale = 3–8) with a 30 mg/kg I.V. bolus can significantly increase survival and the recovery of speech (Giannotti et al., 1984). However, although these data are suggestive, flaws in the study prevent it from being definitive. For instance, there were three groups of patients: a placebo group, a low dose methylprednisolone group (100 mg I.V. bolus plus 8 days of tapered maintenance dosing), and a high dose group (30 mg/kg I.V. bolus plus tapered dosing). There was clearly no difference between the placebo and low dose groups, but only when these two groups were combined post-hoc and statistically compared to the high dose group could the significant effect on survival and speech recovery be demonstrated. Furthermore, the effect was only demonstrable if those patients over 40 years of age were excluded, also post-hoc. Thus, an additional trial would be required to determine the efficacy of methylprednisolone in brain injury.

Studies of the efficacy of the 21-aminosteroid U74006F in acute head injury have also demonstrated the ability of the compound to improve early neurological recovery and survival of head-injured mice (Hall et al., 1988). Administration of an I.V. dose of U74006F produced a significant improvement in the 1 hr post-injury neurological status (grip test score) over a broad range (0.003–30 mg/kg). A 1 mg/kg I.V. dose given within 5 min and again at 1.5 hrs after a severe injury, in additional to improving early recovery, also increased the 1-week survival to 78.6% compared to 27.3% in vehicle-treated mice ($p < .02$).

Additional experiments have been conducted in severely head-injured cats to assess the effects of U74006F on brain energy metabolites (Dimlich et al., 1990). A 1 mg/kg I.V. dose adminstered at 30 min post-injury, plus a second 0.5 mg/kg dose 2 hrs later, resulted in an improved metabolic profile within the injured hemisphere measured at 4 hrs. Most notably, U74006F significantly reduced post-traumatic lactic acid accumulation in both the cerebral cortex and subcortical white matter. This biochemical effect suggests an improved maintenance of cerebral blood flow in the injured brain. As noted above, U74006F effectively reduces progressive post-traumatic ischemia development in experimental cat spinal cord injury (Hall, 1988; Hall et al., 1989), which may also provide the explanation for the reduction of post-traumatic lactate levels in the injured brain.

Summary

This chapter has reviewed the current state of knowledge regarding the occurrence and possible role of oxygen radical generation and lipid peroxidation in experimental models of acute CNS injury. Although much work remains, four criteria that are logically required to establish the pathophysiological importance of oxygen radical reactions have been met, at least in part. First, oxygen radical generation and lipid peroxidation appear to be early biochemical events subsequent to CNS trauma. Second, a growing body of direct or circumstantial evidence suggests that oxygen radical formation and lipid peroxidation are linked to pathophysiological processes such as secondary post-traumatic ischemia, edema, axonal conduction failure, failure of energy metabolism, and anterograde (Wallerian) degeneration. Third, there is a striking similarity between the pathology of blunt mechanical injury to CNS tissue and that produced by chemical induction of peroxidative injury. Fourth and most convincing is the repeated observation that compounds that inhibit lipid peroxidation or scavenge oxygen radicals can block post-traumatic pathophysiology, reduce neuronal damage, and promote functional recovery and survival in experimental studies.

Nevertheless, the significance of oxygen radicals and lipid peroxidation ultimately depends on whether it can be demonstrated that early application of effective anti–free radical or antiperoxidative agents can promote survival and neurological recovery after CNS injury and stroke in humans. The results of the NASCIS II clinical trial, which have shown that an antioxidant dosing regimen with methylprednisolone begun within 8 hrs after spinal cord injury can significantly enhance chronic neurological recovery, strongly supports the significance of lipid peroxidation as post-traumatic degenerative mechanism. However, ongoing Phase III trials with the more selective and effective antioxidant U74006F (tirilazad mesylate) will give a more clear-cut answer as to the therapeutic importance of inhibition post-traumatic free radical reactions in the injured central nervous system.

References

Anderson DK, Waters TR, Means ED (1988a): Pretreatment with alpha tocopherol enhances neurologic recovery after spinal cord compression injury. *J Neurotrama* 6:61–68.

Anderson DK, Braughler JM, Hall ED, Waters TR, McCall JM, Means ED (1988b): Effects of treatment with U-74006F on neurological recovery following experimental of spinal cord injury. *J Neurosurg* 69:562–567.

Anderson DK, Hall ED, Braughler JM, McCall JM, Means ED (1991): Effect of delayed administration of U74006F (tirilazad mesylate) on recovery of locomotor function following experimental spinal cord injury. *J Neurotrauma* 8:187–192.

Anderson DK, Means ED (1983): Lipid peroxidation in spinal cord. $FeCl_2$ induction and protection with antioxidants. *Neurochem Path* 1:249–264.

Anderson DK, Means ED (1985): Iron-induced lipid peroxidation in spinal cord: protection with mannitol and methylprednisolone. *J Free Rad Biol Med* 1:59–64.

Anderson DK, Means ED (1985): Pathophysiological mechanisms in acute spinal cord trauma: effects of decompartmentalized iron on cellular membranes. In: *Trauma of the Central Nervous System*, RG Dacey, HR Winn, RW Rimel, JA Jane, eds. New York: Raven Press.

Anderson DK, Saunders RD, Demediuk P, Dugan LL, Braughler JM, Hall ED, Means ED, Horrocks LA (1985): Lipid hydrolysis and peroxidation in injured spinal cord: partial protection with methylprednisolone or vitamin E and selenium. *CNS Trauma* 2:257–267.

Audus KL, Guillot FL, Braughler JM (1991): Evidence for 21-aminosteroid association with the hydrophobic domains of brain microvessel endothelial cells. *Free Rad Biol Med* 11:361–371.

Blight AR (1983): Axonal physiology of chronic spinal cord injury in the cat: intracellular recording *in vitro. Neuroscience* 10:1471–1486.

Bracken MB, Shepard MJ, Collins WF, Holford TR, Young W, Baskin DS, Eisenberg HM, Flamm ES, Leo-Summers L, Maroon J, Marshall LF, Perot PL, Piepmeier J, Sonntag VKH, Wagner FC, Wilberger JE, Winn HR (1990): A randomized controlled trial of methylprednisolone or naloxone in the treatment of acute spinal cord injury. *N Engl J Med* 322:1405–1411.

Braughler JM, Chase RL, Neff GL, Day JS, Hall ED, Sethy VH, Lahti RA (1988): A new 21-aminosteroid antioxidant lacking glucocorticoid activity stimulates ACTH secretion and blocks arachidonic acid release from mouse pituitary tumor (AtT-20) cells. *J Pharmacol Exp Ther* 244:423–427.

Braughler JM, Hall ED (1983): Lactate and pyruvate metabolism in injured cat spinal cord before and after a single large intravenous dose of methylprednisolone. *J Neurosurg* 59:256–261.

Braughler JM, Hall ED (1989): Central nervous system trauma and stroke: I. Biochemical considerations for oxygen radical formation and lipid peroxidation. *Free Rad Biol Med* 6:289–301.

Braughler JM, Hall ED, Means ED, Waters T, Anderson DK (1987): Evaluation of an intensive methylprednisolone sodium succinate dosing regimen in experimental spinal cord injury. *J Neurosurg* 67:102–105.

Braughler JM, Pregenzer JF (1989): The 21-aminosteroid inhibitors of lipid peroxidation: reactions with lipid peroxyl and phenoxyl radicals. *Free Rad Biol Med* 7:125–130.

Braughler JM, Pregenzer JF, Chase RL, McCall JM, Jacobsen EJ (1987): Novel 21-aminosteroids as potent inhibitors of iron-dependent lipid peroxidation. *J Biol Chem* 262:10438–10440.

Chan PH, Longar S, Fishman RA (1985): Oxygen-free radicals: potential edema mediators in brain injury. In: *Brain Edema* Inaba Y, Klatzo I, Spatz M, ed. Tokyo: Springer-Verlag.

Clendenon NR, Allen H, Gordon WA, Bingham WG (1978): Inhibition of $Na^{+}+K^{+}-$ activated ATPase activity following experimental spinal cord trauma. *J Neurosurg* 49:563–568.

Demopoulos HB, Flamm ES, Pietronigro DD, Seligman ML (1980): The free radical pathology and the microcirculation in the major central nervous system disorders. *Acta Physiol Scand* 492(suppl):91–119.

Demopoulos HB, Flamm ES, Seligman ML, Pietronigro DD, Tomasula J, DeCrescito V (1982): Further studies on free radical pathology in the major central nervous system disorders: effects of very high doses of methylprednisolone on the functional outcome, morphology and chemistry of experimental spinal cord impact injury. *Can J Physiol* 60:1415–1424.

Dimlich RVW, Tornheim PA, Kindel RM, Hall ED, Braughler JM, McCall JM (1990): Effects of a 21-aminosteroid (U-74006F) on cerebral metabolites and edema after severe experimental head trauma. In: *Advances in Neurology, Vol 52*, Long D et al., eds. New York: Raven Press.

Ellis EF, Wright KF, Wei EP, Kontos HA (1981): Cyclooxygenase products of arachidonic acid metabolism in cat cerebral cortex after experimental concussive brain injury. *J Neurochem* 37:892–896.

Giannotta SL, Weiss MH, Apuzzo MLJ, Martin E (1984): High dose glucocorticoids in the management of severe head injury. *Neurosurgery* 15:497–501.

Greenwood J (1991): Mechanisms of blood-brain barrier breakdown. *Neuroradiology* 33:95–100.

Hall ED (1985): High-dose glucocorticoid treatment improves neurological recovery in head-injured mice. *J Neurosurg* 62:882–887.

Hall ED (1987): Intensive anti-oxidant pretreatment retards motor nerve degeneration. *Brain Res* 413:175–178.

Hall ED (1988): Effect of the 21-aminosteroid U74006F on post-traumatic spinal cord ischemia. *J Neurosurg* 68:462–465.

Hall ED, Baker T, Riker WF (1977): Glucocorticoid preservation of motor nerve function during early degeneration. *Ann Neurol* 1:263–269.

Hall ED, Braughler JM (1981): Acute effects of intravenous glucocorticoid pretreatment on the *in vitro* peroxidation of cat spinal cord tissue. *Exp Neurol* 73:321–324.

Hall ED, Braughler JM (1982): Effects of intravenous methylprednisolone on spinal cord peroxidation and $Na^{+}+K^{+}$-ATPase activity: dose-response analysis during 1st hour after contusion injury in the cat. *J Neurosurg* 57:247–253.

Hall ED, Braughler JM (1986): Role of lipid peroxidation in post-traumatic spinal cord degeneration: a review. *CNS Trauma* 3:281–293.

Hall ED, Braughler JM, Yonkers PA, Smith SL, Linseman KL, Means ED, Scherch HM, VonVoigtlander PF, Lahti RA, Jacobsen EJ (1991): U78517F, a potent inhibitor of lipid peroxidation with activity in experimental brain injury and ischemia. *J Pharmacol Exp Ther* 258:688–694.

Hall ED, McCall JM, Braughler JM (1988): New pharmacological treatments for spinal cord trauma. *J Neurotrauma* 5:81–89.

Hall ED, Riker WF, Baker T (1983): Beneficial action of glucocorticoid treatment on neuromuscular transmission during early motor nerve degeneration. *Exp Neurol* 79:488–496.

Hall ED, Telang FW (1985): Characteristics of lipid peroxidative conduction block induced by an organic hydroperoxide in axons of isolated frog nerve. *CNS Trauma* 2:161–168.

Hall ED, Travis MA (1988): Inhibition of arachidonic acid-induced vasogenic brain edema by the non-glucocorticoid 21-aminosteroid U74006F. *Brain Res* 451:350–352.

Hall ED, Wolf DL (1984): Methylprednisolone preservation of motor nerve function during early degeneration. *Exp Neurol* 84:715–720.

Hall ED, Wolf DL (1986): A pharmacological analysis of the pathophysiological mechanisms of post-traumatic spinal cord ischemia. *J Neurosurg* 64:951–961.

Hall ED, Wolf DL, Braughler JM (1984): Effects of a single large dose of methylprednisolone sodium succinate on experimental post-traumatic spinal cord ischemia: dose-response and time-action analysis. *J Neurosurg* 61:124–130.

Hall ED, Yonkers PA (1990): Preservation of motor nerve function during early degeneration by the 21-aminosteroid antioxidant U74006F. *Brain Res* 513: 244–247.

Hall ED, Yonkers PA, Andrus PK, Cox JW, Anderson DK (1992): Biochemistry and pharmacology of lipid antioxidants in acute brain and spinal cord injury. *J Neurotrauma*, 9(Suppl. 2):S425–S442.

Hall ED, Yonkers PA, Horan KL, Braughler JM (1989): Correlation between attenuation of post-traumatic spinal cord ischemia and preservation of vitamin E by the 21-aminosteroid U-74006F: evidence for an *in vivo* antioxidant action. *J Neurotrauma* 6:169–176.

Hall ED, Yonkers PA, McCall JM, Braughler JM (1988): Effect of the 21-aminosteroid U74006F on experimental head injury in mice. *J Neurosurg* 68:456–461.

Hsu CY, Halushka PV, Hogan EL, Banik NL, Lee WA, Perot PL (1985): Alteration of thrombone and prostacyclin levels in experimental spinal cord injury. *Neurology* 35:1003–1009.

Jonsson HT, Daniell HB (1976): Altered levels of PGF in cats spinal cord tissue following traumatic injury. *Prostaglandins* 11:51–61.

Kontos HA, Povlishock JT (1986): Oxygen radicals in brain injury. *CNS Trauma* 3:257–263.

Kontos HA, Wei EP, Povlishock JT, Dietrich WD, Mageira DJ, Ellis EF (1980): Cerebral arteriolar damage by arachidonic acid and prostaglandin G_2. *Science* 209:1242–1245.

Kukreja RC, Kontos HA, Hess ML, Ellis EF (1986): PGH synthase and lipoxygenase generate superoxide in the presence of NADH and NADPH. *Circ Res* 59:612–619.

Kurihara M (1985): Role of monoamines in experimental spinal cord injury. Relationship between $Na^{+}+K^{+}$-ATPase and lipid peroxidation. *J Neurosurg* 62:743–749.

McIntosh T, Banbury M, Smith D (1992): The novel 21-aminosteroid U74006F attenuates cerebral edema and improves survival after brain injury in the rat. *Acta Neurochirurg* 51(suppl):329–330.

Milvy P, Kari S, Campbell JB, Demopoulos HB (1973): Paramagnetic species and radical products in cat spinal cord. *Ann NY Acad Sci* 222:1102–1111.

Monyer H, Hartley DM, Choi DW (1990): 21-Aminosteroids attenuate excitotoxic neuronal injury in cortical cell cultures. *Neuron* 5:121–126.

Pietronigro DD, Hovsepian M, Demopoulos HB, Flamm ES (1983): Loss of ascorbic acid from injured feline spinal cord. *J Neurochem* 41:1072–1076.

Saunders RD, Dugan LL, Demediuk P, Meand ED, Horrocks LA, Anderson DK (1987): Effects of methylprednisolone and the combination of alpha tocopherol and selenium on arachidonic acid metabolism and lipid peroxidation in traumatized spinal cord tissue. *J Neurochem* 49:24–31.

Stokes B, Fox P, Hollinden G (1983): Extracellular calcium activity in the injured spinal cord. *Exp Neurol* 80:561–572.

Wei EP, Lamb RG, Kontos HA (1982): Increased phospholipase C activity after experimental brain injury. *J Neurosurg* 56:695–698.

Willmore LJ, Rubin JJ (1984): Effects of antiperoxidants on FeC12-induced lipid peroxidation and focal edema in rat brain. *Exp Neurol* 83:62–70.

Young W (1985): Blood flow, metabolic and neurophysiological mechanisms in spinal cord injury. In: *Central Nervous System Trauma* Becker DB, Povlishock JT, eds. *Status Report.* Bethesda; NIH.

Young W, Flamm ES (1982): Effect of high-dose corticosteroid therapy on blood flow, evoked potentials, and extracellular calcium in experimental spinal injury. *J Neurosurg* 57:667–673.

Young W, Yen V, Blight A (1982): Extra-cellular calcium ionic activity in experimental spinal cord contusion. *Brain Res* 253:105–113.

Zuccarello M, Anderson DK (1989): Protective effect of a 21-aminosteroid on the blood-brain barrier following subarachnoid hemorrhage in rats. *Stroke.* 20:367–371.

Chapter 10

Nitric Oxide as a Mediator of Cerebral Blood-Flow, Synaptic Plasticity, and Superoxide-Mediated Brain Injury

Joseph S. Beckman, Jun Chen, Harry Ischiropoulos, Ling Zhu, Karl A. Conger, and James H. Halsey, Jr

We propose that excessive production of nitric oxide initiated by activation of glutamate receptors after cerebral ischemia potentiates free radical injury to the brain. Nitric oxide, produced by neurons and possibly astrocytes, helps regulate local cerebral blood flow and plays an essential role in synaptic plasticity and normal development of the brain. Nitric oxide itself is a weak oxidizing agent, but after reaction with superoxide (O_2^-), it forms the strong and relatively long-lived oxidant peroxynitrite anion ($ONOO^-$) (Beckman et al., 1990). Peroxynitrite is sufficiently stable even in the presence of physiological concentrations of glutathione and other cellular antioxidants to diffuse for up to several cell diameters. Depending on what peroxynitrite reacts with, it can produce oxidants with the reactivity of hydroxyl radical (HO·), nitrogen dioxide (NO_2), nitronium ion (NO_2^+), and possibly singlet oxygen. Thus, the conversion of nitric oxide to peroxynitrite and related secondary oxidants may provide a common link between glutamate and free radical–mediated injury. Glutamate antagonists and free radical scavengers can reduce infarct volume by approximately the same extent in rat middle cerebral artery occlusion models of stroke (Oh and Betz, 1991). Recently, nitroarginine, a competitive inhibitor of nitric oxide synthesis, has also been shown to reduce infarct volume in a mouse middle cerebral ischemia model (Nowicki et al., 1991). Nitroarginine also can protect neurons in culture from glutamate toxicity (Dawson et al., 1991b). Thus, three major

Oxygen Free Radicals in Tissue Damage
Merrill Tarr and Fred Samson, Editors

therapeutic interventions for the treatment of cerebral ischemic injury—glutamate antagonist, free radical scavengers, and inhibitors of nitric oxide synthesis—may share a common mechanism of action through preventing the formation of peroxynitrite. In this brief chapter we explore some of nitric oxide's functions in brain and how ischemia may subvert the processes controlling the production of nitric oxide. We propose that elevated concentrations of nitric oxide may react with superoxide to form peroxynitrite, which causes the formation of strong oxidants that contribute to cell death.

Nitric Oxide and the Endothelium-Derived Relaxing Factor

The evidence is becoming increasingly strong that nitric oxide is indeed the endothelium-derived relaxing factor (EDRF), the endogenous humoral agent mimicked by nitrovasodilators such as nitroglycerin and nitroprusside (Furchgott and Vanhoutte, 1989; Ignarro, 1990; Moncada et al., 1987; Moncada et al., 1991; Palmer et al., 1987). Endothelium and neurons produce nitric oxide by a calmodulin-activated enzyme that oxidizes arginine and requires biopterin, NADPH, and oxygen (Bredt and Snyder, 1990; Palmer et al., 1988). Vasodilatory agents such as acetylcholine, adenosine triphosphate (ATP), and bradykinin trigger the production of nitric oxide by initiating an influx of calcium into endothelium. Nitric oxide is a hydrophobic gas and will freely diffuse from endothelium into surrounding muscle cells and platelets. Nitric oxide activates soluble guanylate cyclases by binding to the heme cofactor. The cyclic guanosine monophosphate (cGMP) produced by guanylate cyclase activated by nitric oxide promotes relaxation of vascular smooth muscle and inhibits both platelet–platelet aggregation and platelet adhesion to endothelium (Busse et al., 1987). Thus, nitric oxide mediates both vasorelaxation as well as modulation of the antithrombotic properties of endothelium. Nitric oxide has a half-life of only a few seconds because it reacts with oxygen eventually to form nitrite (NO_2^-) and nitrate (NO_3^-). Thus, it is a short-acting messenger that can diffuse for no more than roughly a few hundred micrometers before being inactivated.

Neurons Produce Nitric Oxide by a Process Stimulated by Excitory Amino Acids

The first indirect evidence that the brain might produce nitric oxide came when a soluble factor required to activate guanylate cyclase in neuroblastoma cells was shown to be arginine (Deguchi and Yoshioka, 1982). Garthwaite et al. showed that when cerebellar neurons are treated with the excitory amino acid antagonist N-methyl-D-aspartic acid (NMDA), they produce an EDRF-like factor that stimulates brain cGMP formation (East and Garthwaite, 1991; Garthwaite et al., 1988; Southam et al., 1991). Shibuki and Okado have directly measured nitric oxide produced by cerebellar slices when electrically stimulated (Shibuki, 1990; Shibuki and Okada, 1991). Bredt and Snyder have subsequently purified a 160-kDa protein from rat cerebellum that appears to be the nitric oxide synthase (Bredt and Snyder, 1990), and the brain enzyme has been prepared by other groups as well (Mayer et al., 1990; Schmidt et al., 1991). The brain and endothelial enzymes are both activated by calmodulin and inhibited by calmodulin inhibitors. However, the principal isoform to the endothelial enzyme is associated with membranes (Pollock et al., 1991), whereas the brain enzyme is soluble.

The brain nitric oxide synthase has recently been cloned and its sequence was revealing (Bredt et al., 1991). One segment of the sequence bears a striking resemblance to cytochrome P-450 reductase, containing binding regions for flavin adenine dinucleotide, FAD, flavin mononucleotide, FMN, and reduced nictonamide adenine dinucleotide phosphate, NADPH. Cytochrome P-450 reductase oxidizes NADPH by two electrons, which are temporarily stored on the flavin cofactors and transferred, one electron at a time, to cytochrome P-450. This facility for transferring single electrons is essential because the oxidation of arginine requires a total of five electrons. The actual oxidation of arginine to nitric oxide appears to involve biopertin, a cofactor used in phenylalanine and tyrosine hydroxylases to introduce hydroxyl groups on to phenyl rings. It appears that the oxidation of arginine involves two steps catalyzed by the same active site, producing N-hydroxyarginine as an intermediate (Stuehr et al., 1991; Wallace and Fukuto, 1991). Presumably, N-hydroxyarginine remains bound to the active site and then is oxidized by a second step to release nitric oxide and citrulline. Thus, the stoichiometry of the overall reaction appears to require 1 mol of arginine, 1.5 mol of NADPH, and 2 mol of molecular oxygen to produce 1 mol of nitric oxide plus 1 mol of citrulline and 2 mol of water. The second electron from the "left over" half mole of NADPH may be stored in the flavin cofactors for the production of a second nitric oxide.

Activated neutrophils and macrophages also produce nitric oxide from arginine, but the enzyme differs from the endothelial or neuronal enzymes (Marletta et al., 1988). The macrophage enzyme is not activated by calcium and is approximately 20 kDa smaller than the brain enzyme, with the sequence needed to recognize calmodulin missing. The brain sequence contains several phosphorylation sites, which might be important in regulation of both the brain and macrophage enzyme. In macrophages, nitric oxide may serve as an additional cytotoxic agent enhancing the toxicity of oxygen radicals produced.

The macrophage nitric oxide synthase is generally not present in resting macrophages and requires an induction period of several hours after stimulation with bacterial lipopolysaccharide or cytokines including interferon, interleukin, and tumor necrosis factor. A wide variety of cell types are capable of producing the calmodulin-independent inducible nitric oxide synthase in response to a lipopolysaccharide challenge (Moncada et al., 1991). The profound hypotension resulting from septic shock appears to be reversed by *N*-methylarginine. Corticosteroids block the production of this isozyme (Radomski et al., 1990).

Localization in Brain

The distribution of nitric oxide synthase is restricted to a relatively small percentage of neurons, constituting about 2% in cortex (Bredt et al., 1990). However, these neurons are present in all brain regions and are highly branched such that no region is far removed from a source of nitric oxide. Regionally, the cerebellum and olfactory bulb are the richest sources of nitric oxide synthase, but the regional variation in enzyme activity is only about eightfold (East and Garthwaite, 1991; Förstermann et al., 1990).

Recently, it has been suggested that nitric oxide synthase may be a major form of NADPH diaphorase (Hope et al., 1991). The high specific localization of nitric oxide synthase in neurons closely matches that of NADPH diaphorase (Dawson et al., 1991a). NADPH diaphorase is a histochemical marker enzyme of unknown function, detected by the reduction of nitroblue tetrazolium salts to insoluble blue formazam precipitates in the presence of NADPH. It appears the NADPH diaphorase neurons may be selectively spared in Huntington's chorea and possibly resistant to brief ischemic insults (Uemura et al., 1990). Partial denaturation of the nitric oxide synthase, such as might occur during freezing or weak formalin fixation, could easily expose its flavin cofactors and thereby allow the direct reduction of nitroblue tetrazolium.

Glutamate as a Neurotransmitter

The amino acid glutamate is an excitatory neurotransmitter linked to the process of long-term synaptic potentiation, the most likely candidate for the neuronal basis of learning (Brown et al., 1988). Glutamate also has a role in neuropathology (Mayer and Miller, 1990; Seisjo and Bengtsson, 1989). Three classes of glutamate-activated channels have been distinguished by their sensitivities to the agonists quisqualate, kainate, and NMDA. Quisqualate and kainate activate fast sodium and potassium channels, whereas the NMDA receptor controls a postsynaptic, voltage-dependent slow calcium channel. The opening of the calcium channel requires the binding of glutamate plus glycine, normally considered to be an inhibitory amino acid. In addition, the membrane must be partially depolarized. The voltage dependence of the NMDA receptor channel requires physiological concentrations of magnesium, which block the NMDA channel at normal membrane potentials by binding in the channel (Mayer et al., 1984). When the membrane is partially depolarized, magnesium no longer prevents calcium from entering the neuron.

The unique voltage-dependence of the NMDA channel initiates the process of long-term potentiation (Brown et al., 1988). Repeated patterns of stimulation have been shown to cause a prolonged decrease in the firing threshold of the affected neuron. Long-term potentiation is prevented by antagonists of the NMDA receptor, such as high magnesium concentrations or MK-801, and by removal of calcium from the medium.

Nitric Oxide and Long-Term Depression in Cerebellum

The role of nitric oxide in synaptic plasticity was first established in long-term depression of Purkinje cell discharge in cerebellum (Shibuki and Okada, 1991). Simultaneous electrical stimulation of both parallel fibers and climbing fibers in cerebellar slices at a moderate frequency for 3 to 5s will cause a long-term depression in the firing rate of Purkinje cells. Shibuki (1990) invented a simple electrochemical probe for measuring nitric oxide by sealing a platinum wire polarized to 0.9 V inside of a glass micropipet covered with a thin hydrophobic membrane (chloroneoprene). Because nitric oxide is a hydrophobic gas, it can diffuse across the membrane and be detected by measuring the current generated by its oxidation to nitric oxide. When both climbing and parallel fibers are simultaneously stimulated, nitric oxide concentrations increased to 70 nM within seconds. Blockade of nitric oxide synthesis with methylarginine (a competitive subtrate inhibitor of arginine) blocked the synthesis of nitric oxide and the induction of long-term depression. Addition of nitroprusside, a synthetic source of nitric oxide, allowed induction of long-term depression in cerebellum slices incubated with methylarginine.

Nitric Oxide and Long-Term Potentiation

More recently, several groups have reported that nitric oxide is important for long-term potentiation in the hippocampus. Inhibitors of nitric oxide synthase as well as hemoglobin block in the induction of long-term potentiation in hippocampal slices (Böhme et al., 1991; Shulman and Madison, 1991). One difficulty that remains to be resolved is that the immunocytochemical localization reported by Bredt and Snyder (Bredt et al., 1990) shows no localization of nitric oxide synthase in the neurons involved in long-term potentiation. Possibly, other isoforms of the brain nitric oxide synthase exist that are not recognized by the existing antibodies. For example, astrocytes in culture produce a calcium-regulated EDRF-like factor in culture (Murphy et al., 1990), yet astrocytes were not found to be stained (Bredt et al., 1990).

Synaptic Plasticity

Hebb (1949) postulated that learning requires a mechanism for signaling from the postsynaptic cell to the presynaptic terminals such that rapidly firing synapses would be strengthened while less active synapses would be destabilized. Williams et al. (1989) and Gally et al. (1990) have recently proposed that nitric oxide may serve as the Hebbian mediator of learning (Fig. 1). When dendritic processes have become sufficiently depolarized by the rapid synaptic stimulation to activate the NMDA receptor, calcium entry through the NMDA receptor will initiate release of nitric oxide that diffuses to the surrounding synapses. Nitric oxide would cause changes that strengthen rapidly firing presynaptic terminals while destabilizing less active synapses. The effects on synapses could be mediated through the action of soluble guanylate cyclases or nitric–activated protein ADP-ribosyltransferases (Brune and Lapetina, 1990), both of which are found in synaptosomes.

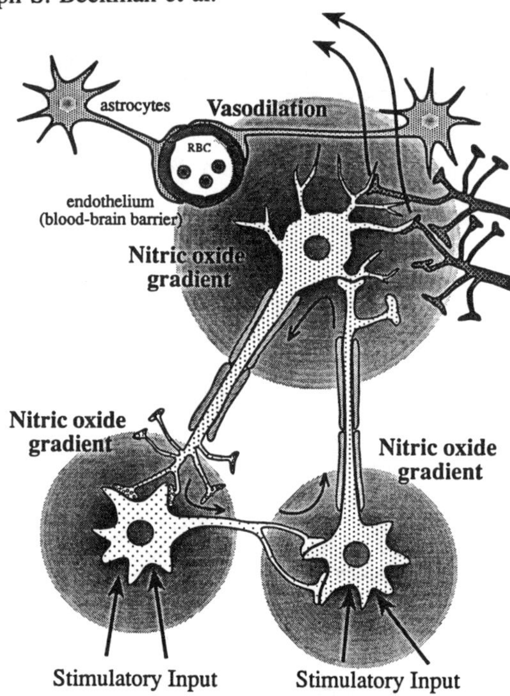

Figure 1 Neuronal nitric oxide as a transduction mechanism in brain. Increased neuronal activity will cause depolarization allowing calcium entry through the NMDA receptor. Nitric oxide produced by neurons activated by the NMDA receptor can diffuse to surrounding synapses to affect synaptic stability and to resistance vessels to increase local cerebral blood flow. Synapses that have not been firing rapidly during the exposure to nitric oxide may form weaker connections and eventually disconnect.

Nitric Oxide and the Control of Cerebral Blood Flow

Gally et al. (1990) also hypothesized that nitric oxide may contribute to the local increase of cerebral blood flow in response to neuronal activation by dilating local resistance vessels. We have recently shown that inhibition of nitric oxide synthesis *in vivo* with the competitive inhibitor, *N*-nitroarginine, causes a 40% decrease in cerebral blood flow in rats when infused at a rate of 30 mg/kg·hr (Fig. 2). The overall average cerebral blood flow was 95 ± 3.6 ml·100 g^{-1}·min^{-1} and average systemic pressure was 110 ± 10 mm Hg before nitroarginine infusion. At the high dosage of nitroarginine, the reduction of cerebral blood flow could not be reversed by arginine. When the dosage of nitroarginine was reduced to a 5-min intravenous (I.V.) infusion of 5 mg·kg^{-1}, the reduction of cerebral blood flow (CBF) occurred without increasing systemic pressure and could be reversed with infusion of 30 mg·kg^{-1} of arginine. Nitroarginine causes the most potent cerebral vasoconstriction of any systemic agent known and at lower concentrations than will cause systemic hypertension. The increase in cerebral vascular resistance could be reversed with arginine, the substrate for nitric oxide synthesis. Thus, nitric oxide has a major influence on controlling resistance vessels in the cerebral circulation, although the role of neuronal versus endothelial sources of nitric oxide needs to be further elucidated.

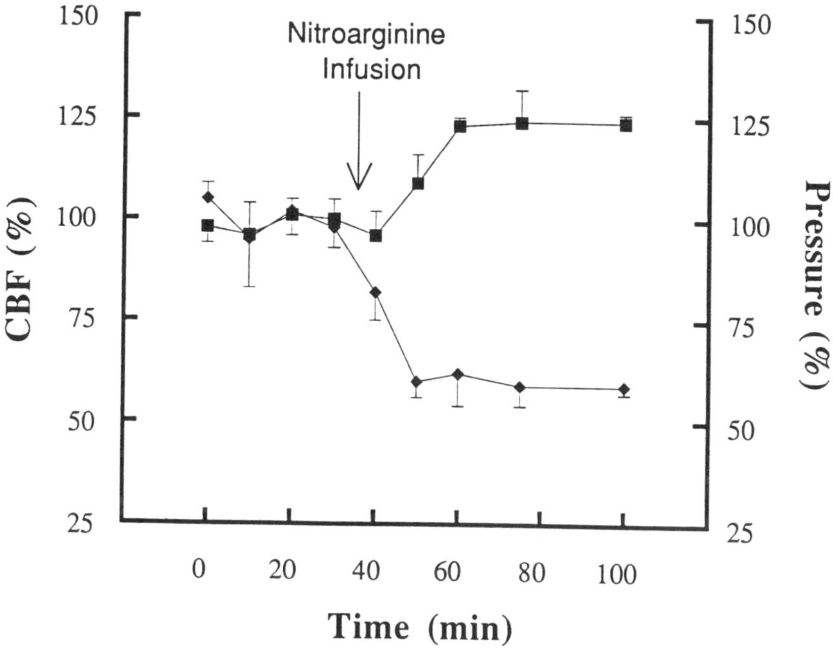

Figure 2 Effect of nitroarginine on resting blood flow in rats. Cerebral blood flow (◇) was measured by hydrogen clearance from 4 cortical electrodes implanted 3 days before the experiments in male Sprague Dawley rats (~300 g). The rats were anesthetized with 1% halothane/60% N_2O/40% O_2 and maintained at 37°C. The hydrochloride salt of nitroarginine (Aldrich) was prepared by dissolving nitroarginine in a minimal volume of 3 N HCl and evaporating under vacuum to dryness. A 6 mg/ml solution in saline was adjusted to pH 7.0 and infused through the femoral vein into the rats at approximately 1.5 ml h^{-1}. Systemic pressure (■) was monitored from a carotid catheter. Cerebral blood flows were averaged for the four electrodes per rat and then reported as the mean ± SEM for the six rats.

Nitric Oxide and Superoxide-Mediated Tissue Injury

Ischemia/reperfusion, inflammation, xenobiotic metabolism, hyperoxic exposure, and other diseases can destroy endothelial barrier function, cause adhesion of platelets, and disrupt vasoregulation. Superoxide dismutase can reduce such injury, which indirectly implicates superoxide anion (O_2^-) in many pathological processes (Fridovich, 1986). While superoxide can be directly toxic, its limited reactivity with many biological molecules raises questions about its toxicity per se (Baum, 1984; Sawyer and Valentine, 1981). Because nitric oxide contains an unpaired electron and is paramagnetic, it rapidly reacts with superoxide to form peroxynitrite anion ($ONOO^-$) in high yield (Blough and Zafiriou, 1985):

$$\cdot NO + O_2^- \Rightarrow ONOO^- \tag{1}$$

The rate constant for this reaction has recently been reported to be 4×10^7 $M^{-1}s^{-1}$ (Saran, 1990), which is about 200-fold slower than the diffusion limit. Superoxide dismutase is well known to stabilize EDRF (Rubanyi and Vanhoutte, 1986), suggesting that the formation of peroxynitrite *in vivo* may be a significant reaction. In alkaline solutions, peroxynitrite anion is relatively stable and can be stored in the freezer for months. However, it has a pK_a of 6.8 at 37°C and the protonated form, peroxynitrous acid (ONOOH), is a reactive oxidant:

$$ONOO^- + H^+ \Leftrightarrow ONOOH \tag{2}$$

Peroxynitrous acid can form an activated intermediate, which acts as an oxidant resembling hydroxyl radical ($\cdot OH$) plus nitrogen dioxide (NO_2):

$$ONOOH \Rightarrow \cdot NO_2 \cdots \cdot OH \tag{3}$$

The activated intermediate can also rearrange to form nitrate directly:

$$\cdot NO_2 \cdots \cdot OH \Rightarrow {}^-NO_3 + H^+ \tag{4}$$

Thus, only 25% of peroxynitrite reacts as hydroxyl radical and nitrogen dioxide under physiological conditions with the remainder recombining to form nitrate (Beckman et al., 1990). However, the formation of hydroxyl radical–like oxidant is only one of several cytotoxic reactions carried out by peroxynitrite. At physiological pH, peroxynitrite anion is sufficiently stable to diffuse over several cell diameters to critical cellular targets before becoming protonated and decomposing. Peroxynitrite initiates lipid peroxidation and reacts directly with sulfhydryl groups at a thousand-fold greater than hydrogen peroxide at pH 7.4 (Radi et al., 1991). In addition, peroxynitrite reacts with transition metals to form a powerful nitrating agent with reactivity suggestive of the nitronium ion (NO_2^+) (Ischiropoulos et al., 1990). Thus, peroxynitrite is a reactive species with a sufficiently long lifetime to be highly toxic.

Ischemia and Peroxynitrite

Many pathophysiological processes including reperfusion of ischemic tissue, acute inflammation, and sepsis might initiate events that stimulate the simultaneous production of nitric oxide and superoxide. The subsequent formation of peroxynitrite could cause tissue injury. For example, ischemia in any organ will allow calcium entry into endothelial cytoplasm because of failure of ionic pumps and opening of ion channels, but endothelium will be unable to support nitric oxide synthesis without oxygen. Reperfusion will allow rapid nitric oxide synthesis by providing the oxygen needed to produce nitric oxide, because the intracellular messenger, Ca^{+2}, and other substrates—arginine and NADPH—would already have been made available by the ischemic insult.

Ischemia also induces intracellular superoxide production by xanthine oxidase, mitochondria, and other sources, which can escape extracellularly through anion channels (Lynch and Fridovich, 1978). Nitric oxide then rapidly reacts with superoxide both intracellularly and in the vascular lumen to form peroxynitrite. Because the rate of peroxynitrite formation depends on the product of superoxide and nitric oxide concentrations, each 10-fold increase in the concentrations of nitric oxide and superoxide results in a hundredfold amplification in the rate of peroxynitrite formation.

Perhaps the most puzzling effect of superoxide dismutase is its ability to protect ischemic tissue in experimental models when injected into the circulation just before reperfusion. The protective action of superoxide dismutase may in part be due to preventing the decomposition of nitric oxide by scavenging superoxide, which would help maintain normal vasodilation and inhibit thrombosis. We propose that superoxide dismutase also protects endothelium *in vivo* by preventing the formation of peroxynitrite that decomposes to form potent, cytotoxic oxidants.

Recently, high dosages of superoxide dismutase were found to be less protective than moderate dosages in isolated, perfused rat hearts (Bernier et al., 1989; Omar and McCord, 1990; Omar et al., 1990). The higher superoxide dismutase dosages have no effect on control rat hearts. Peroxynitrite contains the superoxide moiety and thus should be able to fit into the active site pocket from peroxynitrite, which might account for the apparent toxicity of high superoxide dismutase concentrations *in vivo.*

Peroxynitrite Reacts with Superoxide Dismutase

We have observed that peroxynitrite reacts with metal ions and with the active site of superoxide dismutase to produce the highly reactive and toxic nitronium ion (NO_2^+), which was detected by nitration of the sole tyrosine on bovine Cu, Zn superoxide dismutase. Copper in the active site of superoxide dismutase was necessary for this reaction. Removal of the copper by reduction with borohydride and dialysis against 50 mM KCN prevented nitrotyrosine formation, whereas restoration of copper to superoxide dismutase restored full superoxide-scavenging and nitrating activity. To account for the essential role of copper in the active site and the subsequent formation of the 3-nitrotyrosine located 18 to 21 Å distal from the active site, we propose that peroxynitrite is attracted by the same electrostatic field that draws superoxide anion into the active site (Getzoff et al., 1983). Once in the active site, peroxynitrite forms a transient cuprous adduct:

$$\mathrm{SOD} - \mathrm{Cu}^{2+} \cdot \cdot {}^{-}\mathrm{OO} - \mathrm{N} = \mathrm{O} \rightarrow \mathrm{SOD} - \mathrm{Cu}^{+1}\mathrm{O}^{-} \cdot \cdot \mathrm{O} = \mathrm{N}^{+} = \mathrm{O}$$

This intermediate complex can donate a nitronium to phenolics to form nitrophenols. After releasing hydroxide ion, native superoxide dismutase is regenerated:

$$\mathrm{SOD} - \mathrm{Cu}^{+1}\mathrm{O}^{-} \cdot \cdot \mathrm{O} = \mathrm{N}^{+} = \mathrm{O} + \mathrm{phenol} \rightarrow \mathrm{SOD} - \mathrm{Cu}^{+2} + \mathrm{OH}^{-} + \mathrm{NO_2}{-}\mathrm{phenol}$$

In the absence of other phenolic compounds, a second order collision between the superoxide dismutase–peroxynitrite intermediate and a second superoxide dismutase molecule results in nitration of the sole tyrosine at residue 108. Cu,Zn superoxide dismutase acts catalytically and is not inactivated as a result of its reaction with peroxynitrite. Superoxide dismutase catalyzes the nitration of a wide range of phenolics, including tyrosines in lysozyme and histone, at a reaction rate of $10^5 M^{-1} s^{-1}$. This reaction is about 50 times faster than the fastest peroxynitrite reaction we have determined to date (the oxidation of cysteine) (Radi et al., 1991). The implications from these data are (a) exogenous superoxide dismutase may also catalyze toxic reactions in ischemic tissue and (b) superoxide dismutase–nitrating reactions are specific and sensitive probes for measuring peroxynitrite *in vivo*. Nitric oxide produced by macrophages in the presence of high extracellular superoxide dismutase still formed peroxynitrite, which then reacted with superoxide dismutase to generate a reactive nitrating agent resembling nitronium ion. Peroxynitrite may be formed in regions not accessible to superoxide dismutase, but could diffuse to superoxide dismutase. The reaction of superoxide dismutase with peroxynitrite is approximately a thousand-fold slower than with superoxide. Therefore, higher concentrations of superoxide dismutase will continue to accelerate the peroxynitrate reaction, whereas superoxide scavenging will have been limited by escape from cells.

Superoxide Dismutase can be a Specific Probe for Peroxynitrite

The reaction of superoxide dismutase with peroxynitrite could lead to a highly useful and specific probe for peroxynitrite *in vivo*. We inactivated >99% of the superoxide scavenging activity of superoxide dismutase by cleaving one histidine in the active site with H_2O_2 and by adding phenylglyoxal to the arginine 141. Both of these modifications inhibit superoxide dismutation (Beyer et al., 1987), but do not significantly affect the reaction with peroxynitrite, providing a modified superoxide dismutase suitable for trapping peroxynitrite.

Macrophages Produce Peroxynitrite

Considerable evidence suggests that the production of nitric oxide from *L*-arginine contributes to the cytotoxicity of macrophages (Granger et al., 1988; Stuehr and Nathan, 1989). Early reports found that fluid from tumors containing activated macrophages were depeleted of arginine. Resting macrophages do not significantly produce nitric oxide, but the pathway is inducible by treatment with lipopolysaccharide, interferon, or interleukin (Iyengar et al., 1987; Marletta et al., 1988). Immunostimulated murine macrophages have been shown to induce cytostasis (Hibbs et al., 1987), to inhibit the mitochondrial respiration of target tumor cells (Hibbs et al., 1988), and to cause microbiostasis by an *L*-arginine–dependent pathway (Green et al., 1990). Activated macrophages also produce superoxide (Johnston et al., 1978), suggesting that peroxynitrite could contribute to cytotoxicity. Inhibitors of nitric oxide synthesis increase the amount of superoxide detectable from stimulated macrophages, whereas superoxide dismutase increases the amount of nitric oxide released from these macrophages (Albina et al., 1989). Thus, it appears that peroxynitrite may be produced by activated macrophages. Activated neutrophils also produce nitric oxide (McCall et al., 1989), but the production of nitric oxide relative to superoxide remains to be demonstrated.

We have used the superoxide dismutase–catalyzed nitration of a tyrosine analog to measure peroxynitrite production from activated rat alveolar macrophages, observing 0.1 nmol peroxynitrite$\cdot$min$^{-1}\cdot 10^{-6}$ cells. The rate of nitration was the same whether native Cu, Zn superoxide dismutase or the phenylglyoxyl-H_2O_2 modified superoxide dismutase (which is >99% inhibited with respect to its superoxide-scavenging activity) was used. Thus, nitric oxide effectively competes with superoxide dismutase for superoxide in a biological system where nitric oxide can react with superoxide in regions inaccessible to superoxide dismutase.

Three other independent but indirect estimates of macrophage peroxynitrite formation were consistent with the superoxide dismutase–based measurement: (a) the stable decomposition products of nitric oxide, nitrite (NO_2^-), and nitrate (NO_3^-), accumulated at a rate of 0.10 ± 0.01 nmol$\cdot 10^6$ cells$^{-1}\cdot$ min^{-1} in activated macrophages, (b) inhibition of nitric oxide synthesis with methylarginine increased the amount of superoxide detected by the superoxide dismutase–inhibitable cytochrome *c* reduction by 0.12 ± 0.02 nmol$\cdot 10^6$ cells$^{-1}\cdot$min^{-1}, and (c) the percentage of nitrate relative to nitrite increased from 30% to 67% after phorbol ester treatment. Nitric oxide in dilute solution without a source of superoxide predominantly decays to produce nitrite (Ignarro, 1990). Peroxynitrite decomposes principally to nitrate in buffer, although it can also yield nitrite after reaction with other molecules (Hughes et al., 1971).

Excitatory Neurotransmitters in Cerebral Ischemia

Low concentrations of glutamate (~100 μM) induce swelling and cell death of cultured neurons. Selective blockade of NMDA receptors or depeletion of Ca^{2+} from the culture medium protects neurons from low concentrations of glutamate (Goldberg et al., 1987). Antagonists of NMDA also prevent neuronal death in cell culture resulting from short durations of hypoxia or starvation (Rothman, 1984). Hypoxic stress depolarizes membranes and causes synaptic vesicles to discharge glutamate (Novelli et al., 1988; Simon et al., 1984), thereby activating NMDA receptors and allowing intracellular Ca^{+2} accumulation. Remarkably, Park et al. (1988) found that the NMDA antagonist, MK-801, can reduce cortical infarct volume by 50% in cats undergoing occlusion of the middle cerebral artery (MCA). The mechanism whereby inhibition of NMDA receptors affects infarct volume remains unknown, but may involve rescuing marginally perfused penumbra regions of the ischemic lesion.

Superoxide and Cerebral Injury

Superoxide also contributes to infarct development after focal cerebral ischemia. We have shown that infarct volume in a focal MCA stroke model in the rat was reduced by the combination of superoxide dismutase plus catalase (Beckman et al., 1988; Liu et al., 1989). Both superoxide dismutase and catalase were conjugated to polyethylene glycol to increase their circulating half-lives and interactions with endothelial cells (Beckman et al., 1988). Imaizuma et al. (1990) have demonstrated liposome-entrapped superoxide dismutase also reduces infarct volume in a similar MCA occlusion model in the rat. The use of antioxidants such as superoxide dismutase or catalase as therapeutic agents has implicated oxygen radicals in cerebral injury resulting from trauma (Kontos and Wei, 1986), hypertension (Kontos, 1985), cold edema (Ando et al., 1989; Chan et al., 1987), intraventricular hemorrhage (Ment et al., 1985), and ischemia (Araki et al., 1992; Cao et al., 1988). Transgenic mice overexpressing superoxide dismutase are also protected against cold edema (Chan et al., 1991). Superoxide dismutase also reduces neuronal death in culture resulting from brief glucose deprivation (Saez et al., 1987) and from hypoxia (Kinoshita et al., 1991). Cultured neurons derived from transgenic mice overexpressing superoxide dismutase are protected from glutamate toxicity (Chan et al., 1990). Low molecular weight antioxidants also protect cultured neurons from glutamate toxicity (Miyamoto et al., 1989). The sources of superoxide in cerebral ischemia are unknown, but could be generated by arachidonic acid metabolism (Kukreja et al., 1986), mitochondria (Takeuchi et al., 1991; Turrens et al., 1991), and perhaps circulating xanthine oxidase (Yokoyama et al., 1990).

Nitric Oxide and Cerebral Injury

The protective effects of superoxide dismutase in cerebral injury seems to parallel many protective effects associated with inhibition of the NMDA receptor. We have previously suggested that nitric oxide produced by neurons could be the common link (Beckman, 1990, 1991). We hypothesize that the synaptic discharge of excitatory neurotransmitters activate NMDA receptors in marginally perfused regions of ischemic brain, which causes maximal synthesis of nitric oxide by neurons when reperfusion occurs. Indeed, high nitric oxide concentrations may be partially responsible for the reactive hyperemia that frequently accompanies reperfusion. Derangements in oxidative metabolism increases production of superoxide during reperfusion. The reaction of superoxide, itself a mild reducing agent, with nitric oxide will produce the far more damaging peroxynitrite anion.

Recently, several groups have reported that nitric oxide may be an important agent in brain injury. Dawson et al. (1991b) found that both methyl-and nitroarginine protect cultured cortical neurons from glutamate toxicity. However, there have been negative reports as well (Demerlè-Pallardy et al., 1991). Relatively low dosages of nitroarginine were shown to reduce cortical infarct volume in a mouse MCA occlusion model of stroke (Nowicke et al., 1991). Peroxynitrite has also been implicated in acute immune-stimulated lung injury (Mulligan et al., 1991) and in myocardial ischemia (Matheis et al., 1992).

Early Activation of Microglia by Ischemia and Brain Injury

As shown earlier, we have established that rat alveolar macrophages produce substantial amounts of peroxynitrite. Microglia are widely distributed cells in brain that can differentiate into macrophages and are important for the clearing of cellular debris after injury. They are observed in large numbers several days after injury. However, recent histological observations indicate that microglia may be activated in the first hours of ischemia and brain trauma (Gehrmann et al., 1992). Activated microglia produce superoxide (Colton and Gilbert, 1987) and also may be able to express the inducible form of the nitric oxide synthase. Thus, peroxynitrite might be produced by the activation of microglia.

Acknowledgement. This work was supported by the National Institutes of Health (NS 24338) and from a Grant-in-Aid from the American Heart Association. Part of this work was preformed while Joseph S. Beckman was an Established Investigator of the American Heart Association.

References

Albina J, Mills C, Henry WL Jr, Caldwell M (1989): Regulation of macrophage physiology by L-arginine. Role of the oxidative L-arginine deiminase pathway. *J Immunol* 143:3641–3646.

Ando Y, Inoue M, Hirota M, Morino Y, Araki S (1989): Effect of superoxide dismutase derivative on cold-induced brain edema. *Brain Res* 477:286–291.

Araki N, Greenberg J, Uematsu D, Sladky J, Reivich M (1992): Effect of superoxide dismutase on intracellular calcium in stroke. *J Cereb Blood Flow Metab* 12:43–52.

Baum RM (1984): Superoxide theory of oxygen toxicity is center of heated debate. *Chem Engin News* April 9, 1984:20–28.

Beckman JS (1990): Ischemic injury mediator. *Nature (Lond)* 345: 27–28.

Beckman JS (1991): The double edged role of nitric oxide in brain function and superoxide-mediated pathology. *J Devl Physiol* 15:53–59.

Beckman JS, Beckman TW, Chen J, Marshall PM, Freeman BA (1990): Apparent hydroxyl radical production from peroxynitrite: implications for endothelial injury by nitric oxide and superoxide. *Proc Natl Acad Sci USA* 87:1620–1624.

Beckman JS, Liu TH, Hogan EL, Lindsay SL, Freeman BA, Hsu CY (1988): Evidence for a role of oxygen radicals in cerebral ischemic injury. In: *Cerebrovascular Diseases*, Ginsberg WDD, ed. New York: Raven Press.

Beckman JS, Minor RM Jr, White CJ, Repine J, Rosen GM, Freeman BA (1988): Superoxide dismutase and catalase conjugated to polyethylene glycol increases endothelial enzyme activity and oxidant resistance. *J Biol Chem* 263:6584–6802.

Bernier M, Manning A, Hearse D (1989): Reperfusion arrhythmias: dose-related protection by anti-free radical interventions. *Am J Physiol* 256:H1344-H1352.

Beyer W Jr, Fridovich I, Mullenbach GT, Hallewell R (1987): Examination of the role of arginine-143 in the human copper and zinc superoxide dismutase by site-specific mutagenesis. *J Biol Chem* 262:11182–11187.

Blough NV, Zafiriou OC (1985): Reaction of superoxide with nitric oxide to form peroxonitrite in alkaline aqueous solution. *Inorg Chem* 24:3504–3505.

Böhme G, Bon C, Stutzmann JM, Doble A, Blanchard JC (1991): Possible involvement of nitric oxide in long-term potentiation. *Eur J Pharmacol* 199:379–381.

Bredt D, Hwang P, Glatt C, Lowenstein C, Reed R, Snyder S (1991): Cloned and expressed nitric oxide synthase structurally resembles cytochrome P-450 reductase. *Nature* 351:714–718.

Bredt DS, Snyder HS (1990): Isolation of nitric oxide synthetase, a calmodulin-requiring enzyme. *Proc Natl Acad Sci USA* 87:682–685.

Brown TH, Chapman PF, Kairiss EW, Keenan CL (1988): Long-term synaptic potentiation. *Science* 242:724–728.

Brune B, Lapetina EG (1990): Properties of a novel nitric oxide-stimulated ADP-ribosyltransferase. *Arch Biochem Biophys* 279:286–290.

Busse R, Luckhoff A, Bassenge E (1987): Endothelium-derived relaxant factor inhibits platelet activation. *Naunyn-Schmiedeberg's Arch Pharmacol* 336:566–571.

Cao W, Carney JM, Duchon A, Floyd RA, Chevion M (1988): Oxygen free radical involvement in ischemia and reperfusion injury to brain. *Neurosci Lett* 88:233–238.

Chan P, Chu L, Chen S, Carlson E, Epstein C (1990): Reduced neurotoxicity in transgenic mice overexpressing human copper-zinc-superoxide dismutase. *Stroke* 21(suppl III):III-80–82.

Chan P, Yang G, Chen S, Carlson E, Epstein C (1991): Cold-induced brain edema and infarction are reduced in transgenic mice overexpressing CuZn-superoxide dismutase. *Ann Neurol* 29:482–486.

Chan PH, Longar S, Fishman RA (1987): Protective effects of liposome-entrapped superoxide on post-traumatic brain edema. *Ann Neurol* 21:540–547.

Colton A, Gilbert D (1987): Production of superoxide by a CNS macrophage, the microglia. *FEBS Lett* 223:284–288.

Dawson T, Bredt D, Fotuhi M, Hwang P, Snyder S (1991a): Nitric oxide synthase and neuronal NADPH diaphorase are identical in brain and peripheral tissues. *Proc Natl Acad Sci USA* 88:7797–7801.

Dawson V, Dawson T, London E, Brent D, Snyder S (1991b): Nitric oxide mediates glutamate neurotoxicity in primary cortical cultures. *Proc Natl Acad Sci USA* 88:6368–6371.

Deguchi T, Yoshioka M (1982): L-Arginine identified as an endogenous activator for soluble guanylate cyclase from neuroblastoma cells. *J Biol Chem* 257:10147–10151.

Demerlè-Pallardy C, Lonchampt M-O, Chabrier P-E, Braquet P (1991): Absence of implication of L-arginine/nitric oxide pathway in neuronal cell injury induced by L-glutamate or hypoxia. *Biochem Biophys Res Commun* 181:456–464.

East S, Garthwaite J (1991): NMDA receptor activation in the rat hippocampus induces cyclic GMP formation through the *L*-arginine-nitric oxide pathway. *Neurosci Lett* 123:17–19.

Föstermann U, Gorsky L, Pollock J, Schmidt H, Heller M, Murad F (1990): Regional distribution of EDRF/NO-synthesizing enzyme(s) in rat brain. *Biochem Biophys Res Commun* 168:727–732.

Fridovich I (1986): Biological effects of the superoxide radical. *Arch Biochem Biophys* 247:1–11.

Furchgott RF, Vanhoutte PM (1989): Endothelium-derived relaxing and contracting factors. *FASEB J* 3:2007–2018.

Gally JA, Montague PR, Reeke GN Jr, Edelman GM (1990): The NO hypothesis: possible effects of a short-lived, rapidly diffusible signal in the development and function of the nervous system. *Proc Natl Acad Sci USA* 87:3547–3551.

Garthwaite J, Charles SL, Chess-Williams R (1988): Endothelium-derived relaxing factor release on activation of NMDA receptors suggests role as intercellular messenger in the brain. *Nature (Lond)* 336:385–388.

Gehrmann J, Bonnekoh P, Miyazawa T, Hossman K, Kreutzberg G (1992): Immunocytochemical study of an early microglial activation in ischemia. *J Cereb Blood Flow Metab* 12:257–269.

Getzoff ED, Tainer JA, Weiner PK, Kollman PA, Richardson JS, Richardson DC (1983): Electrostatic recognition between superoxide and copper, zinc superoxide dismutase. *Nature (Lond)* 306:287–290.

Goldberg MP, Weiss JH, Pham P-C, Choi DW (1987): *N*-methyl-*D*-aspartate receptors mediate hypoxic neuronal injury in cortical culture. *J Pharmacol Exp Ther* 24:784–791.

Granger D, Hibbs J Jr, Perfect J, Durak D (1988): Specific amino acid (L-arginine) requirement for the microbiostatic activity of murine macrophages. *J Clin Invest* 81:1129–1136.

Hebb DO (1949): *The organization of behavior*. New York: Wiley.

Hibbs J Jr, Taintor R, Vavrin Z (1987): Macrophage cytotoxicity: Role of *L*-arginine deminiase and imino nitrogen oxidation to nitrite. *Science* 235:473–235.

Hibbs J Jr, Taintor R, Vavrin Z, Rachlin E (1988): Nitric oxide: a cytotoxic activated macrophage effector molecule. *Biochem Biophys Res Commun* 157:87–94.

Hope B, Michael G, Knigge K, Vincent S (1991): Neuronal NADPH diaphorase is a nitric oxide synthase. *Proc Natl Acad Sci USA* 88:2811–2814.

Hughes M, Nickline H, Sackrule W (1971): The chemistry of peroxynitrites with nucleophiles in alkali, and other nitrite producing reactions. *J Chem Soc A* 3722–3725.

Ignarro LJ (1990): Biosynthesis and metabolism of endothelium-derived nitric oxide. *Annu Rev Pharmacol Toxicol* 30:535–560.

Imaizumi S, Wollworth V, Fishman RA, Chan PH (1990): Liposome-entrapped superoxide dismutase reduces cerebral infarction in cerebral ischemia in rats. *Stroke* 21:1312–1317.

Ischiropoulos H, Chen J, Tsai JM, Martin J, Smith C, Beckman J (1990): Peroxynitrite (ONOO-) reacts with superoxide dismutase to give the reactive nitronium ion. *Free Rad Biol Med* 9(S1):131 Abstract.

Iyengar R, Stuehr D, Marletta M (1987): Macrophage synthesis of nitrite, nitrate, and N-nitrosamines: precursors and role of the respiratory burst. *Proc Natl Acad Sci USA* 84:6369–6373.

Johnston R Jr, Godzik C, Cohn Z (1978): Increased superoxide anion production by immunologically activated and chemically elicited macrophages. *J Exp Med* 22:115–127.

Kinoshita A, Yamada K, Kohmura E, Hayakawa T (1991): Human recombinant superoxide dismutase protects cultured neurons against hypoxic injury. *Pathobiology* 59:340–344.

Kontos HA (1985): Oxygen radicals in cerebral vascular injury. *Circ Res* 57:508–516.

Kontos HA, Wei EP (1986): Superoxide production in experimental brain injury. *J Neurosurg* 64:803–807.

Kukreja RC, Kontos HA, Hess ML, Ellis EF (1986): PGH synthase and lipoxygenase generate superoxide in the presence of NADH or NADPH. *Circ Res* 59:612–619.

Liu TH, Beckman JS, Freeman BA, Hogan EL, Hsu CY (1989): Polyethylene glycol-conjugated superoxide dismutase and catalase reduce ischemic brain injury. *Am J Physiol* 256:H589–H593.

Lynch RE, Fridovich I (1978): Permeation of erythrocyte stroma by superoxide radical. *J Biol Chem* 253:4697–4699.

Marletta MA, Yoon PS, Iyengar R, Leaf CD, Wishnok JS (1988): Macrophage Oxidation of L-arginine to nitrite and nitrate: nitric oxide is an intermediate. *Biochemistry* 27:8706–8711.

Matheis G, Sherman M, Buckberg G, Haybron D, Young H, Ignarro L (1992): Role of *L*-arginine-nitric oxide pathway in myocardial reoxygenation injury. *Am J Physiol* 262:H616–H620.

Mayer B, John M, Böhme E (1990): Purification of a Ca^{+2}/calmodulin-dependent nitric oxide synthase from porcine cerebellum. Cofactor-role of tetrahydrobiopterin. *FEBS Lett* 277:215–219.

Mayer ML, Miller RJ (1990): Excitatory amino acid receptors, second messengers and regulation of intracellular Ca^{+2} in mammalian neurons. *Trends Pharmacol Sci* 11:254–260.

Mayer ML, Westbrook GL, Gurthie PE (1984): Voltage-dependent block by Mg^{+2} of NMDA responses in spinal cord neurons. *Nature (Lond)* 309:261–277.

McCall TB, Boughton-Smith NK, Palmer RMJ, Whittle BJR, Moncada S (1989): Synthesis of nitric oxide from L-arginine by neutrophils. *Biochem J* 261:293–296.

Ment LA, Steward WB, Duncan CC (1985): Superoxide dismutase protects in a beagle puppy model of neonatal intraventricular hemorrhage. *J Neurosurg* 62:563–567.

Miyamoto M, Murphy T, Schnaar R, Coyle J (1989): Antioxidants protect against glutamate-induced cytotoxicity in a neuronal cell line. *J Pharmacol Exp Ther* 250:1132–1140.

Moncada S, Herman AG, Vanhoutte PM (1987): Endothelium-derived relaxing factor is identified as nitric oxide. *Trends Pharmacol Sci* 8:365–368.

Moncada S, Palmer R, Higgs E (1991): Nitric oxide: physiology, pathophysiology, and pharmacology. *Pharmacol Rev* 43:109–142.

Mulligan M, Hevel J, Marletta M, Ward P (1991): Tissue injury caused by deposition of immune complexes is *L*-arginine dependent. *Proc Natl Acad Sci USA* 88:6338–6342.

Murphy S, Minor R Jr, Wel G, Harrison D (1990): Evidence for an astrocyte-derived vasorelaxing factor with properties similar to nitric oxide. *J Neurochem* 55:349–351.

Novelli A, Reilly JA, Lysko PG, Henneberry RC (1988): Glutamate becomes neurotoxic with the *N*-methyl-*D*-asptarate receptor when intracellular energy levels are reduced. *Brain Res* 451:205–212.

Nowicki J, Duval D, Poignet H, Scatton B (1991): Nitric oxide mediates neuronal death after focal cerebral ischemia in the mouse. *Eur J Pharmacol* 204:339–340.

Oh S, Betz A (1991): Interaction between ree radicals and excitatory amino acids in the formation of ischemic brain edema in rats. *Stroke* 22:915–921.

Omar B, Gad N, Jordan M, Striplin S, Russell W, Downey J, McCord J (1990): Cardioprotection by Cu,Zn-superoxide dismutase is lost at high doses in the reoxygenated heart. *Free Rad Biol Med* 9:465–471.

Omar BA, McCord J (1990): The cardioprotective effect of Mn-superoxide dismutase is lost at high doses in the postischemic isolate rabbit heart. *Free Rad Biol Med* 9:473–478.

Palmer RMJ, Ashton DS, Moncada S (1988): Arginine is the source of endothelial-derived nitric oxide. *Nature (Lond)* 333:664–666.

Palmer RMJ, Ferrige AG, Moncada S (1987): Nitric oxide release accounts for the biological activity of endothelium-derived relaxing factor. *Nature (Lond)* 327:523–526.

Park CK, Nehls DG, Graham DI, Teasdale GM, McCullough J (1988): Focal cerebral ischemia in the cat: treatment with the glutamate antagonist MK-801 after induction of ischemia. *J Cereb Blood Flow Metab* 8:757–762.

Pollock J, Förstermann J, Mitchell J, Warner T, Schmidt H, Nakane M, Murad F (1991): Purification and characterization of particulate endothelium-derived relaxing factor synthase from cultured and native bovine aortic endothelial cells. *Proc Natl Acad Sci USA* 88:10480–10484.

Radi R, Beckman JS, Bush KM, Freeman BA (1991): Sulfhydryl oxidation by peroxynitrite: the cytotoxic potential of superoxide and nitric oxide. *J Biol Chem* 266:4244–4250.

Radomski M, Palmer R, Moncada S (1990): Glucocorticoids inhibit the expression of an inducible, but not the constitutive, nitric oxide synthase in vascular endothelial cells. *Proc Natl Acad Sci USA* 87:10043–10047.

Rothman SM (1984): Synaptic release of excitatory amino acid neurotransmitter mediates anoxic neuronal death. *J Neurosci* 4:1884–1891.

Rubanyi GM, Vanhoutte PM (1986): Superoxide anions and hyperoxia inactive endothelium-derived relaxing factor. *Am J Physiol* 250:H822–H827.

Saez JS, Kessler JA, Bennett MVL, Spray DC (1987): Superoxide dismutase protects cultured neurons against death by starvation. *Proc Natl Acad Sci USA* 84:3056–3059.

Saran M (1990): Reaction of NO with O_2^-. Implications for the action of endothelium-derived relaxing factor (EDRF). *Free Rad Res Commun* 10:221–226.

Sawyer DT, Valentine J (1981): How super is superoxide? *Acct Chem Res* 14:393–400.

Schmidt H, Pollock J, Nakane M, Gorsky L (1991): Purification of a soluble isoform of guanylyl cyclase-activating factor synthase. *Proc Natl Acad Sci USA* 88:365–369.

Seisjo BK, Bengtsson F (1989): Calcium fluxes, calcium antagonists, and calcium-related pathology in brain ischemia, hypoglycemia, and spreading depression: a unifying hypothesis. *J Cereb Blood Flow Metab* 9:127–140.

Shibuki K (1990): An electrochemical microprobe for detecting nitric oxide release in brain tissue. *Neurosci Res* 9:69–76.

Shibuki K, Okada D (1991): Endogenous nitric oxide release required for long-term synaptic depression in the cerebellum. *Nature (Lond)* 349:326–329.

Shulman E, Madison D (1991): A requirement for the intercellular messenger nitric oxide in long-term potentiation. *Science* 254:1503–1506.

Simon RP, Swan JH, Griffins T, Meldrum BS (1984): Blockade of *N*-methyl-*D* aspartate receptors may protect against ischemic damage in the brain. *Science* 226:850–885.

Southam E, East S, Garthwaite J (1991): Excitatory amino acid receptors coupled to the nitric oxide/cyclic GMP pathway in rat cerebellum during development. *J Neurochem* 56:2072–2081.

Stuehr D, Kwon N, Nathan C, Griffin O, Feldman P, Wiseman J (1991): N^W-hydroxy-*L*-arginine is an intermediate in the biosynthesis of nitric oxide from *L*-arginine. *J Biol Chem* 266:6259–6263.

Stuehr DJ, Nathan C (1989): Nitric oxide. A macrophage product responsible for cytostasis and respiratory inhibition in tumor cells. *J Exp Med* 169:1543–1555.

Takeuchi Y, Morii H, Tamura M, Hayaishi O, Watanabe Y (1991): A possible mechanism of mitochondrial dysfunction during cerebral ischemia: inhibition of mitochondrial respiration activity by arachidonic acid. *Arch Biochem Biophys* 289:33–38.

Turrens J, Beconi M, Barilla J, Chavez U, McCord J (1991): Mitochondrial generation of oxygen radicals during reoxygenation of ischemic tissues. *Free Rad Res Commun* 12:681–689.

Uemura Y, Kowall N, Beal M (1990): Selective sparing of NADPH-diaphorase somatostatin-neuropeptide Y neurons in ischemic gerbil striatum. *Ann Neurol* 27:620:625.

Wallace GC, Fukuto J (1991): Synthesis and bioactivity of N^W-hydroxyarginine: a possible intermediate in the biosynthesis of nitric oxide from arginine. *J*

Med Chem 34:1746–1748.

Williams JH, Errington ML, Lynch MA, Bliss TVP (1989): Arachidonic acid induces a long-term activity-dependent enhancement of synaptic transmission in the hippocampus. *Nature (Lond)* 341:739–742.

Yokoyama Y, Beckman JS, Wheat J, Cash TG, Freeman BA, Parks DA (1990): Circulating xanthine oxidase: potential mediator of ischemic injury. *Am J Physiol* 258:G564–G570.

Chapter 11

Contributions of the Physical Properties of Neuroprotective Agents to their Efficacy as Inhibitors of Lipid Peroxidation

Kenneth L. Audus

The chemistry and role of oxygen radicals and lipid peroxidation in brain damage has been extensively reviewed (Aust et al., 1985; Halliwell and Gutteridge, 1989; Siesjo et al., 1989). Collectively, the literature implicates oxygen radicals and the process of lipid peroxidation in central nervous system (CNS) tissue damage resulting from head and spinal cord injury, trauma, inflammation, and ischemic cerebrovascular disorders (Braughler and Hall, 1989). A causative role for oxygen radicals has also been suggested for other human diseases including cancer, multiple sclerosis, Parkinson's disease, autoimmune diseases, senile dementia and the aging process (Gutteridge, 1987), and atherosclerosis (Henning and Chow, 1988; Henry, 1991). Accordingly, current drug research includes design and development of neuroprotective agents with appropriate antioxidant chemistry.

Accumulating literature suggests that the efficacy of a neuroprotective agent may include beneficial physical interactions between the molecule and the cell membrane. Researchers have noted that a relatively lipophilic nature, facilitating intercalation into cell membranes (as well as crossing the blood–brain barrier), appears to be an important factor for an antioxidant directed at oxygen radical–mediated CNS cell or tissue injuries (Braughler et al., 1989; Demopoulos et al., 1972, 1982; Hall et al., 1987; Janero et al., 1988; Janero and Burghardt, 1989; Kaneko et al., 1991). Moreover, many of the effective lipid peroxidation inhibitors have

Oxygen Free Radicals in Tissue Damage
Merrill Tarr and Fred Samson, Editors

been observed to stabilize cell membranes physically, influencing membrane permeability and other membrane-dependent functions (Audus et al., 1991; Demopolous et al., 1972, 1982; Hall et al., 1987; Spetzler and Hadley, 1989). Consequently, design and development of effective neuroprotective agents might require consideration of substances that have not only appropriate antioxidant chemistry and favorable membrane intercalating properties, but also certain other beneficial physical interactions with cell membranes. The focus of this chapter is to highlight properties of several neuroprotective molecules, aside from antioxidant chemistry, that may contribute to future drug design and development.

Cell Membrane, Lipid Peroxidation, and Drug Development

The primary cellular target sites of lipid peroxidation are unsaturated phospholipids of the plasma and organelle membranes (Mead, 1976; Tappel 1975, 1978). Peroxidation of membrane phospholipids alters bilayer integrity, influencing membrane-dependent functions such as permeability (transcellular and paracellular), enzyme activities, and receptors (Divakaran and Wiggins, 1987; Dobretsov et al., 1977; Ernster et al., 1982; Ivanov, 1985; Koster and Slee, 1980; Leibovitz and Johnson, 1971; Smolen and Shohet, 1974).

Oxidation of membrane-associated proteins may also occur during lipid peroxidation and contribute to the overall breakdown in membrane-dependent functions (Konat and Wiggins, 1985; Konat et al., 1986). Oxygen radicals are believed to be involved in altering the surface charge of endothelia, for example, by depolymerizing the glycocalx, perhaps altering the permselectivity of the vasculature (Inauen et al., 1990). In addition, the cytoskeleton, which includes microfilaments associated with the plasma membrane, retracts cells under oxidant stress. In the vasculature, oxidant-induced cell retraction increases permeability of the vessels to albumin and water. Phalloidin, a bicyclic peptide that binds to and stabilizes the cytoskeleton, prevents oxidant-induced endothelial cell retraction, supporting the role of the cytoskeleton in regulating cell functions under oxidative stress (Alexander et al., 1988; Phillips et al., 1989). Damage to such membrane-associated proteins may originate directly from reactions with oxygen radicals or from reactions with lipid peroxides and/or their decomposition products (Divakaran and Wiggins, 1987).

Natural membrane antioxidants include the lipophilic substances, vitamin E, and β-carotene (Gutteridge, 1987; Palossa and Krinsky, 1991). By contrast, hydrophilic substances such as vitamin C, transferrin, lactoferrin, extracellular superoxide dismutase, ceruloplasmin, albumin, urate, and glucose are extracellular fluid antioxidants. Enzymes, catalase, glutathione peroxidase, and superoxide dismutases generally provide intracellular antioxidant protection. Vitamin E represents the most important endogenous inhibitor of lipid peroxidation (Gutteridge, 1987).

Partitioning into cellular lipid domains may be critical for the action of potential therapeutic agents directed at lipid peroxidation. Lipid radicals are formed within cell membranes. Therefore, the lipophilic nature of interior domains of cell membranes requires that agents directed at inhibition of lipid peroxidation be relatively lipophilic. This point has been demonstrated experimentally by Braughler et al. (1987) with the iron chelator, desferroxiamine. Compared to vitamin E, desferroxiamine was determined to be a weak inhibitor of iron-dependent lipid peroxidation. However, on coupling desferrioxamine in inhibiting lipid peroxidation was essentially equivalent to vitamin E (Braughler et al., 1987).

Design and development of agents directed at injuries mediated by oxygen radicals will likely include multiple mechanisms of action. At the cellular level, it appears that combined therapies, those that inhibit iron delocalization/free radical formation and control calcium flux across membranes, are among those strategies with the greatest potential (Hall and Braughler, 1989; Siesjo et al., 1989). Summarized in Table 1 are some of the potential mechanisms through which antioxidants may influence oxygen–radical–mediated cell damage. Some or all of these mechanisms have been observed as potential contributors to the therapeutic efficacy of several drugs or drug groups generally classified as neuroprotective agents.

Membrane Interactions Characteristic of Neuroprotective Agents

The following agents are representative of a group of antioxidants with membrane interactions suggested as contributors to their roles as inhibitors of lipid peroxidation.

Vitamin E. Generally, tocopherols penetrate into membranes as the unsaturation of phospholipids increases. In addition, vitamin E influences

Table 1. Some potential mechanisms of action for inhibitors of lipid peroxidation

Membrane intercalculation and control of free radicals
Membrane stabilization and permeability regulation
Regulation of metabolic rate
Shielding fatty acids
Control of phospholipase
Cytoskeletal stabilization

cholesterol distribution into membranes (Diplock, 1982). Typically, dramatically increasing membrane levels of vitamin E requires considerable time, reducing the application of vitamin E as a treatment. However, elevated dietary vitamin E has been shown to provide some measure of protection against lipid peroxidation in animal models surveyed in the literature (Hall and Braughler, 1989).

In recent studies, Kaneko et al. (1991) have looked at vitamin E and three analogs with respect to lipophilicity and inhibition of lipid peroxidation. Their results suggest that vitamin E and analogs having both significant antioxidant activity and a relatively lipophilic nature were the most effective in protecting endothelial cells from lipid–peroxide induced injury.

Significant interactions of vitamin E with phospholipids coincide with the antioxidant effects of the molecule (Ohyashiki et al., 1986; Patel et al., 1991). Several reports confirm that vitamin E stabilizes membranes in a manner analogous to cholesterol (Bisby and Birch, 1989; Massey et al., 1982; Ohyashiki et al., 1986; Steiner, 1981). In some phospholipids, vitamin E has actually been shown to be more effective than cholesterol in modifying membrane structure (Massey et al., 1982). Possibly through modifying the physical state of membranes, vitamin E may modulate cell membrane transport. Patel et al. (1991) have recently demonstrated that vitamin E–induced alterations in the physical state of the plasma membranes of pulmonary endothelia modulates membrane-dependent functions including 5-hydroxytryptamine uptake. Although not fully appreciated at present, the definition of the physiological role of vitamin E as a protector of cell functions should perhaps include both the physical and antioxidant chemical properties of the molecule (Divakaran and Wiggins, 1987; Patel et al., 1991).

Glucocorticoids. Methylprednisolone has been shown to protect CNS cell membranes from lipid peroxide–induced injury (Braughler, 1985). The precise mechanism of action of this class of compounds is unknown. Demopoulos et al. (1972, 1982) have hypothesized that high dose steroid inhibition of lipid peroxidation may involve multiple actions, including control of free radicals, membrane intercalation, physical stabilization of membranes, shielding of membrane fatty acids from free radicals, and inhibiting phospholipase A_2, as briefly reviewed below.

Steroid intercalation and stabilization of membranes and subsequent modulation of membrane permeability has been established (Bangham et al., 1965; Kiss et al., 1990; Young and Flamm, 1982). Recent works with methylprednisolone and a nonglucocorticoid steroid analog, U-72099E, has further demonstrated that antioxidant properties were unrelated to either the glucocorticoid activity or classic receptor-mediated actions of the steroid (Braughler, 1985; Hall et al., 1987). The latter study indicated that U-72099E was a more effective antioxidant, with the improvement attributed to an increased lopophilicity (Hall et al., 1987). Based in part on these collective observations, a link between membrane intercalation/stabilization and regulation of permeability has been proposed to contribute to the action of steroids and nonsteroid analogs in lipid peroxide–induced cell damage (Hall et al., 1987).

Investigation of the antioxidant of methylprednisolone has revealed evidence suggestive of the physical shielding of membrane fatty acids from free radicals. Braughler (1985) observed that glucocorticoids appeared to inhibit lipid peroxidation at a step before conjugated diene formation, the intermediate step in lipid peroxidation (Buege and Aust, 1958). These results were consistent with the hypothesis of Demopoulos et al. (1972, 1980), who proposed that the steroid molecules (e.g., methylprednisolone, dexamethasone) may insert within archways formed by unsaturated fatty acids whose double bonds were in a *cis* configuration (Braughler, 1985). The insertion of the steroid into the unsaturated membrane fatty acids, in effect, may physically block initiation of oxygen radical chain reactions by shielding target hydrogens in the divinyl methane structures (Braughler, 1985). Similar observations and conclusions were made with U-72099E (Hall et al., 1987).

Mentz et al. (1980) conducted a study in a cell free system and concluded that glucocorticoids may also modulate phospholipase A_2 directly. The influence of glucocorticoids on enzyme function through membrane stabilization was essentially ruled out in their experimental system. However, the possibility remains that steroids may influence the enzyme through other indirect mechanisms (e.g., inhibition of prostaglandin release) (Mentz et al., 1980).

21-Aminosteroids. The 21-aminosteroid (i.e., lazaroids) were synthesized as more potent nonsteroidal antioxidant mimics of methylprednisolone. These agents have been observed to be substantially more potent than methylprednisolone as inhibitors of lipid peroxidation and are devoid of glucocorticoid activity (Braughler et al., 1989). Recently, using fluorescence labeling techniques, we have shown that 21-aminosteroids, U-74500A and U-74006F, effectively distribute into brain microvessel endothelial cell (BMEC) membranes (Audus et al., 1991), a major target for oxygen radicals (Hall et al., 1989). These agents differ in antioxidant action. The former agent (U-74500A) behaves more like an iron chelator, whereas U-74006F is a free radical scavenger in the manner of vitamin E. Currently, U-74006F is in clinical trials (Braughler et al., 1989).

The association of 21-aminosteroids with BMEC membranes was established in suspended cells that were labeled with either diphenylhexatriene or trimethylammonium-diphenylhexatriene, fluorescent probes for the hydrophobic domains of membranes. The cationized probe, trimethylammonium-diphenylhexatriene, localizes at the lipid–water interface and can be used to measure the lipid order (average molecular packing) changes at the surface of cells or in the glycerol side-chain regions of BMECs (Cranney et al., 1983; Engel and Prendergast, 1981; Prendergast et al., 1981). This probe remains on the surface of viable endothelia for up to 4 hrs (Sheridan and Block, 1988). Diphenylhexatriene has been used to measure the average molecular packing order of the deeper core lipid or hydrophobic regions of membranes throughout a cell (Cranney et al., 1983; Pottel et al., 1983; van Blitterswijk et al., 1981). Studies of the effects of certain agents on the fluorescence parameters of these probes can indicate which membrane domain (i.e., superficial versus deep hydrophobic core distribution) the agent is likely to affect or associate.

The fluorescence anisotropy and lifetimes of diphenylhexatriene in labeled BMECs were determined after exposure to lazaroids and other antioxidants. As shown in Fig. 1, U-74500A and U-74006F increased the fluorescence anisotropy of diphenylhexatriene significantly over a 5 to 70 μM concentration range. These drugs were less effective in altering the fluorescence parameters of the cationized diphenylhexatriene probe (Audus et al., 1991). The fluorescence anisotropy of diphenylhexatriene and trimethylammonium-diphenylhexatriene was unaffected by solvents for the drugs (e.g., ethanol and phosphate-buffered saline). Concentrations of 21-aminosteroids noted to distribute into dephenylhexatriene-labeled BMEC membranes correlate well with concentrations effective in inhibiting lipid peroxidation (Audus et al., 1991). Similar observations were made in Fig. 1 with a nonsteroidal lazaroid, U-78517F, a 2-methylamino-chromans. This compound combines the amino functionality of a 21-aminosteroid with the antioxidant ring portion of vitamin E. The result was a more potent antioxidant (Hall et al., 1990).

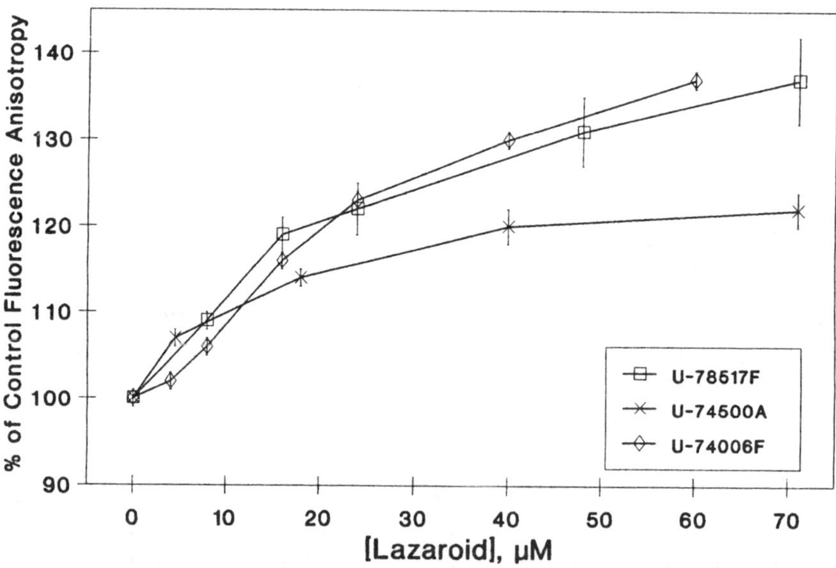

Figure 1 Concentration-dependent effect of selected lazaroids on the fluorescence anisotropy of diphenylhexatriene as a measure of membrane order in labeled bovine brain microvessel endothelial cells. The control anisotropy was 0.162 ± 0.004 at 37°C (Audus et al., 1991).

Changes in the fluorescence anisotropy and lifetimes of diphenylhexatriene in the BMEC membranes provided an indication of either interactions of the drug with either the environment immediately surrounding the probe or directly with the probe itself (Audus et al., 1991). In the absence of significant fluorescence lifetime changes, the fluorescence anisotropy of diphenylhexatriene reflects a change in lipid packing order of membranes. An increase in fluorescence anisotropy would be interpreted as an increase in packing order. Conversely, a decrease in fluorescence anisotropy would be interpreted as a disordering of lipid packing (Sklar, 1984). Analyzed on the basis of accompanying fluorescence lifetime changes, the dose-dependent increase in anisotropy suggsted that lazaroids may stabilize BMEC membranes to some degree (Audus et al., 1991). These observations were consistent with the earlier work of Braughler et al. (1988), who demonstrated that 21-aminosteroids block the release of arachidonic acid from injured cell membranes, a manifestation of the membrane stabilizing effects of these agents.

Another finding from this study was the localization of lazaroid effects to predominantly diphenylhexatriene-labeled membrane domains. The hydrocarbon core localization of lazaroids in cell membranes has been corroborated by another laboratory (Raub et al., 1991). The lack of effects of the lazaroids on trimethylammonium-diphenylhexatriene–labeled domains was similar to findings with anesthetic drugs, which also fail to alter superficially labeled membrane domains (Harris and Bruno, 1985).

As a comparison, Fig. 2 illustrates the concentration-dependent effect of other antioxidants, ascorbate, cholesterol, and butylated hydroxytoluene (BHT), on diphenylhexatriene fluorescence anisotropy in BMEC membranes. Ascorbate, the extracellular antioxidant (Gutteridge, 1987), as might be expected, had no effect on diphenylhexatriene anisotropy or lifetimes in BMEC membranes (Audus et al., 1991). In addition to inhibiting lipid peroxidation (Demopoulos et al., 1980; Smolen and Shohet, 1974), both cholesterol and BHT have been shown to stabilize membranes physically (Shertzer et al., 1991; Smolen and Shohet, 1974; van Ginkel et al., 1989). The biophysical interpretation of BHT's significant but opposite effect on diphenylhexatriene fluorescence anisotropy in the BMEC membrane remains to be solved. However, recent work by Shertzer et al. (1991) suggests that at low concentrations of BHT, subtle increases in membrane order can be detected, as monitored by increases in diphenylhexatriene in red blood cells. The increased membrane order correlates with membrane stabilization.

The 21-aminosteroids and 21-methylaminochromans control cell calcium (Hall et al., 1990). Uncontrolled cytosolic calcium levels result in depletion of cell energy resources and activation of lipases and proteases (Battaini et al., 1988; Naylor, 1983). Therefore, the lazaroids may also influence control of the metabolic rate of damaged cells through modulation of membrane calcium permeability (Hall et al., 1987, 1990). Retention of U-74500A, U74006F, and U78517F in the BMEC membranes here and in other cell membrane systems (Raub et al., 1991) has been shown to be effectively nonreversible unless serum albumin was present. The high

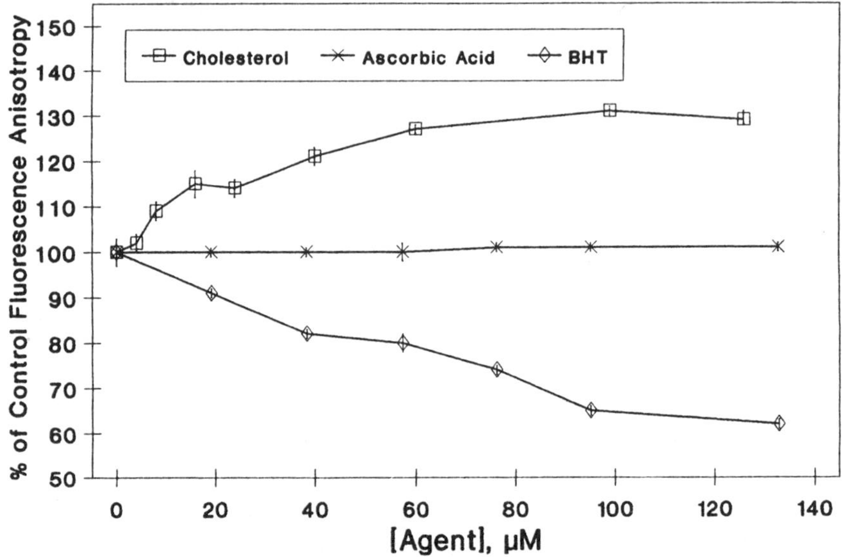

Figure 2 Concentration-dependent effect of selected antioxidants on the fluorescence anisotropy of diphenylhexatriene as a measure of membrane order in labeled bovine brain microvessel endothelial cells. The control anisotropy was 0.162 ± 0.004 at 37°C (Audus et al., 1991).

affinity of lazaroids for cell membranes would by implication present the possibility of low brain uptake by these agents (Raub et al., 1991). Conversely, membrane retention may favor a sustained action that extends to control of membrane stabilization, permeability regulation, and other functions.

Calcium antagonists. As mentioned above, control of cell calcium in oxidant-induced injury is critical for survival. Uncontrolled levels of cell calcium impairs cellular energy metabolism and activates proteases and lipases that may damage intracellular organelles. Overload of calcium results in a complete shutdown of adenosine triphosphate (ATP) production and eventually cell death (Battaini et al., 1988; Naylor, 1983). Therefore, pharmacological agents that regulate cell calcium may protect against cell damage. Perhaps not surprising then, calcium antagonists have been investigated as potential tissue protectants in oxidative injury.

The more specific calcium channel antagonists (e.g., verapamil) have little or reduced effectiveness as protectants against oxidative injury (Braughler et al., 1985; Janero et al., 1988). This observation suggests that membrane control of calcium permeability in oxidative injury may be related to other mechanisms (e.g., activation of phospholipases and degradation of phospholipids) as opposed to substantial disturbances in membrane calcium channels (Braughler et al., 1985). The reasons for calcium antagonist effectiveness in oxidative injury can apparently be attributed primarily to their ability to act as antioxidants (Goncalves et al., 1991; Henry, 1991; Janero and Burghardt, 1989; Janero et al., 1988). While lipophilicity has been a factor for calcium antagonist activity as antioxidants, the antioxidant activity of dihydropyridines depends more on the apparent radical scavenging capacity of their chemical structure rather than on apparent lipophilicity (Goncalves et al., 1991). On the other hand, the antioxidant activity of other calcium antagonists (e.g., diphenylpiperazines, alkylamines) coincides with the individual molecule's relative liposolubility (Goncalves et al., 1991; Henry, 1991). As for the other antioxidants discussed here, calcium antagonists with significant liposolubility also appear to stabilize membranes and, perhaps, transmembrane calcium flux (Robak and Duniec, 1986). Thus, protection of membranes by calcium antagonists may also be due to multiple mechanisms including both antioxidant chemistry and a physical stabilization of the cell membranes.

Barbiturates. Spetzler and Hadley (1989) have reviewed in detail the potential utility of barbiturate anesthesia for temporary ischemic conditions. The anti-ischemic acitivity of this class of compounds has been related primarily to membrane stabilizing activity, preserving ion pumps, and membrane permeability (Shapiro, 1985). Effectively, barbiturates depress cell energy demands and thus the continuous need for a supply of essential metabolic substrates that may be disrupted in ischemia. Unlike the other agents here, the barbiturates are not as well known as free radical scavengers or lipid peroxidation inhibitors. However, Demopoulos et al. (1980) have shown that some barbiturates (e.g., thiopental, methohexital) are free radical scavengers.

Summary and Conclusions

Aside from appropriate antioxidant chemistry, the potential exists that other physical interactions between effective inhibitors of lipid peroxidation and cell membranes may be desirable. Considering some of the neuroprotective agents above, the role of certain drug–membrane interactions have been observed and credited with contributions to the action of the of the individual agents. As the knowledge of oxygen radical and lipid peroxidation chemistry at the cell–molecular level becomes better understood, perhaps the significance of certain membrane interactions will be established. Certainly continued investigation of the agents above and the potential utility of other strategies including NMDA antagonists (Benveniste, 1991; Buchan, 1990), angiotensin-converting enzyme inhibitors (Mak et al., 1990), xanthine oxidase inhibitors (Halliwell and Gutteridge, 1986; Roy and McCord, 1983), polyethylene glycol–conjugated superoxide dismutase (Haun et al., 1991), and demethylthiourea (Parker et al., 1985) in oxidant-induced injuries will contribute further to our understanding of appropriate needs in neuroprotective drug development.

Acknowledgments. This work was supported by the Upjohn Company, Kalamazoo, Michigan, and The Center for Biomedical Research, Higuchi Biosciences Center, The University of Kansas.

References

Alexander JS, Hechtman HB, Shepro D (1988): Phalloidin enhances endothelial barrier function and reduces inflammatory permeability *in vitro. Microvasc Res* 35:308–315.

Audus KL, Guillot FL, Braughler JM (1991): Evidence for 21-aminosteroid association with the hydrophobic domains of brain microvessel endothelial cells. *Free Rad Biol Med* 11:361–371.

Aust SD, Morehouse LA, Thomas CE (1985): Role of metals in oxygen radical reactions. *J Free Rad Biol Med* 1:3–25.

Bangham AD, Standish MM, Weissman G (1965): The action of steroids and streptolysin S on the permeability of phospholipid structures to cations. *J Mol Biol* 13:253–259.

Battaini F, Govoni S, Trabucchi M, Paoletti R (1988): Calcium antagonists in tissue protection. *Pharmacol Ther* 39:385–388.

Benveniste H (1991): The excitotoxin hypothesis in relation to cerebral ischemia. *Cerebr Blood Flow Metab* 3:213–245.

Bisby RH, Birch JS (1989): A time-resolved fluorescence anisotropy study of bilayer membranes containing α-tocopherol. *Biochem Biophys Res Commun* 158:386–391.

Braughler JM (1985): Lipid peroxidation-induced inhibition of gamma-aminobutyric acid uptake in rat brain synaptosomes: Protection by glucocorticoids. *J Neurochem* 44:1282–1288.

Braughler JM, Hall ED (1989): Central nervous system trauma and stroke. I. Biochemical considerations for oxygen radical formation and lipid peroxidation. *Free Rad Biol Med* 6:289–301.

Braughler JM, Pregenzer JF, Chase RL, Duncan LA, Jacobsen EJ, McCall JM (1987): Novel 21-amino steroids as potent inhibitors or iron-dependent lipid peroxidation. *J Biol Chem* 262:10438–10440.

Braughler JM, Chase RL, Neff GL, Day JS, Yonkers PA, Hall ED, Sethy VH, Lahti RA (1988): A new 21-aminosteroid antioxidant lacking glucocorticoid activity stimulates ACTH secretion and blocks arachidonic acid release from mouse pituitary tumor (AtT-10) cells. *J Pharmacol Exp Ther* 244:423–427.

Braughler JM, Hall ED, Jacobsen EJ, McCall JM, Means ED (1989): The 21-aminosteroids: potent inhibitors of lipid peroxidation for the treatment of central nervous system trauma and ischemia. *Drugs of the Future* 14:143–152.

Buchan AM (1990): Do NMDA antagonists protect against cerebral ischemia: Are clinical trials warranted? *Cerebr Blood Flow Metab* 2:1–26.

Buege JA, Aust SD (1958): Microsomal lipid peroxidation. *Methods Enzymol* 52:302–310.

Cranney M, Cundall RB, Jones GR, Richards JT, Thomas EW (1983): Fluorescence lifetime quenching studies on some interesting diphenylhexatriene membrane probes. *Biochim Biophys Acta* 735:418–425.

Demopoulos HB, Flamm ES, Pietronigro DD, Seligman MI (1972): Molecular aspects of membrane structure in cerebral edema. In: *Steroids in Brain Edema*, Reulen HJ, Schurman K, eds. New York/Vienna: Springer-Verlag.

Demopoulos HB, Flamm ES, Pietronigro DD, Seligman M (1980): The free radical pathology and the microcirculation in the major central nervous system disorders. *Acta Physiol Scand Suppl* 492:91–120.

Demopoulos HB, Flamm ES, Seligman ML, Pietronigro DD, Tomasula J, DeCrescito V (1982): Further studies on free-radical pathology in the major central nervous system disorders: Effect of very high doses of methylprednisolone on the functional outcome, morphology, and chemistry of experimental spinal cord impact injury. *Can J Physiol Pharmacol* 60:1415–1424.

Diplock AT (1982): The modulating influence of vitamin E in biological membrane unsaturated phospholipid metabolism. *Acta Vitaminol Enzymol* 4:303–309.

Divakaran P, Wiggins RC (1987): Tocopherol in brain metabolism and disease: A review. *Metab Brain Dis* 2:1–12.

Dobretsov GE, Borschevskaya TA, Petrov VA, Vladimirov YA (1977): The increase of phospholipid bilayer rigidity after lipid peroxidation. *FEBS Lett* 84:125–128.

Engel LW, Prendergast FG (1981): Values for and significance of order parameters and "cone angles" of fluorophore rotation in lipid bilayers. *Biochemistry* 20:7338–7345.

Ernster L, Nordenbrand K, Orrenius S (1982): Microsomal lipid peroxidation: Mechanisms and some biomedical implications. In: *Lipid Peroxidation in Biology and Medecine*, Yagi, K, ed. New York: Academic Press.

Goncalves T, Carvalho AP, Oliveira CR (1991): Antioxidant effect of calcium antagonists on microsomal membranes isolated from different brain areas. *Eur J Pharmacol* 204:315–322.

Gutteridge JMC (1987): The role of oxygen radicals in tissue damage and ageing. *Pharm J* 239:401–406.

Hall ED, Braughler JM (1989): Central nervous system trauma and stroke. II. Physiological and pharmacological evidence for involvement of oxygen radicals and lipid peroxidation. *Free Rad Biol Med* 6:303–313.

Hall ED, McCall JM, Chase RL, Yonkers PA, Braughler JM (1987): A nonglucocorticoid steroid analog of methylprednisolone duplicates its high-dose pharmacology in models of central nervous system trauma and neuronal membrane damage. *J Pharmacol Exp Ther* 242:137–142.

Hall ED, Pazara KE, Braughler JM (1990): Nonsteroidal lazaroid U78517F in models of focal and global ischemia. *Stroke* 21(supp III):83–87.

Halliwell B, Gutteridge JMC (1986): Oxygen free radicals and iron in relation to biology and medecine: some problems and concepts. *Arch Biochem Biophys* 246:501–514.

Halliwell B, Gutteridge JMC (1989): Free radicals in biology and medecine, 2nd ed. Oxford: Clarendon Press.

Harris RA, Bruno P (1985): Membrane disordering by anesthetic drugs: relationship to synaptosomal sodium and calcium fluxes. *J Neurochem* 44:1274–1281.

Haun SE, Kirsch JR, Helfaer MA, Kubos KL, Traystman RJ (1991): Polyethylene glycol-conjugated superoxide dismutase fails to augment brain superoxide dismutase activity in piglets. *Stroke* 22:655–659.

Henning B, Chow CK (1988): Lipid peroxidation and endothelial cell injury: Implications in athersclerosis. *Free Rad Biol Med* 4:99–106.

Henry PD (1991): Antiperoxidative actions of calcium antagonists and atherogenesis. *J Cardiovasc Pharmacol* 18(suppl I):S6–S10.

Inauen W, Payne DK, Kvietys PR, Granger DN (1990): Hypoxia/Reoxygenation increases the permeability of endothelial cell monolayers: Role of oxygen radicals. *Free Rad Biol Med* 9:219–223.

Ivanov II (1985): A relay model of lipid peroxidation in biological membranes. *J Free Rad Biol Med* 1:247–253.

Janero DR, Burghardt B (1989): Antiperoxidant effect of dihydropyridine calcium antagonists. *Biochem Pharmacol* 38:4344–4348.

Janero DR, Burghardt B, Lopez R (1988): Protection of cardiac membrane phospholipid against oxidative injury by calcium antagonists. *Biochem Pharmacol* 37:4197–4203.

Kaneko T, Nakano S, Matsuo M (1991): Protective effect of vitamin E on linoleic acid hydroperoxide-induced injury to human endothelial cells. *Lipids* 26:345–348.

Kiss C, Balazs M, Keri-Fulop I (1990): Dexamethasone decreases membrane fluidity of leukemia cells. *Leukemia Res* 14:221–225.

Konat G, Wiggins RC (1985): Effects of reactive oxygen species on myelin membrane proteins. *J Neurochem* 45:1113–1118.

Konat G, Gantt G, Gorman A, Wiggins RC (1986): Peroxidative aggregation of myelin membrane proteins. *Metab Brain Dis* 1:177–186.

Koster JF, Slee RG (1980): Lipid peroxidation of rat liver microsomes. *Biochim Biophys Acta* 620:489–499.

Leibovitz ME, Johnson MC (1971): Relation of lipid peroxidation to loss of cations trapped in liposomes. *J Lipid Res* 12:662–670.

Mak IT, Freedman AM, Dickens BF, Weglicki WB (1990): Protective effects of sulfhydryl-containing angiotensin converting enzyme inhibitors against free radical injury in endothelial cells. *Biochem Pharmacol* 40:2169–1275.

Massey JB, She HS, Pownall HJ (1982): Interaction of vitamin E with saturated phospholipid bilayers. *Biochem Biophys Res Commun* 106: 842–847.

Mead JF (1976): Free radical mechanisms of lipid damage and consequences for cellular membranes. In: *Free radicals in biology*, vol. 1, Pryor WA, ed. New York: Academic Press.

Mentz P, Giebler C, Forster W (1980): Evidence for a direct inhibitory effect of glucocorticoids on the activity of phospholipase A_2 as a further possible mechanism of some actions of steroidal anti-inflammatory drugs. *Pharmacol Res Commun* 12:817–827.

Naylor WG (1983): The role of calcium in myocardial ischemia and cell death. In: *Calcium channel blocking agents in the treatment of cardiovascular disorders*, Stone PH, Antman EM, eds. Mount Kisco, NY: Futura Publishing.

Ohyashiki T, Ushiro H, Mohri T (1986): Effects of α-tocopherol on the lipid peroxidation and fluidity of porcine intestinal brush-border membranes. *Biochim Biophys Acta* 858:294–300.

Palozza P, Krinsky NI (1991): The inhibition of radical-initiated peroxidation of microsomal lipids by both α-tocopherol and β-carotene. *Free Rad Biol Med* 11:407–414.

Parker NB, Berger EM, Curtis WE, Linas SL, Repine JE (1985): Hydrogen peroxide causes dimethylthiourea consumption while hydroxyl radical causes dimethylsulfoxide consumption *in vitro*. *J Free Rad Biol Med* 1:415–419.

Patel JM, Sekharam M, Block ER (1991): Vitamin E distribution and modulation of the physical state and function of pulmonary endothelial cell membranes. *Exp Lung Res* 17:707–723.

Phillips PG, Lum H, Malik AB, Tsan M-F (1989): Phallacidin prevents thrombin-induced increases in endothelial permeability to albumin. *Am J Physiol* 257:C562–C567.

Pottel H, van der Meer W, Herreman W (1983): Correlation between the order parameter and the steady-state fluorescence anisotropy of 1,6-diphenyl-1,3,5-hexatriene and an evaluation of membrane fluidity. *Biochim Biophys Acta* 730:181–186.

Prendergast FG, Haugland RP, Callahan PJ (1981): 1-[4-(Trimethylamino)phenyl]-6-phenyl hexa-1,3,5-triene: Synthesis, fluorescence properties and use as a fluorescence probe of lipid bilayers. *Biochemistry* 20:7333-7338.

Raub TJ, Ho NFH, Barsuhn CL (1991): Transcellular permeability of a highly lipophilic, protein bound antioxidant via lipid bilayer diffusion in a cell culture model. *Pharm Res* 8:S-128.

Robak J, Duniec Z (1986): Membrane activity, antioxidant, antiaggregatory, and antihemolytic properties of four calcium channel blockers. *Pharmacol Res Commun* 18:1107–1117.

Roy RS, McCord JM (1983): Superoxide and ischemia: Conversion of xanthine dehydrogenase to xanthine oxidase. In: *Oxy radicals and their scavenger systems*, Greenwald R, Cohen G, eds. New York: Elsevier Science.

Shapiro HM (1985): Barbiturates in brain ischaemia. *Br J Anaesth* 57:82–95.

Sheridan NP, Block ER (1988): Plasma membrane fluidity measurements in intact endothelial cells: Effect of hyperoxia on fluorescence anisotropies of 1-[4-(trimethylamino)phenyl]-6-phenyl hexatriene. *J Cell Physiol* 134:117–134.

Shertzer HG, Bannenberg GL, Rundgren M, Moldeus P (1991): Relationship of membrane fluidity, chemoprotection, and the intrinsic toxicity of butylated hydroxytoluene. *Biochem Pharmacol* 42:1587–1593.

Siesjo BK, Agardh C-D, Bengtsson F (1989): Free radicals and brain damage. *Cerebrovasc Brain Metab Rev* 1:165–211.

Sklar LA (1984): Fluorescence polarization studies of membrane fluidity: Where do we go from here? In: *Biomembranes*, Morris K, Manson LA, eds. New York: Plenum Press.

Smolen JE, Shohet SB (1974): Permeability changes induced by peroxidation in liposomes prepared from human erythrocyte lipids. *J Lipid Res* 15:273–280.

Spetzler RT, Hadley MN (1989): Protection against cerebral ischemia: The role of barbiturates. *Cerebrovasc Brain Metab Rev* 1:212–229.

Steiner M (1981): Vitamin E changes the membrane fluidity of human platelets. *Biochim Biophys Acta* 640:100–105.

Tappel AL (1975): Lipid peroxidation and fluorescent molecular damage to membranes. In: *Pathobiology of cell membranes*, Vol 1, Trump BF, Arstila AU, eds. New York: Academic Press.

Tappel AL (1978): Protection against free radical lipid peroxidation reactions. *Adv Exp Med Biol* 97:111-131.

van Blitterswijk WJ, van Hooven RP, van der Meer BW (1981): Lipid structural order parameters (reciprocal of fluidity) in biomembranes derived from steady-state fluorescence polarization measurements. *Biochim Biophys Acta* 644:323–332.

van Ginkel G, van Langen H, Levine YK (1989): The membrane fluidity concept revisited by polarized fluorescence spectroscopy on different model membranes containing unsaturated lipids and sterols. *Biochimie* 71:23–32.

Young W, Flamm ES (1982): Effects of high-dose corticosteroid therapy on blood flow, evoked potentials, and extracellular calcium in experimental spinal injury. *J Neurosurg* 57:667–673.

Chapter 12

Oxidative Stress in the Pathogenesis of Postischemic Ventricular Dysfunction (Myocardial "Stunning")

Marcel Zughaib, Xiao Ying Li, Mohamed O. Jeroudi, Craig J. Hartley and Roberto Bolli

Since its first description in 1982 by Braunwald and Kloner (1982), extensive research has focused on the phenomenon of myocardial "stunning" or postischemic dysfunction, and on the elucidation of its mechanisms. Only a handful of other areas of cardiology in the 1980s have generated so much interest and experimental work in such a short time span. As is often the case with newly described pathophysiological entities, several pathogenetic hypotheses were initially advanced (for review, see Bolli, 1990). At the time of this writing, the "oxyradical" and the "calcium" hypotheses appear to be the most plausible explanations for stunning (for review, see Bolli, 1990).

The evidence for a pathogenetic role of cytotoxic oxygen-derived free radicals in myocardial stunning has been accumulated through extensive work in several laboratories. Since its first formulation in the mid-1980s (Gross et al., 1986; Myers et al., 1985; Przyklenk and Kloner, 1986), more than 50 full-length papers have been written on the oxyradical hypothesis of stunning. Unlike many other areas, the results of these studies have been consistent in supporting a pathogenetic role of oxygen metabolites in postischemic dysfunction—indeed, a rare example of concordance among different laboratories (Bolli, 1990). This is in striking contrast to the controversy that surrounds the role of oxygen radicals in myocardial infarction (Bolli, 1991) and emphasizes the concept that results obtained

Oxygen Free Radicals in Tissue Damage
Merrill Tarr and Fred Samson, Editors

in models of *irreversible* myocardial injury (i.e., myocardial infarction) should not be extrapolated to models of *reversible* myocardial injury (i.e., myocardial stunning). Myocardial infarction and myocardial stunning are two completely different pathophysiological entities, and there is no reason to suspect that they share a common pathogenesis. Although there is strong evidence for an involvement of oxygen radicals in stunning, there is little evidence for an involvement of these species in infarction (Bolli, 1991).

This chapter examines the foundations and the validity of the evidence supporting the oxyradical hypothesis of stunning, and hence sheds some light on its underlying pathophysiology, the associated controversies, and possible clinical implications. In doing so, an attempt will be made to reconcile the oxyradical hypothesis with other leading hypotheses in the framework of a unifying pathogenetic scheme. Finally, future research directions will be highlighted.

Historical Background

As mentioned above, the oxyradical hypothesis of myocardial stunning is less than a decade old. However, the evolution of our understanding of its mechanism and inherent properties has been extremely rapid, due to the extensive investigational interest it has generated. Generally speaking, this process has occurred through three phases that have followed a logical sequence.

The first phase was the one that begot the hypothesis that oxygen-derived free radicals might play a role in postischemic dysfunction. To test this hypothesis, indirect evidence was sought by investigating whether antioxidants enhance recovery of function of the stunned myocardium. Several studies demonstrated that this was the case, and thus provided the foundation for the oxyradical hypothesis (reviewed in Bolli, 1990). This approach was nevertheless limited by the unproven assumptions that oxygen-derived free radicals were produced, and that antioxidants improved contractile recovery specifically through their inhibition and/or scavenging of free radicals.

Further insights into the question were provided by the work done in the second phase. Here, the role of free radicals was substantiated by directly measuring their production in the stunned myocardium (Bolli et al., 1988b; Bolli et al., 1989b, c). These studies demonstrated that free radicals are *indeed* produced in stunned myocardium and that antioxidant therapy does *indeed* inhibit their production, thus providing the missing link for the results of the first phase and removing concerns about a possible nonspecific effect of the antioxidant agents.

The third phase, which is just under way, proposes to tackle the yet to be answered fundamental question: how does the reperfusion-related brief oxidative stress cause a prolonged depression of contractility in the absence of myocyte death? The answer evidently lies at the molecular level. Elucidation of this problem will be one of the major challenges in the 1990s.

Experimental Models of Stunning

There currently exists a wide array of experimental settings of myocardial stunning. However, they may not share the same pathophysiology and pathogenesis and therefore one has to remember that observations made in a particular model may not be generalized to other models. This concept was detailed in a recent publication (Bolli, 1990).

The various models of stunning are all based on an oxygen supply/demand mismatch and can be categorized as follows:

A. Stunning secondary to decreased O_2 supply: the ischemia/reperfusion models.

1. *Regional ischemia*

   a. Single, completely reversible ischemic episode (e.g., occlusion of $<$ 20 min in the canine model)

   b. Multiple completely reversible ischemic episodes (e.g, successive 5–10 min occlusions/reperfusion)

   c. Single, partly irreversible episode resulting in subendocardial infarction (e.g., occlusion of $>$ 20 min but $<$ 3 hrs in the canine model)

2. *Global ischemia*

   a. Isolated heart preparations

   b. *In vivo* cardioplegic arrest (e.g., cardiac arrest with bypass circulation)

B. Stunning secondary to increased O_2 demand: with normal or increased coronary flow that nevertheless cannot match demands (e.g., exercise-induced ischemia).

Table 1 details the possible pathogenetic mechanisms associated with each of the above-described models. There is strong evidence for a pathogenetic role of oxygen-derived free radicals in models of regional and global ischemia that do not result in subendocardial infarction, but not in models of ischemia resulting in subendocardial infarction or in models of exercise-induced ischemia (Table 1). In view of this heterogeneity of models of myocardial stunning, this chapter specifies the exact model alluded to in each of the discussions to follow.

Table 1. Classification of myocardial stunning and evidence for the various mechanisms proposed in experimental animals

	Evidence for a pathogenic role of:			
Experimental setting	Oxygen radicals	Sarcoplasmic reticulum dysfunction	Calcium overload	Reduced calcium sensitivity
Stunning due to decreased blood flow				
Regional ischemia				
1. Single, completely reversible ischemic episode	++	?	?	–
2. Multiple, completely reversible ischemic episodes	+	+	?	–
3. Single, partly irreversible ischemic episode (subendocardial infarction)	±	?	?	?
Global ischemia				
4. Isolated heart in vitro	+	?	+	+
5. Cardioplegic arrest in vivo	+	?	?	?
Stunning due to increased O_2 demands				
6. Exercise-induced ischemia	–	?	?	?

+ published studies support this mechanism; ++ published studies from multiple laboratories consistently support this mechanism; evidence is also available in conscious animal preparations; – published studies do not support this mechanism; ± published studies are conflicting; ? No data are available.

Reproduced (with minor changes) with permission of the *American Heart Association* from *Circulation* (1990): 82:723–738.

Oxyradicals in Stunning Secondary to One Reversible Episode of Regional Ischemia

A large section of this chapter is devoted to the discussion of this model, in view of its relatively large contribution to the body of evidence supporting the oxyradical hypothesis. In the dog, this model entails a 15-min coronary occlusion, which is well established to produce no irreversible myocardial injury.

Effect of antioxidants in open-chest animals. In 1984, we and others postulated that hydrogen peroxide (H_2O_2), superoxide ($\cdot O_2^-$), and the hydroxyl ($OH\cdot$) radicals contribute to the pathogenesis of myocardial stunning. Accordingly, a number of laboratories, including ours, designed experimental protocols to test this hypothesis. We used a dog model in which the left anterior descending artery is occluded for 15 min and then reperfused; this results in a reversible myocardial dysfunction that lasts hours to days (Bolli et al., 1988a). However, when superoxide dismutase (SOD) and catalase (Fig. 1) were administered, a significant enhancement of the recovery of contractile function was observed (Myers et al., 1985). Similar results were obtained by other investigators in the same (dog) model (Gross et al., 1986; Murry et al., 1989; Przyklenk and Kloner, 1986), as well as in rabbits (Koerner et al., 1991). Both $\cdot O_2^-$ and H_2O_2 contribute to the free radical–induced myocardial injury, as shown by the finding (Jeroudi et al., 1990) that neither SOD nor catalase exerted a protective effect when they were administered separately. Again, similar results were obtained by Koerner et al. (1991) in rabbits and by Buchwald et al. (1989a) in pigs.

The question then was whether $\cdot O_2^-$ and H_2O_2 exerted a direct cytotoxic effect on cardiac myocytes, or rather required the formation of the more reactive hydroxyl radical ($\cdot OH$) (Halliwell and Gutteridge, 1984) through the iron-catalyzed Haber-Weiss reaction. To address this problem, one would have to dissociate the effects of $\cdot OH$ from those of $\cdot O_2^-$ or H_2O_2 through the use of a potent $\cdot OH$ scavenger that ideally would not react with the other species. Accordingly, we investigated the agents dimethylthiourea (DMTU) (Bolli et al., 1987a) and *N*-2-mercaptopropionyl glycine (MPG) (Bolli et al., 1989c; Myers et al., 1986a), both of which are potent $\cdot OH$ radical scavengers, with no *in vitro* activity on either $\cdot O_2^-$ or H_2O_2 (Bolli et al., 1987a, 1989c). Both agents produced attenuation of postischemic dysfunction, suggesting that $\cdot OH$ plays an important pathophysiologic role, and that the salutary effects of SOD and catalase (Gross et al., 1986; Jeroudi et al., 1990; Koerner et al., 1991; Murry et al., 1989; Myers et al., 1985; Przyklenk and Kloner, 1986) may be related, at least in part, to the prevention of $\cdot OH$ generation. It was therefore logical to investigate whether the iron chelator desferrioxamine, through its inhibition of $\cdot OH$ formation (Halliwell and Gutteridge, 1984), also reduced the severity of myocardial stunning. This was indeed found to be the case (Bolli et al., 1987b; Farber et al., 1988), thus further corroborating the role of $\cdot OH$ in stunning.

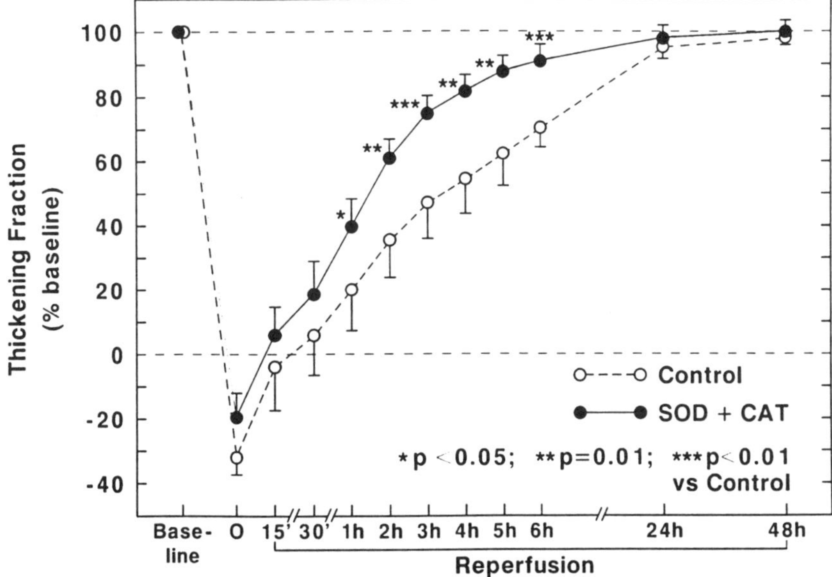

Figure 1 Systolic thickening fraction in the ischemic/reperfused region 5 min after coronary occlusion (O) and at selected times after reperfusion in control (*dashed line with open circles*) and SOD and catalase-treated (*continuous line with solid circles*) conscious dogs. Thickening fraction is expressed as percent of baseline values. The data pertain to all of the conscious dogs studied (controls, n =19; treated, n = 21). Data are mean ± SEM. Reproduced with permission of *The American Heart Association* from *Circulation Research* (1991): 69:731–747.

A word of caution is warranted at this point. The above discussion should not lead one to believe that ·OH is the only mediator of oxyradical injury. H_2O_2, although relatively unreactive in itself, is important in this context as a precursor of ·OH (Halliwell and Gutteridge, 1984). Furthermore, there is evidence that $\cdot O_2^-$ can exert direct toxicity *in vitro* (Schrier and Hess, 1988, reviewed in DiGuiseppi and Fridovich, 1982); the fact that SOD has to be added to catalase to attenuate stunning *in vivo* (Jeroudi et al., 1990; Koerner et al., 1991) suggest that $\cdot O_2^-$ may cause toxic effects that are not mediated by ·OH. Hence, each of the three oxygen metabolites appears to contribute to the pathogenesis of postischemic dysfunction.

To summarize, the evidence that oxygen-derived free radicals play a crucial role in the development of myocardial stunning is overwhelming and has been uniformly reported by several independent investigators (Bolli et al., 1987a,b; Bolli et al., 1989c; Dage et al., 1991; Farber et al., 1988; Gross et al., 1986; Jeroudi et al., 1990; Koerner et al., 1991; Myers et al., 1985; Myers et al., 1986a; Murry et al., 1989; Przyklenk and Kloner, 1986). Nevertheless, the relative individual importance of H_2O_2, $\cdot O_2^-$, and $\cdot OH$ remains to be determined.

Effect of antioxidants in conscious animals. The above-mentioned animal experiments (Bolli et al., 1987a,b; Bolli et al., 1989c; Dage et al., 1991; Farber et al., 1988; Gross et al., 1986; Jeroudi et al., 1990; Koerner et al., 1991; Myers et al., 1985; Myers et al., 1986a; Murry et al., 1989; Przyklenk and Kloner, 1986) were performed in open-chest models and under general anesthesia. These models suffer from major limitations due to their unphysiologic nature; therefore, one should be careful in extrapolating those results to the clinical setting. The sources of artifacts believed to plague the open-chest, anesthetized preparations are the trauma of surgery, anesthesia itself, the lack of integrity of the chest wall, the unstable (and usually low) body temperature, and various neurohumoral and metabolic perturbations. This concept was corroborated by a recently published study (Triana et al., 1991) where striking differences between the open-chest and conscious dog models were observed. In that study (Triana et al., 1991), both groups of dogs were subjected to a 15-min coronary occlusion. Even after controlling for hemodynamic variables (preload, afterload, double product, etc.), temperature, acid–base status, and collateral flow and size of the ischemic areas at risk, myocardial stunning was found to be greatly exaggerated in the open-chest model and the depression of contractility was about double that seen in the conscious model. Even more importantly, in recent unpublished experiments we have found that the magnitude of cumulative radical production is much greater in the open-chest compared with the conscious dog: 1245 $\pm$ 267 vs. 345 $\pm$ 154 arbitrary units per gram of tissue (U/g), 2359 $\pm$ 463 vs. 715 $\pm$ 254 U/g, and 16007 $\pm$ 3224 vs. 3109 $\pm$ 702 U/g over the first 5, 10, and 180 min of reperfusion, respectively.

What are the implications of these findings? In our opinion, open-chest animal experiments should not be extrapolated to the human clinical setting without adequate investigation and testing in conscious animal models, which are free from the problems alluded to above. Two recent studies published from our laboratory were specifically aimed at addressing these concerns. In the first (Triana et al., 1991), SOD and catalase were found to attenuate myocardial stunning in conscious unsedated dogs. The degree of protection was directly proportional to the severity of ischemia (and thus inversely proportional to collateral flow; Fig. 2). The second study (Sekili et al., 1991) demonstrated a beneficial effect of a ·OH scavenger (MPG), and an inhibitor of ·OH generation (desferrioxamine), administered separately on myocardial stunning; this study demonstrated for the first time the importance of the hydroxyl radical in the pathogenesis of postischemic dysfunction independently from the possible artifacts generated by the open-chest preparations.

In summary, the evidence for a pathogenetic role for oxygen-derived free radicals in myocardial stunning, initially limited to open-chest animal models (Bolli et al., 1987a, 1989c; Dage et al., 1991; Gross et al., 1986; Jeroudi et al., 1990; Koerner et al., 1991; Myers et al., 1985; Myers et al., 1986a; Przyklenk and Kloner, 1986), has been recently extended to the more physiologic and clinically relevant conscious animal model (Bolli et al., 1991a; Sekili et al., 1991; Triana et al., 1991; Zughaib et al., 1991). Thus, the oxyradical hypothesis appears to be valid in the most physiologic animal model available—the conscious animal.

Systolic thickening fraction in the ischemic/reperfused region 5 min after coronary occlusion (O) and at selected times after reperfusion in subset I (*upper panel*), subset II (*middle panel*), and subset III (*lower panel*) of the conscious dogs. Control dogs are represented by the dashed line with open circles; SOD and catalase-treated dogs are represented by the continuous line with solid circles. Thickening fraction is expressed as percent of baseline values. Data are mean $\pm$ SEM. Subset I was formed by dogs with collateral flow $< 10\%$ of nonischemic zone flow (NZF); subset II included animals with collateral flow between 10% and 30% of NZF; subset III consisted of dogs with collateral flow $> 30\%$ of NZF. In subset I, transmural ischemic zone flow averaged 0.07 ± 0.02 ml/min/g in controls (n = 7) and 0.09 ± 0.01 ml/min/g in treated animals (n = 11) ($4.1 \pm 1.1\%$ and $5.9 \pm 0.9\%$ of NZF, respectively). In subset II, ischemic zone flow was 0.25 ± 0.04 ml/min/g in controls (n = 7) and 0.31 ± 0.03 ml/min/g in treated dogs (n = 5) ($15.7 \pm 2.0\%$ and $23.0 \pm 2.6\%$ of NZF, respectively). In subset III, ischemic zone flow averaged 0.52 ± 0.06 ml/min/g in controls (n = 5) and 0.63 ± 0.04 ml/min/g in treated dogs (n = 5) ($43.2 \pm 4.2\%$ and $41.6 \pm 4.9\%$ of NZF, respectively). Reproduced with permission of *The American Heart Association* from *Circulation Research* (1991): 69:731–747.

See Figure 2 next page

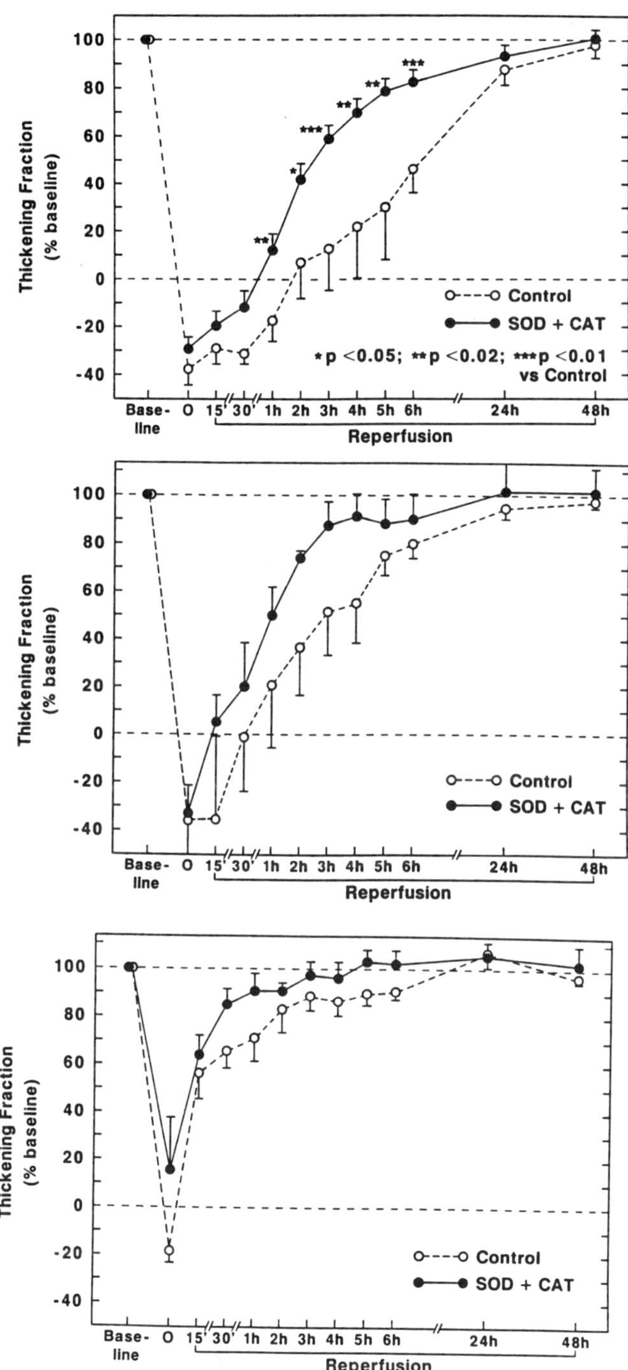
Thickening Fraction
(% baseline)
100
80
60
40
20
0
-20
-40
Control
SOD + CAT
*p <0.05; **p <0.02; ***p <0.01
vs Control
Base-
line
O
15'
30'
1h
2h
3h
4h
5h
6h
24h
48h
Reperfusion

*Where are the free radicals generated in stunned myocardium?*The potential sources of free radicals in the setting of postischemic dysfunction are multiple, and several of them remain controversial. The most important are (a) the enzyme xanthine oxidase, (b) the arachidonate cascade, and (c) the activated neutrophil. Other possibilities include autoxidation of catecholamines and other compounds, damage to the mitochondrial electron transport chain, and accumulation of reducing equivalents.

Xanthine oxidase is the enzyme that catalyzes the oxidation of hypoxanthine to xanthine and to uric acid. In the process, superoxide $\cdot O_2^-$ is generated from O_2. We (Charlat et al., 1987) have demonstrated attenuation of stunning in the dog with the use of the xanthine oxidase inhibitor allopurinol; at the same time, we demonstrated that allopurinol inhibited the increased cardiac production of urate that was observed in control dogs during ischemia and early reperfusion. Others (Holzgrefe and Gibson, 1988; Puett et al., 1987) obtained similar results with oxypurinol. Although these results suggest an important role for xanthine oxidase as a contributor to myocardial stunning in the dog, their applicability to humans remains in doubt because the presence of the enzyme in human hearts is still being debated (Eddy et al., 1987; Grum et al., 1989; Huizer et al., 1987; Jarasch et al., 1986; Muxfeldt and Schaper, 1987).

In contrast to these studies, Werns et al. (1989) have recently reported that two xanthine oxidase inhibitors (oxypurinol and amflutizole) failed to attenuate stunning in a 15-min coronary occlusion model in dogs. The reason(s) for this discrepancy is(are) unclear, although one cannot rule out the possibility that different degrees of inhibition of xanthine oxidase led to different results. In any case, the finding that antioxidants attenuate the severity of myocardial stunning in the rabbit (Dage et al., 1991; Koerner et al., 1991), a species believed to lack myocardial xanthine oxidase, suggests that xanthine oxidase is not the sole or the major source of the oxygen radicals responsible for postischemic dysfunction.

Leukocytes are known to produce free radicals and have been thought to be major mediators of irreversible injury and vascular plugging (Engler et al., 1986; Engler, 1987; Lucchesi and Mullane, 1986). However, they do not appear to contribute to myocardial stunning which, by definition, is a reversible injury (see above). Although postischemic dysfunction after a 15-min coronary occlusion has been reported to be prevented (Engler and Covell, 1987) or alleviated (Westlin and Mullane, 1989) through granulocyte depletion by leukopak filtration, other studies have arrived at opposite results. No beneficial results were observed by depleting neutrophils through antiserum (O'Neill et al., 1989; Shea et al., 1987) or with filtration (Jeremy and Becker, 1989). Similarly, dextran (which inhibits leukocyte adherence to endothelium (Kerber et al., 1989), nafazatrom (which inhibits the production of leukotrienes (O'Neill et al., 1987), and antibodies to the adhesion-promoting Mo1 glycoprotein (Schott et al., 1989) failed to attenuate stunning. Furthermore, recent studies (Go et al., 1988) have shown that the stunned myocardium has a decreased neutrophil content and that myeloperoxidase activity (a marker of neutrophil content) was not different from nonischemic myocardium. Moreover, myocardial stunning can be produced in models that are deprived of circulating neutrophils, such as the isolated heart preparation. Taken together, these studies do not support a role for leukocytes in the pathogenesis of myocardial stunning (reversible injury). After the recent report (Juneau et al., 1991) that virtually complete neutrophil depletion fails to mitigate stunning in the pig, there is now a general consensus that granulocytes do not play a prominent role is this phenomenon.

Another potential source of free radicals may be the cyclooxygenase pathway that metabolizes arachidonic acid. Nevertheless, there exists some indirect evidence to the contrary as indomethacin, used in doses that were sufficient to inhibit the cardiac production of TXB_2 and 6-keto-$PGF_{2\alpha}$, failed to alleviate myocardial stunning (Farber and Gross, 1990).

In summary, the sources of oxygen radicals in the stunned myocardium produced by a single reversible episode of ischemia remain uncertain, mainly because of the complexity of the intact animal models, although xanthine oxidase may play a role in the dog. The weight of the evidence does not support any role for leukocytes or the arachidonate cascade. Further research is warranted to clarify the issue and explore other potential sources such as the autoxidation of catecholamines and other substances, the accumulation of reducing equivalents, and the mitochondrial electron transport chain.

Time-frame of oxyradical-mediated insult. In the complex sequence of events encompassing ischemia and reperfusion, it is important to identify the exact timing of the injury mediated by free radicals. Studies from our laboratory have specifically addressed this problem. Using the antioxidant MPG, we have demonstrated attenuation of postischemic dysfunction when infusion of the drug was initiated before ischemia or 1 min before reflow, but not when it was started 1 min after reperfusion

(Bolli et al., 1989c) (Fig. 3). Comparable results were obtained with the iron chelator desferrioxamine (Bolli et al., 1990b) (Fig. 4), which prevents the formation of hydroxyl radicals, and with the spin trap α-phenyl *N-tert*-butyl nitrone (Bolli et al., 1988b). Other investigators (Westlin and Mullane, 1988) have reported attenuation of stunning with captopril (an angiotensin-converting enzyme inhibitor with free radical scavenging properties) when treatment was started 2 min before reperfusion. In the rabbit model, antioxidant therapy commenced immediately before reperfusion was likewise beneficial (Koerner et al., 1991).

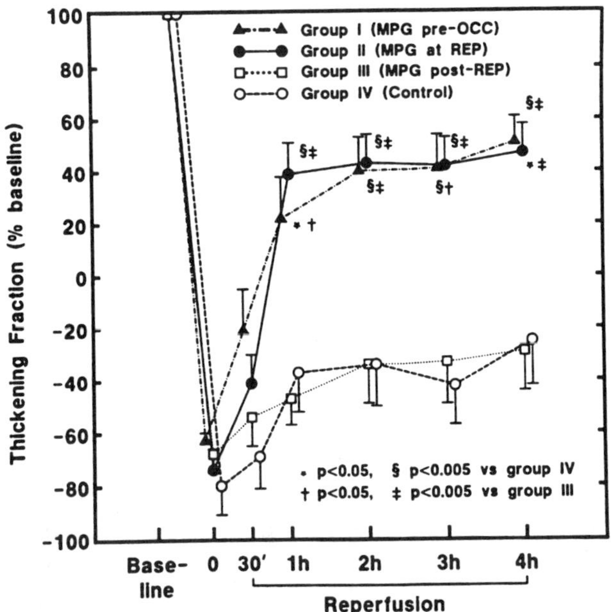

Figure 3 Systolic thickening fraction in the ischemic/reperfused region 5 min after coronary occlusion (O) and at selected times after reperfusion in the following groups: Group I (MPG infusion started 15 min before ischemia, n = 8), group II (MPG started 1 min before reperfusion, n = 9), group III (MPG infusion started 1 min after reperfusion, n = 10), and group IV (controls, n = 10). Thickening fraction is expressed as percent of baseline values. Data are mean ± SEM. MPG attenuated postischemic dysfunction to a similar extent irrespective of whether the infusion was started before ischemia or just before reperfusion; however, infusion started 1 min after reperfusion was ineffective, suggesting that the critical radical-mediated injury occurs in the first few moments after reperfusion. MPG, *N*-2-mercaptopropionyl glycine. Reproduced with permission of *The American Heart Association* from *Circulation Research* (1989): 65:607–622.

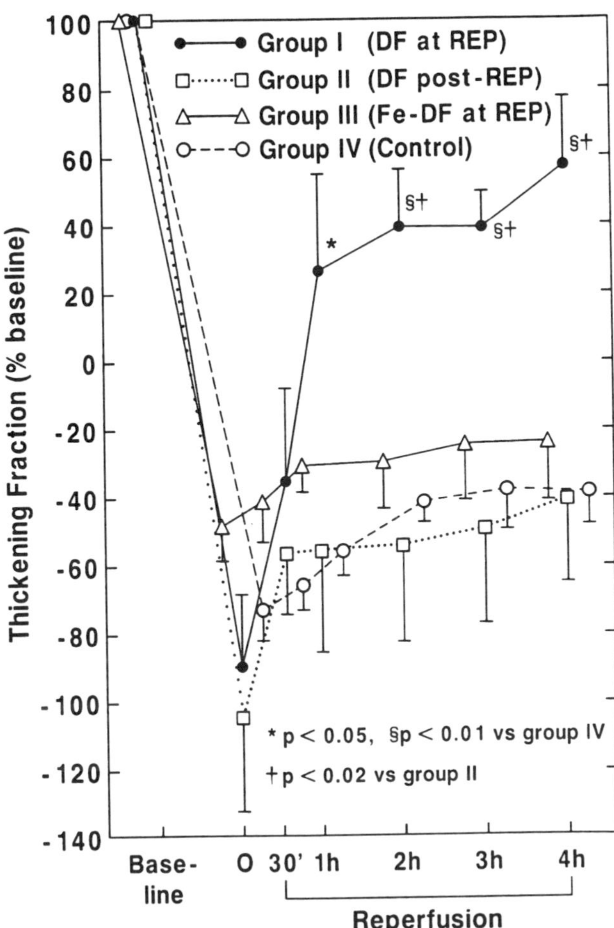

Figure 4 Systolic thickening fraction in the ischemic/reperfused region 5 min after coronary occlusion (O) and at selected times after reperfusion in the following groups: Group I [desferrioxamine (DF) administration started 2 min before reperfusion (REP), n = 8], group II (DF administration started 1 min after reperfusion, n = 6), group III [iron-loaded desferrioxamine (Fe-DF) administration started 2 min before reperfusion, n = 7], and group IV (controls, n = 9). Thickening fractions is expressed as percent of baseline values. Data are mean ± SEM. Desferrioxamine attenuated postischemic dysfunction when started just before reperfusion; however, administration started 1 min after reperfusion was ineffective, suggesting that the critical iron-mediated injury occurs in the first few moments after reperfusion. Reproduced with permission from the *American Journal of Physiology* (1990): 259:H1901–H1911.

In further support of the concept that the early minutes of reflow constitute the time-frame during which oxyradical injury occurs, direct measurements of free radicals (Bolli et al., 1988b, 1989b,c) and of lipid peroxidation products (Romaschin et al., 1987) both demonstrated a burst that coincided with the first few moments after reflow. Furthermore, inhibition of free radical production is only salutary when it takes place during this initial burst, and not when it begins 5 min after reperfusion (Bolli et al., 1989c) (Fig. 5).

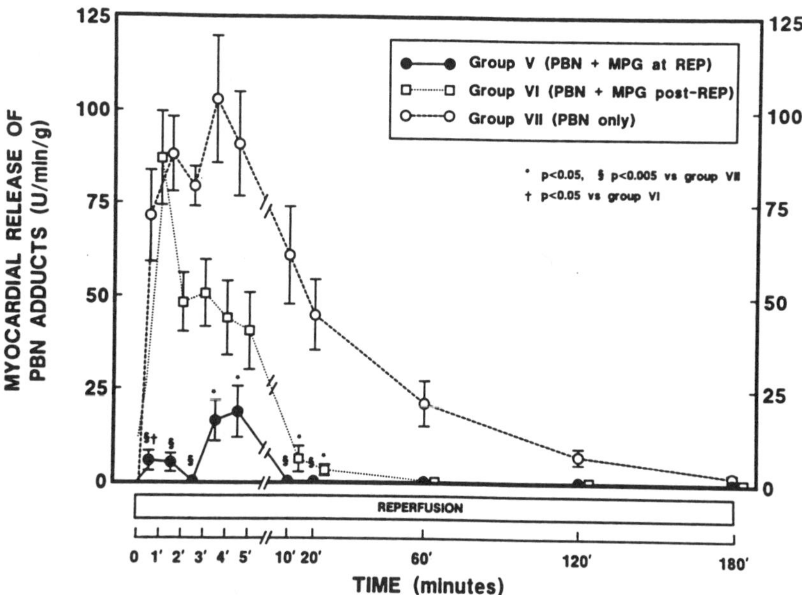

Figure 5 Time course of myocardial release of PBN adducts in group V (MPG infusion started 1 min reperfusion, (n = 5)), group VI (MPG infusion started 1 min after reperfusion, (n = 5)), and group VII (controls, n = 6). Data are mean ± SEM. All groups received PBN by the intracoronary route. Infusion of MPG started 1 min before reperfusion markedly suppressed production of free radicals in the stunned myocardium. Infusion of MPG started 1 min after reperfusion did not affect the initial production of free radicals and produced a delayed suppression that became evident by 10 min of reflow. Since only MPG given as in group V attenuated postischemic dysfunction (whereas MPG given as in group VI did not) (see Fig. 1), these data suggest that the free radicals important in myocardial stunning are those generated immediately after reperfusion. MPG, *N*-2-mercaptopropionyl glycine; PBN, α-phenyl *N-tert*-butyl nitrone. Reproduced with permission of *The American Heart Association* from *Circulation Research* (1989): 65:607–622.

In summary, it appears from the above-presented data that the free radicals that play an important role in the pathogenesis of postischemic dysfunction are those that are produced immediately after reflow. Thus, myocardial stunning appears to be a form of "reperfusion injury."

There are important therapeutic implications that emerge from the above discussion. In the setting of spontaneous or active (therapeutic) reperfusion, antioxidant therapy could be beneficial even if begun after the onset of ischemia, as long as it is started before reperfusion. However, if begun after reperfusion, then its impact may be minimal. It is crucial to remember, however, that postischemic dysfunction also has a component resulting from the injury incurred during ischemia (Hearse, in press), and that antioxidant therapy may or may not modulate that component.

Direct evidence for the oxyradical hypothesis. It may be argued, and rightfully so, that the bulk of the above-discussed evidence supporting the oxyradical hypothesis (Bolli et al., 1987a,b; Bolli et al., 1989c; Dage et al., 1991; Farber et al., 1988; Gross et al., 1986; Jeroudi et al., 1990; Koerner et al., 1991; Myers et al., 1985; Myers et al., 1986a; Murry et al., 1989; Przyklenk and Kloner, 1986; Triana et al., 1991) is indirect and therefore inconclusive, and that objective proof should rest on the direct demonstration and quantitation of free radicals in the setting of myocardial stunning.

Free radical generation has been demonstrated in the *in vitro* isolated rat or rabbit heart models (reviewed in Davies, 1989). However, data obtained *in vitro* may not necessarily be applicable to the *in vivo* situation. Therefore, we have used electron paramagnetic resonance (EPR) spectroscopy and the spin trap α-phenyl *N-tert*-butyl nitrone (PBN) to detect and quantitate the production of free radicals in our intact animal model (15-min coronary occlusion in open-chest dogs) (Bolli and McCay, 1990). The venous effluent blood returning from the area that was rendered ischemic was sampled; analysis of the plasma by EPR spectroscopy revealed EPR spectra characteristic of PBN adducts (Bolli et al., 1988b) (Fig. 6). The myocardial production of radicals during ischemia was barely detectable and was followed by a burst at reperfusion that peaked between 2 and 4 min (Fig. 7). After this initial peak, production decreased but did not stop, persisting at low levels up to 3 hr after reflow (Bolli et al., 1988b) (Fig. 7).

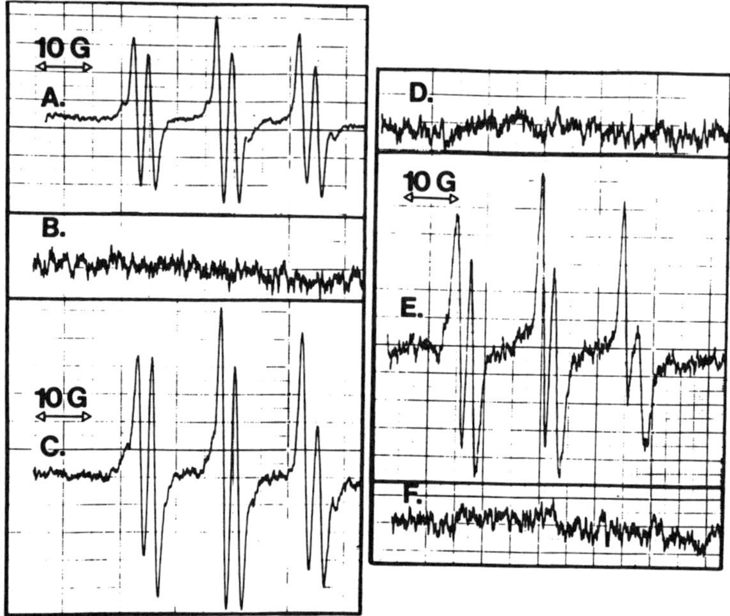

Figure 6 Representative electron paramagnetic resonance (EPR) spectra of PBN radical adducts detected in the coronary venous effluent blood in open-chest dogs. Dogs underwent a 15-min coronary occlusion followed by reperfusion. The spin trap, α-phenyl *N-tert*-butyl nitrone (PBN), was infused intracoronarily beginning 5 min before occlusion and continuing until 10 min after reperfusion (group I), beginning 20 s before reperfusion and continuing until 10 min after reperfusion (group II), or beginning 30 min after reperfusion and continuing until 40 min after reperfusion (group III). The average rate of infusion was 4.4 mg/min. Blood samples were obtained from a vein draining the ischemic/reperfused region and the plasma lipids were extracted by the Folch procedure and analyzed by EPR spectroscopy. Shown in this figure are signals from plasma obtained: **A**: 3 min after reperfusion in a dog from group I (PBN infusion started 5 min before ischemia) (a_N = 14.75, $a^H \beta = 2.69$ G; gain, 5×10^5); **B**: at corresponding time after start of PBN infusion in a control dog (gain, 1×10^6); **C**: 5 min after reperfusion in a dog in group II (PBN infusion started 20 s before reperfusion) (a_N = 14.77, a^H $\beta = 2.69$ G; gain, 5×10^5); **D**: at corresponding time after start of PBN in a second control dog (gain, 1×10^6); **E**: 35 min after reperfusion in a dog in group III (PBN infusion started 30 min after reperfusion) (a_N = 15.00, a^H $\beta = 2.78$ G; gain, 2×10^6); **F**: at corresponding time after start of PBN in a third control dog (gain, 1×10^6). The signals observed are characteristic of radical adducts of PBN. These results demonstrate that free radicals are generated in the stunned myocardium in the intact dog. The spectrometer settings were as follows: microwave power, 19.7 mW; modulation amplitude, 1 G; time constant, 1.25 s; scan range, 100 G; and scan time, 8 min. All spectra were recorded at room temperature (26°C). Reproduced with permission from the *Journal of Clinical Investigation* (1988): 82:476–485.

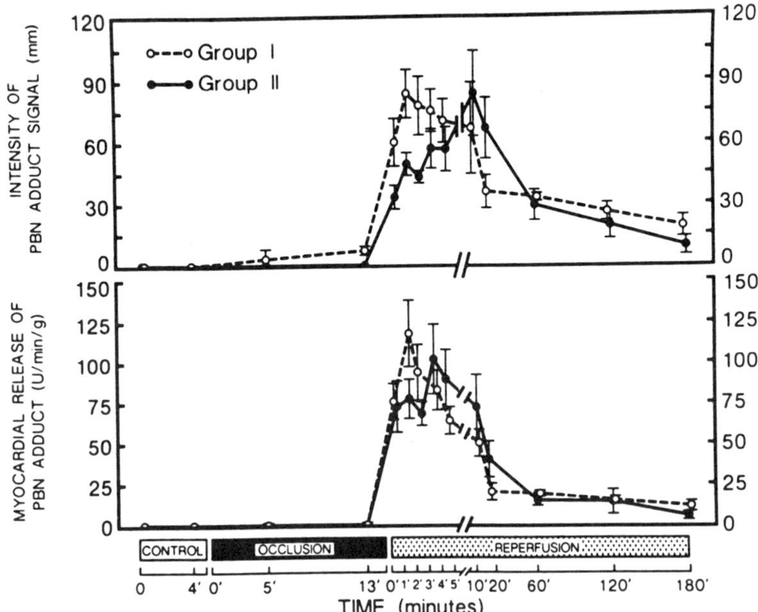

Figure 7 **Upper panel**: Intensity of the electron paramagnetic resonance spectroscopy signals detected in the coronary venous effluent blood in groups I and II. **Lower panel**: Time course of myocardial release of PBN adducts in groups I and II. Data are mean ± SEM. Open-chest dogs underwent a 15-min coronary occlusion followed by reperfusion. Group I received intracoronary PBN starting 5 min before occlusion and continuing until 10 min after reperfusion (n = 5); group II received intracoronary PBN starting at reperfusion and ending 10 min thereafter (n = 5). Reperfusion after 15 min of ischemia was associated with a burst of free radical generation, which abated by 10 min but continued up to 3 hr after reflow. PBN, α-phenyl *N-tert*-butyl nitrone. Reproduced with permission from *The Journal of Clinical Investigation* (1988): 82:476–485.

The amplitude of this reperfusion-related free radical production was directly proportional to the severity of the preceding hypoperfusion (Fig. 8): the more pronounced the ischemia, the greater the magnitude of the subsequent free radical generation, and by inference, of the concomitant "reperfusion injury." This would imply that interventions aimed at reducing the severity of ischemia would also lessen the severity of reperfusion-associated injury. Furthermore, we found that PBN itself attenuated myocardial stunning, which suggests that the very radicals that were trapped by it are important pathogenetic contributors (Bolli et al., 1988b). The identity of these radicals remains unknown. We believe they are a mixture of carbon- and oxygen-centered species (e.g., alkyl and alkoxyl radicals), both of which are known to be trapped by PBN (Bolli and McCay, 1990) and to be produced during free radical–initiated lipid peroxidation (Thompson and Hess, 1986). Other investigators have also demonstrated the presence of free radicals associated with regional myocardial stunning *in vivo* (Leiboff et al., 1988).

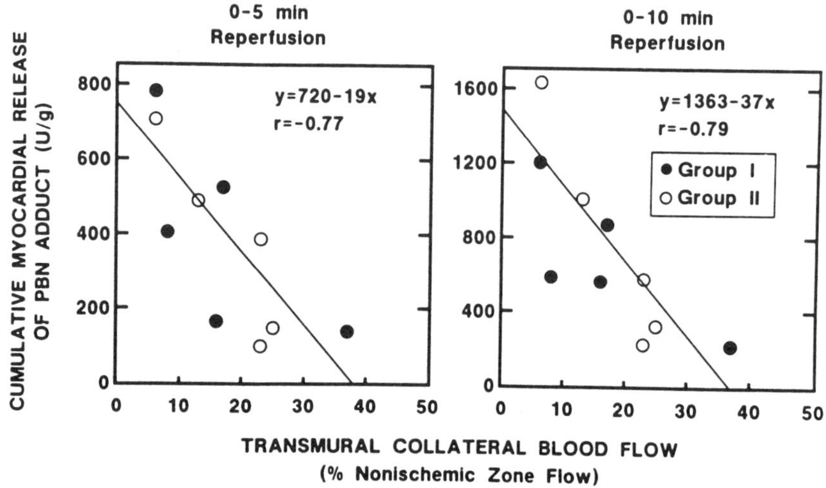

Figure 8 Relationship between mean transmural collateral blood flow to the ischemic region during coronary occlusion (horizontal axis) and total cumulative myocardial release of PBN adducts during the first 5 min (*left*) and 10 min (*right*) of reperfusion. Collateral flow is expressed as percentage of simultaneous nonischemic zone flow; adduct release is expressed in arbitrary units per gram of myocardium. Solid circles represent dogs in group I (PBN given before ischemia); open circles represent dogs in group II (PBN given at reperfusion). In both groups, the myocardial production of PBN adducts after coronary reperfusion was linearly and inversely related to collateral flow during the antecedent occlusion. These data suggest that the severity of the ischemic injury is the major determinant of the severity of the subsequent reperfusion injury. PBN, α-phenyl *N-tert*-butyl nitrone. Reproduced with permission from *The Journal of Clinical Investigation* (1988): 82:476–785.

In a series of subsequent studies, we have investigated the impact of three different antioxidants (SOD plus catalase, MPG, and desferrioxamine) on the generation of free radicals. All three therapeutic regimens blunted the generation of PBN adducts (Figs. 5, 9, 10), hence confirming them to be the by-products of the univalent pathway of reduction of oxygen; at the same time, all three regimens attenuated myocardial stunning (Bolli et al., 1989b, c; Bolli et al., 1990). Again, these beneficial effects were observed when the infusion was started before reperfusion.

When MPG or desferrioxamine was started 1 min after reflow none of the above salutary effects occurred (Figs. 3–5). The fact that three different agents supressed PBN adduct production consistently in the same dose regimens in which they attenuated myocardial stunning suggests a cause–effect relationship between free radical generation and contractile dysfunction.

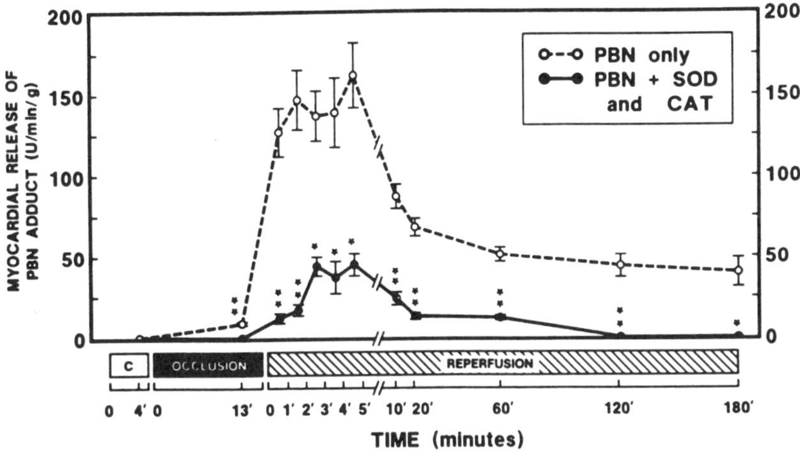

Figure 9 Time course of myocardial release of PBN adducts in group I (PBN only, n = 6) and group II (PBN plus superoxide dismutase and catalase, n = 6). Data are mean ± SEM. Administration of superoxide dismutase and catalase markedly inhibited production of free radicals in the stunned myocardium after 15 min of regional ischemia. $^{*}p < .05$; $^{**}p < .01$ vs. group I. PBN, α-phenyl N-*tert*-butyl nitrone. Reproduced from *Proceedings of the National Academy of Sciences of the USA* 1989;86:4695–4699.

We have also investigated PBN-adduct formation in the more physiological setting of the conscious dog model, also using a 15-min occlusion (Zughaib et al., 1991). The time course of free-radical production was similar to that seen in open-chest preparations (Bolli and McCay, 1990; Bolli et al., 1988b; Bolli et al., 1989b, c; Bolli et al., 1990; Leiboff et al., 1988); however, as mentioned above, its magnitude was less pronounced at every time point. The most important conclusions to be derived from these data (Zughaib et al., 1991) are that (a) free radicals are produced with ischemia/reperfusion independently of the unphysiologic conditions that plague the open-chest preparation (Triana et al., 1991) and (b) the production of free radicals is exaggerated in open-chest as compared with conscious animals. In another recent investigation (Bolli et al., 1991a), we have used phenylalanine in our open-chest and conscious canine models and demonstrated the formation of ortho-, meta-, and para-tyrosine

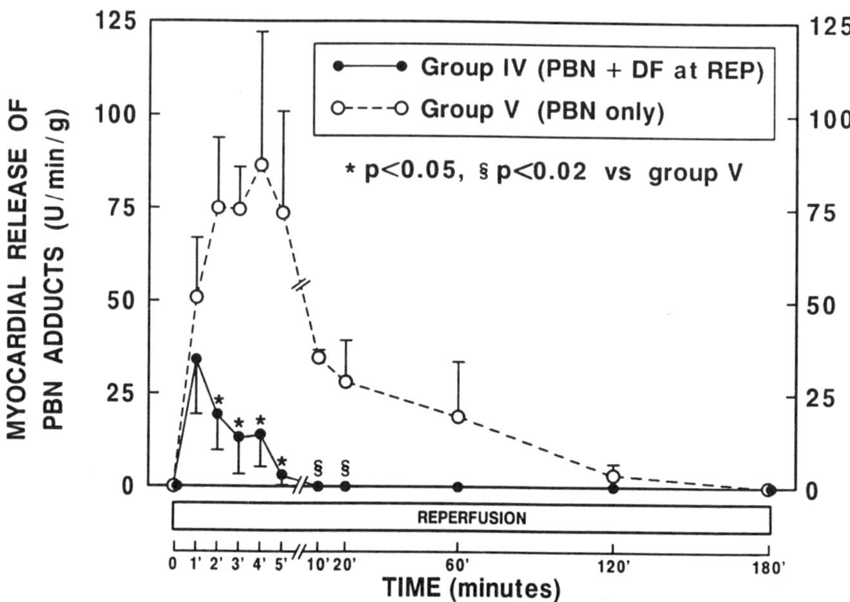

Figure 10 Time course of myocardial release of PBN adducts in group V [desferrioxamine (DF) administration started 2 min before reperfusion (REP), n = 4] and group VI (controls, n = 5). Data are mean ± SEM. Both groups received PBN by the intracoronary route. Infusion of desferrioxamine started 2 min before reperfusion markedly suppressed production of free radicals in the stunned myocardium. Since desferrioxamine given as in group V attenuated postischemic dysfunction (see Fig. 8), these data suggest that iron-mediated free radical reactions occurring immediately after reperfusion contribute to myocardial stunning. Reproduced with permission from the *American Journal of Physiology* (1990): 259:H1901–H1911.

in the reperfused myocardial region in the first few minutes of reflow. Since these compounds are all hydroxylated derivatives of phenylalanine formed by its reaction with the hydroxyl radical, these results provide additional evidence for the generation of hydroxyl radicals in stunned myocardium. Importantly, the similarity of the results obtained with two utterly different techniques [spin trapping (Bolli and McCay 1990; Bolli et al., 1988b; Bolli et al., 1989b,c; Bolli et al., 1990; Leiboff et al., 1988; Zughaib et al., 1991) and aromatic hydroxylation (Bolli et al., 1991a)] further supports the validity of the oxyradical hypothesis of postischemic dysfunction.

In summary, several studies have provided direct evidence for a pathogenetic role for reactive oxygen species in myocardial stunning (Bolli and McCay 1990; Bolli et al., 1991a; Bolli et al., 1989b, c; Bolli et al., 1990b; Leiboff et al., 1988b; Zughaib et al., 1991). The following points have been demonstrated: (a) reperfusion that follows reversible myocardial ischemia is associated with generation of free radicals, (b) these radicals are produced in both open-chest and conscious dog models (i.e., they are not artifacts associated with the former), (c) the radicals are produced by univalent reduction of oxygen, and (d) the radicals are necessary for the occurrence of stunning since inhibition of their production enhances recovery of function.

Other Models of Stunning

The studies discussed heretofore used a 15-min coronary occlusion. There is now a general consensus with respect to the pathogenetic contribution of reactive oxygen species in this model. Table 2 summarizes a

Table 2. Effect of antioxidants on myocardial "stunning" after a coronary occlusion $\leq$ 15 min

Authors	Antioxidant	Preparation	Protection
Myers et al., 1985	SOD + CAT	Open-chest dog	Yes
Przyklenk and Kloner, 1986	SOD + CAT	Open-chest dog	Yes
Gross et al., 1986	SOD + CAT	Open-chest dog	Yes
Murry et al., 1989	SOD + CAT	Open-chest dog	Yes
Bolli et al., 1989b	SOD + CAT	Open-chest dog	Yes
Jeroudi et al., 1990	SOD + CAT	Open-chest dog	Yes
	SOD alone		No
	CAT alone		No
Triana et al., 1991	SOD + CAT	Conscious dog	Yes
Koerner et al., 1991	SOD + CAT	Open-chest rabbit	Yes
	MPG		Yes
Buchwald et al., 1989a	SOD alone	Open-chest pig[a]	No
Myers et al., 1986a	MPG	Open-chest dog	Yes
Bolli et al., 1989c	MPG	Open-chest dog	Yes
Bolli et al., 1987a	DMTU	Open-chest dog	Yes
Bolli et al., 1987b	Desferrioxamine	Open-chest dog	Yes
Farber et al., 1988	Desferrioxamine	Open-chest dog	Yes
Charlat et al., 1987	Allopurinol	Open-chest dog	yes
Bolli et al., 1988b	PBN	Open-chest dog	Yes
Bolli et al., 1989b	PBN	Open-chest dog	Yes
Dage et al., 1991	Probucol	Open-chest rabbit	Yes
Buchwald et al., 1989b	α-Tocopherol	Open-chest pig[a]	No

SOD, superoxide dismutase; CAT, catalase; MPG, *N*-2-mercaptopropionyl glycine; DMTU, dimethylthiourea; PBN, α-phenyl *N-tert*-butyl nitrone.

[a] The duration of occlusion was 8 min, not 15 min.

Reproduced (with minor changes) with permission from *Cardiovas Drug Ther* (1991): 5:249–268.

number of independent investigations demonstrating antioxidant-induced improvement in contractile function after a single 15-min occlusion.

Is there a role for oxyradicals in other models of stunning? In a model of ten 5-min coronary occlusions interspersed with 10-min periods of reflow, we (*unpublished data*) observed that each occlusion/reperfusion cycle was associated with a burst of free-radical generation that was suppressed by MPG (therefore implicating the presence of ·OH). MPG again enhanced the recovery of myocardial function (Triana et al., 1990). These results suggest that oxyradicals contribute to the pathogenesis of stunning after multiple brief coronary occlusions.

In *in vitro* models of global ischemia (isolated hearts), antioxidants consistently attenuate contractile dysfunction (Ambrosio et al., 1987a; Ambrosio et al., 1987b; Casale et al., 1983; Menasche et al., 1986; Myers et al.,1986b; Schlafer et al., 1982a,b; Ytrehus et al., 1987). These models, however, probably represent a mixture of reversible (stunning) and irreversible cell injury (infarction); therefore, their relevance to the problem of stunning is uncertain (Bolli et al., 1990). *In vivo* models of global ischemia abound in the surgical literature. The oxyradicals have been shown to contribute to stunning after cardioplegic arrest (Gardner, 1988; Illes et al., 1989; Johnson et al., 1987; Stewart et al., 1983), and recent observations suggest a role of oxyradicals in the contractile dysfunction observed after cardiac surgery in patients (Ferrari et al., 1990).

Data on the role of oxyradicals in the setting of prolonged coronary occlusions (> 20 min but < 3 hr) resulting in an admixutre of stunning and infarction are more conflicting. As depicted in Table 3, SOD with or without catalase failed to attenuate postischemic dysfunction after 60 min (Asinger et al., 1988), 90 min (Nejima et al., 1989), and 120 min (Przyklenk and Kloner, 1989; Patel et al., 1990) of coronary occlusion in both open-chest anesthetized (Przyklenk and Kloner, 1989; Patel et al., 1990) and conscious unsedated dogs (Asinger et al., 1988; Nejima et al., 1989). On the other hand, when cell permeant antioxidants were tested (oxypurinol, *N*-acetylcysteine, Trolox), myocardial stunning was attenuated independently of any reduction of infarct size in closed-chest dogs subjected to 90-min occlusion and 24-hr reflow (Puett et al., 1987; Forman et al., 1988) and in open-chest pigs with 45-min occlusion and 72-hr reflow (Klein et al., in press). In the latter model, no enhancement of functional recovery was seen with pretreatment with α-tocopherol and ascorbate (Klein et al., 1989) (Table 3). The fact that short-term administration of antioxidants fails to mitigate the myocardial stunning that is associated with infarction suggests that the pathogenesis of this form of stunning may be different from that of stunning associated with brief, reversible ischemia/reperfusion (Tables 2-3). It is possible that only antioxidants with cell-permeant properties may be effective in the setting of prolonged ischemia associated with infarction.

Table 3. Effect of antioxidants on myocardial "stunning" after a coronary occlusion > 20 min

Authors	Antioxidant	Preparation	Duration of ischemia	Protection
Asinger et al., 1988	SOD + CAT	Conscious dog	1hr	No
Vanhaecke et al., 1988	SOD	Closed-chest dog	1hr	No
Nejima et al., 1989	SOD + CAT	Conscious dog	90 min	No
Przyklenk and Kloner, 1989	SOD + CAT	Open-chest dog	2 hr	No
Patel et al., 1990	SOD	Closed-chest dog	2hr	No
Puett et al., 1987	Oxypurinol	Closed-chest dog	90 min	Yes
Forman et al., 1988	*N*-acetylcysteine	Closed-chest dog	90 min	Yes
Klein et al., 1991	Trolox	Open-chest pig	45 min	Yes
Klein et al., 1989	α-Tocopherol + ascorbate	Open-chest pig	45 min	No

SOD, superoxide dismutase; CAT, catalase.

Reproduced with permission (with minor changes) from *Cardiovasc Drugs Ther* (1991) 5:249–268.

Myocardial stunning that is induced by exercise (i.e., stunning due to increased demands rather than decreased supply) was reported not to be affected by SOD and catalase (Homans et al., 1988).

In summary, there exists strong evidence supporting a pathogenetic role of oxyradicals in stunning induced by multiple brief occlusion/reperfusion cycles and by global ischemia *in vitro* as well as *in vivo*. Data in the setting of subendocardial infarction are more controversial, mainly due to the quasi-impossibility of separating the independent contributions of stunning and infarction to myocardial dysfunction after prolonged ischemia. There is no evidence for a role of oxyradicals in exercise-induced stunning.

Correlation between Oxidative Stress and Severity of Antecedent Ischemia

The following lines of evidence indicate that the magnitude of free radical generation is directly related to the severity of the preceding ischemic insult: (a) there is a positive linear relationship between the production of free radicals (as measured directly by spin trapping and EPR spectroscopy or aromatic hydroxylation) and the magnitude of the preceding hypoperfusion (Fig. 8), (b) in conscious dogs, the benefits derived from antioxidant therapy are more pronounced in animals with more severe ischemia (Triana et al., 1991): as collateral flow increases, the enhancement of contractile recovery effected by antioxidants decreases (Fig. 2), and (c) the postischemic depression of myocardial contractility is closely coupled to the degree of hypoperfusion such that even small differences in collateral blood flow during occlusion are associated with significant differences in recovery of function after reflow. This was demonstrated in our laboratory (Bolli et al., 1988a, 1989a) in conscious dogs. In open-chest dogs this relationship is weaker (Gross et al., 1986), probably because of the artifacts from which this preparation suffers (e.g., surgical trauma and anesthesia—see above).

The foregoing observations can be summarized as follows: there is a direct positive correlation between the severity of ischemia and the magnitude of the oxidative stress associated with subsequent reperfusion. As both increase, so will the benefit derived from antioxidant therapy. This concept has important pathophysiological, experimental and therapeutic implications.

1. *Pathophysiologically*, it conforms to the previous proposition (Thompson and Hess, 1986) that ischemia "primes" the myocardium for the oxidative stress incurred upon reperfusion. Certainly, the fact that production of free radicals is inversely related to collateral flow supports this view. The mechanisms through which ischemia modulates oxidative stress may include perturbation of the mitochondrial electron transport chain, the accumulation of reducing equivalents, the activation of the arachidonate cascade, the release of catecholamines and of chemotactic factors for neutrophils, and the conversion of xanthine dehydrogenase to xanthine oxidase. In a sense, the degree of ischemic injury appears to predetermine the degree of "reperfusion injury" (Bolli et al., 1988b).

2. *Methodologically*, evaluation of antioxidant agents should be restricted to animals with low collateral perfusion during coronary occlusion. Because antioxidants are more effective in animals with more severe ischemia, use of animals with low collateral flow would obviate the need for the larger sample sizes that would otherwise be required to show a difference.

3. *Therapeutically*, one would expect that the most effective way to reduce postischemic myocardial dysfunction is the attenuation of the antecedent

ischemia. Furthermore, antioxidant therapies would be expected to be most useful in patients with more severe ischemia and, thus, more severe postischemic dysfunction.

Finally, it is important to recognize that the severity and the duration of ischemia are not equivalent concepts. The severity is defined mainly by the magnitude of collateral flow (and hence of hypoperfusion) and is a major determinant of the intensity of oxidative stress. Whether the duration of ischemia is similarly related to oxidative stress is unknown. There is some evidence to the contrary in dogs: the beneficial effects of antioxidant therapy on recovery of contractility appear to decrease as the duration of ischemia is prolonged (Bolli, 1991b), possibly because of the admixture of myocardial stunning and infarction.

Effect of Oxyradicals on Cardiac Function

Evidence for negative inotropic effects. Evidence for a myocardial depressant action of oxygen radicals has been obtained both *in vitro* and *in vivo.* A decrease in myocardial contractility and adenosine triphosphate (ATP) levels have been uniformly observed when H_2O_2 or free radical–generating solutions were administered to isolated rabbit interventricular septa (Burton et al., 1984), papillary muscles (Schrier and Hess, 1988), or whole hearts (Blaustein et al., 1986; Goldhaber et al., 1988; Jackson et al., 1986; Miki et al., 1988; Shattock et al., 1982; Ytrehus et al., 1986); this decrease in contractility and ATP mimics the changes that occur in *in vivo* models of stunning. These deleterious effects were mitigated by catalase or ·OH scavengers but not by SOD (Blaustein et al., 1986; Burton et al., 1984; Jackson et al., 1986; Miki et al., 1988), strongly implying that ·OH or its precursor H_2O_2 may represent the important offending species. On the other hand, a direct toxicity of $\cdot O_2^-$ cannot be excluded because SOD but not catalase was demonstrated to be beneficial in some cases (Schrier and Hess, 1988). An *in vivo* correlate of these *in vitro* observations was provided by recent studies (Przyklenk et al., 1990) demonstrating a significant depression of myocardial contractility in open-chest dogs in which a free radical–generating system (xanthine oxidase, purine metabolites, and iron-loaded transferrin) was infused into a coronary vein.

In summary it is now clear that reactive oxygen species exert negative inotropic effects both *in vitro* and *in vivo*, although the exact culprits remain to be identified. There is substantial evidence for a major role of ·OH, congruent with previous *in vivo* studies (Bolli et al., 1987a,b; Bolli et al.,

1989c; Farber et al., 1988; Myers et al., 1986a). However, $\cdot O_2^-$ may also be important, either in itself (Jeroudi et al., 1990) or as a precursor of ·OH.

Postulated mechanisms. The mechanism for the depression of contractile function secondary to oxyradical-mediated cellular injury remains to be elucidated. Both cellular proteins and lipids could conceivably represent targets for reactive oxygen species. Oxyradicals are known to denature proteins by oxidation of sulfhydryl groups (Davies, 1987) and to disrupt cellular membranes through peroxidation of their constituent polyunsaturated fatty acids (Thompson and Hess, 1986). The end result would be the disruption of various intracellular organelles and the loss of selective permeability of cell membranes.

Evidence for the occurrence of lipid peroxidation in postischemic myocardium was provided by measurements of one of its by-products, namely hydroxy-conjugated dienes. The tissue concentration of hydroxy-conjugated dienes was found to be increased in stunned (non-necrotic) myocardium during and after a 45-min period of global normothermic ischemia in open-chest dogs, peaking at 5 min of reperfusion (Romaschin et al., 1987). In patients subjected to cardioplegic arrest during coronary bypass surgery, the same conjugated dienes were found to be released in the coronary sinus blood at 3 and 60 min of reflow (Weisel et al., 1989). At the same time, the myocardial concentration of the antioxidant α-tocopherol was found to be reduced (Weisel et al., 1989).

Despite these observations, however, definitive evidence for lipid peroxidation in the 15-min coronary occlusion model of stunning is still lacking. Recently, a 15-min coronary occlusion followed by 30 min of reflow was found to reduce the myocardial non-GSH, non-protein SH pool (composed of cysteine, cysteamine, and other low molecular weight SH groups), whereas the myocardial GSH, GSSG, GSH/GSSH ratio, total SH, and protein SH pools remained unchanged (Lesnefsky et al., 1991). These results suggest that a 15-min coronary occlusion produces a relatively mild oxidative stress whose effects may be subtle and therefore difficult to demonstrate. More importantly, it appears that only the non-GSH, non-protein SH pool is sensitive to this mild oxidative stress and therefore is selectively depleted (Lesnefsky et al., 1991).

The subcellular organelles damaged by oxidative stress remain unknown. Both the sarcoplasmic reticulum and the sarcolemma could be important targets of oxyradical-mediated injury. Exposure of isolated sarcoplasmic reticulum to oxygen radicals results in a decrease in Ca^{2+} uptake and Ca^{2+}, Mg^{2+}-ATPase activity (Rowe et al., 1983; Thompson and Hess, 1986) that is prevented by free radical scavengers (Rowe et al., 1983; Thompson and Hess, 1986). Similar abnormalities were demonstrated in sarcoplasmic reticulum isolated from stunned myocardium (Krause et al., 1989).

Numerous oxyradical-induced sarcolemmal abnormalities have been reported. Oxygen free radicals were shown to interfere with the Na^+-Ca^{2+} exchange (Reeves et al., 1986), calcium transport and Na^+-K^+-ATPase activity (Kaneko et al., 1989a,b; Kramer et al., 1984). Kim and Akera (1987) demonstrated a similar impairment of Na^+-K^+-ATPase activity in reperfused myocardium that was prevented by antioxidants. Impairment of Na^+-K^+-ATPase activity would result in excessive intracellular accumulation of Na^+ leading to activation of Na^+-Ca^{2+} exchange and ultimately in cellular calcium overload.

It is important to stress that the above-described alterations in sarcoplasmic reticulum and sarcolemma as a direct result of free radical–induced damage could represent the link between the "oxyradical" and "calcium" hypotheses of postischemic dysfunction (Bolli, 1990).

Integration of the "Oxyradical" and the "Calcium" Hypotheses

Although this chapter deals with the role of oxyradicals in stunning, it must be stressed that other plausible theories for the pathogenesis of this phenomenon have been developed (see Bolli, 1990 for review). One such theory is the "calcium hypothesis" (Bolli, 1990; Kusuoka and Marban, in press), which has several facets. This theory postulates (a) a reperfusion-induced intracellular Ca^{2+} overload (Marban et al., 1990), (b) an insufficient availability of Ca^{2+} to active contractile elements because of deficient sarcoplasmic reticulum function (Krause et al., 1989), and/or (c) a decreased myofilament responsiveness to calcium (Kusuoka et al., 1987). These perturbations would result in long-lasting depression of contractility, excitation-contraction uncoupling, and reduced force generation for any given $[Ca]_i$, respectively. Because of the central role of Ca^{2+} in modulating cardiac contractility, it seems probable that any or all of the above disturbances of calcium homeostasis could play a pathogenetic role in stunning.

The "oxyradical" and the "calcium" hypotheses of stunning have been developed independently in different experimental settings (Table 1). Despite this, we will attempt to reconcile them in a unifying theory because they are not necessarily contradictory. In fact, there is abundant evidence that these two hypotheses are but two facets of the same phenomenon.

First, the abnormalities described in the "oxygen paradox" and in the "calcium paradox" are strikingly similar and may share a common pathogenesis (Hearse et al., 1978). Second, as discussed in the previous section, oxidative stress can cause profound disturbances of calcium homeostasis. Specifically, oxyradicals induce dysfunction of the sarcoplasmic reticulum (Rowe et al., 1983; Thompson and Hess, 1986) and alteration of transsarcolemmal Ca^{2+} flux (Kaneko et al., 1989a, b; Kim and Akera, 1987; Kramer et al., 1984; Reeves et al., 1986) which would result in uncoupling of excitation and contraction, and in intracellular calcium overload, respectively (Kaneko et al., 1989a, b; Thompson and Hess, 1986). Reoxygenation-induced calcium overload has been found to be preventable by antioxidants (Murphy et al., 1988). On the same lines, SOD and catalase have been found to attenuate the rise of intracellular Na^+, the increased uptake of $^{45}Ca^{2+}$, and the depression of contractility associated with 30 min of ischemia followed by reflow in isolated rabbit hearts (Tani, 1990). Several other investigations support a link between oxidative stress and impaired calcium homeostasis. Exposure of myocytes to free-radical generating solutions results in marked increase in $[Ca]_i$ in cultured cells (Burton et al., 1990a,b) and in isolated rabbit hearts (Corretti et al., 1990). Moreover, Burton et al. (1990a) demonstrated that exposure of myocytes to free radicals causes a rise in conjugated dienes and of [^{3}H] arachidonate, strongly suggesting some degree of peroxidative damage to cell membranes. Oxyradicals could also decrease the sensitivity of the contractile proteins to calcium, despite the presence of this ion in abundance (Kusuoka and Marban, in press). Last, a vicious circle could be produced by the enhancement of production of free radicals by calcium itself through stimulation of the conversion of xanthine dehydrogenase to xanthine oxidase, a reaction that is catalyzed by a calcium-dependent protease (McCord, 1985).

Accordingly, it is our opinion that the "oxyradical" and the "calcium" hypotheses of myocardial stunning represent complementary pathogenetic mechanisms. However, several important gaps in our understanding of this phenomenon remain (Table 1). For instance, the mechanism of oxyradical-mediated contractile dysfunction, of subcellular organelle damage, and of calcium overload are still obscure. Further investigation of these problems will be necessary, and represent an important area for future work. In view of all of the above, and based on the experimental evidence available to date, we propose a unifying paradigm that encompases the present state of our knowledge (Fig. 11). A detailed description of this proposal is provided in the legend to Fig. 11. This proposal attempts to bring the molecular, subcellular, ionic, and mechanical disturbances together in a logical, albeit speculative, framework.

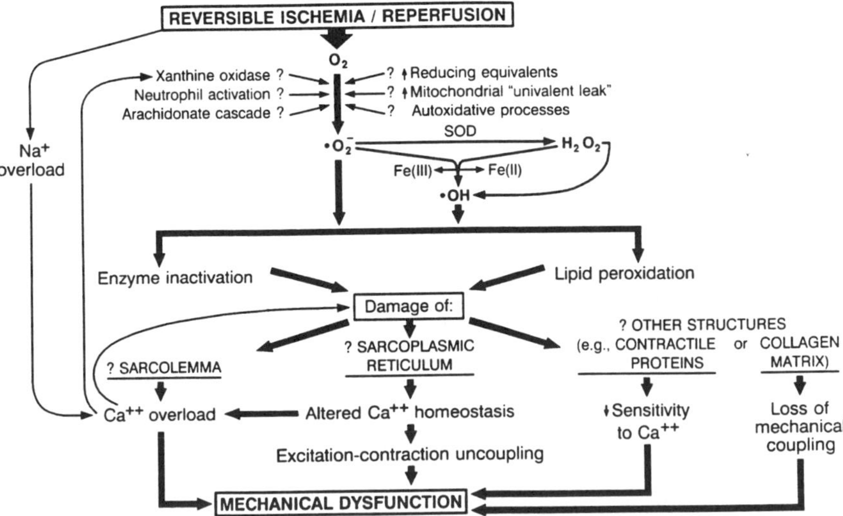

Figure 11 Illustration of the proposed pathogenesis of posischemic myocardial dysfunction. This proposal integrates and reconciles different mechanisms into a unifying pathogenetic hypothesis. Transient reversible ischemia followed by reperfusion could result in increased production of superoxide radicals $\cdot O_2^-$ through several mechanisms, including (a) increased acitivity of xanthine oxidase, (b) activation of neutrophils, (c) activation of the arachidonate cascade, (d) accumulation of reducing equivalents during oxygen deprivation, (e) derangements of the intramitochondrial electron transport system resulting in increased univalent reduction of oxygen, and (f) autoxidation of catecholamines and other substances. Superoxide dismutase (SOD) dismutases $\cdot O_2^-$ to hydrogen peroxide (H_2O_2); the presence of catalytic iron, $\cdot O_2^-$ and H_2O_2 interract in a Haber-Weiss reaction to generate the hydroxyl radical ($\cdot$OH). H_2O_2 can also generate $\cdot$OH in the absence of $\cdot O_2^-$ through a Fenton reaction provided that other substances (such as ascorbate) reduce Fe(III) to Fe(II). $\cdot O_2^-$ and $\cdot$OH attack proteins and polyunsaturated fatty acids, causing enzyme inactivation and lipid peroxidation, respectively.

continued on next page ⟶

In the setting of reversible ischemia, the intensity of this damage is not sufficient to cause cell death, but is sufficient to produce dysfunction of key cellular organelles. Postulated targets of free radicals damage include: (a) the sarcolemma, with consequent loss of selective permeability, impairment of calcium-stimulated ATPase activity and calcium transport out of the cell, and impairment of the Na^+-K^+-ATPase activity. The net result of these perturbations would be increased transarcolemmal calcium influx and cellular calcium overload; (b) the sarcoplasmic reticulum, with consequent impairment of calcium-stimulated ATPase activity and calcium transport. This would result in impaired calcium homeostasis; specifically, decreased calcium sequestration (which would contribute to increase free cystolic calcium) and decreased calcium release during systole (which would cause excitation-contraction uncoupling); (c) Possibly other structures, such as the extracellular collagen matrix (with consequent loss of mechanical coupling) or the contractile proteins (with consequent decreased sensitivity to calcium). At the same time, reversible ischemia/reperfusion could use cellular Na^+ overload due to a) inhibition of sarcolemmal Na^+-K^+-ATPase, and b) acidosis and Na^+-H^+ exchange. This could further exaggerate calcium overload via Na^+-Ca^{2+} exchange. An increase in free cytosolic calcium would activate phospholipases and other degradative enzymes and further exacerbate to the aforementioned key subcellular structures (sarcolemma, sarcoplasmic reticulum, and contractile proteins). Thus, calcium overload could serve to amplify the damage initiated by oxygen radicals. In addition, calcium overload could in itself impair contractile performance and contribute to mechanical dysfunction. It is also possible that the increase in free cystolic calcium could increase oxyradical production by promoting the conversion of xanthine dehydrogenase to xanthine oxidase. The ultimate consequencc of this complex series of perturbations is a reversible depression of contractility. Reproduced with permission of *The Americal Heart Association* from *Circulation* (1990): 82:723–738.

Clinical Implications

In clinical practice, myocardial stunning is a recognized, although incompletely understood phenomenon, that may contribute importantly to the morbidity of patients with coronary artery disease (Bolli, 1991c). Although stunning so far has been reversed with inotropic therapy, it is our belief that this approach is not optimal because it entails *treatment* rather than *prevention.* Furthermore, the prolonged administration of inotropic agents can cause a variety of complications. It appears, therefore, preferable to prevent myocardial stunning from occurring in the first place (e.g., by administering antioxidant therapies). As discussed above, the oxyradical hypothesis implies that selected antioxidants could alleviate postischemic dysfunction if given before reperfusion. Antioxidants like MPG, desferrioxamine, and allopurinol are already being used in patients for other indications; these drugs could soon find new clinical applications in the prevention and/or attenuate of postischemic contractile abnormalities. In particular, both desferrioxamine and superoxide dismutase are being (or will soon be) tested in patients in the setting of pharmacologic (acute myocardial infarction) or surgical (open-heart surgery) coronary reperfusion.

Summary

Although the pathogenesis of myocardial stunning is still incompletely understood, the "oxyradical" and the "calcium" hypotheses are currently the two most plausible theories. The *oxyradical hypothesis* has gained widespread acceptance because of the consistency of the results obtained in different laboratories. Indeed, the role of oxygen radicals in stunning is one of the few areas in which there is relatively little disagreement concerning the pathogenetic contribution of oxygen metabolites to myocardial injury. The evidence supporting the oxyradical hypothesis of stunning is overwhelming and is based on data from several independent investigators that have demonstrated beneficial effects of antioxidants on postischemic dysfunction in the setting of a 15-min coronary occlusion. This evidence has recently gained even more strength when it has been confirmed in the most physiological experimental preparation available, namely, the conscious animal. In addition, spin trapping studies have provided *direct* evidence for the production of oxyradicals in the stunned myocardium and have shown that suppression of oxyradical production by antioxidants is associated with attenuation of postischemic contractile dysfunction. Other observations have also implicated free radicals in the pathogenesis of stunning after global ischemia *in vitro* and *in vivo*. It appears that the oxyradical-mediated injury responsible for stunning occurs in the first few miniutes of reflow, implying that myocardial stunning is a sublethal form of "reperfusion injury."

Considerable evidence also supports the calcium hypothesis. This hypothesis postulates a reperfusion-induced intracellular calcium overload, a reduced sensitivity of the contractile proteins to Ca^{2+}, and/or inadequate release of Ca^{2+} from the sarcoplasmic reticulum resulting in excitation-contraction uncoupling. There is evidence for the first two possibilities in models of postischemic dysfunction alter global ischemia *in vitro*, and for the third possibility in models in postischemic dysfunction after multiple brief episodes of regional ischemia. The importance of these mechanisms in other forms of myocardial stunning remains to be assessed.

The oxyradical theory and the calcium theory are not mutually exclusive and could actually be part of the same pathogenetic scheme. The production of free radicals may disrupt cellular membranes and sarcoplasmic reticulum, leading to calcium overload, which in turn could exacerbate the damage initiated by free radicals. It is hoped that the concepts discussed in this chapter will help to design future research aimed at clarifying the pathophysiology of myocardial stunning and to develop clinically therapeutic strategies.

Acknowledgment. The work described in this chapter was supported in part by NIH Grants HL-43151 and SCOR Grant HL-42267.

References

Ambrosio G, Weisfeldt ML, Jacobus WE, Flaherty JT (1987a): Evidence for a reversible oxygen radical-mediated component of reperfusion injury: reduction by recombinant human superoxide dismutase administered at the time of reflow. *Circulation* 75:282–291.

Ambrosio G, Zweier JL, Jacobus WE, Weisfeldt ML, Flaherty JT (1987b): Improvement of postischemic myocardial function and metabolism induced by administration of desferrioxamine at the time of reflow: the role of iron in the pathogenesis of reperfusion injury. *Circulation* 76:906–915.

Asinger RW, Peterson DA, Elsperger KJ, Homans D (1988): Long-term recovery of LV wall thickening after 1 hour of ischemia is not affected when superoxide dismutase and catalase are administered during the first 45 minutes of reperfusion. *J Am Coll Cardiol* 2:163A (abstract).

Blaustein AS, Schine L, Brooks WW, Fanburg BL, Bing OHL (1986): Influence of exogenously generated oxidant species on myocardial function. *Am J Physiol* 250:H595–H599.

Bolli R (1990): Mechanism of myocardial "stunning." *Circulation* 82:723–738.

Bolli R (1991b): Superoxide dismutase 10 years later: A drug in search of a use. *J Am Coll Cardiol* (in press).

Bolli R, Hartley CJ, Rabinovitz RS (1991c): Clinical relevance of myocardial "stunning." *Cardiovasc Drugs Ther* 5:877–890.

Bolli R, Jeroudi MO, Patel BS, Aruoma OI, Halliwell B, Lai EK, McCay PB (1989c): Marked reduction of free radical generation and contractile dysfunction by antioxidant therapy begun at the time of reperfusion: evidence that myocardial "stunning" is a manifestation of reperfusion injury. *Circ Res* 65:607–622.

Bolli R, Jeroudi MO, Patel BS, DuBose CM, Lai EK, Roberts R, McCay PB (1989b): Direct evidence that oxygen-derived free radicals contribute to postischemic myocardial dysfunction in the intact dog. *Proc Natl Acad Sci USA* 86:4695–4699.

Bolli R, Kaur H, Li XY, Triana JF, Halliwell B (1991a): Demonstration of hydroxyl radical generation in "stunned" myocardium of intact dogs using aromatic hydroxylation of phenylalanine. *FASEB J* 5:A704.

Bolli R, McCay PB (1990): Use of spin traps in intact animals undergoing myocardial ischemia/reperfusion: a new approach to assessing the role of oxygen radicals in myocardial "stunning." *Free Rad Res Commun* 9:169–180.

Bolli R, Patel BS, Hartley CJ, Thornby JI, Jeroudi MO, Roberts R (1989a): Nonuniform transmural recovery of contractile function in the "stunned" myocardium. *Am J Physiol* 257:H375–H385.

Bolli R, Patel BS, Jeroudi MO, Lai EK, McCay PB (1988b): Demonstration of free radical generation in "stunned" myocardium of intact dogs with the use of the spin trap α-phenyl N-*tert*-butyl nitrone. *J Clin Invest* 82:476–485.

Bolli R, Patel BS, Jeroudi MO, Li XY, Triana JF, Lai EK, McCay PB (1990): Iron-mediated radical reactions upon reperfusion contribute to myocardial "stunning." *Am J Physiol* 259:H1901–H1911.

Bolli R, Patel BS, Zhu WX, O'Neill PG, Charlat ML, Roberts R (1987b): The iron chelator desferrioxamine attenuates postischemic ventricular dysfunction. *Am J Physiol* 253:H1372–H1380.

Bolli R, Zhu WX, Hartley CJ, Michael LH, Repine J, Hess ML, Kukreja RC, Roberts R (1987a): Attenuation of dysfunction in the postischemic "stunned" myocardium by dimethylthiourea. *Circulation* 76:458–468.

Bolli R, Zhu WX, Thornby JI, O'Neill PG, Roberts R (1988a): Time-course and determinants of recovery of function after reversible ischemia in conscious dogs. *Am J Physiol* 254:H102–H114.

Braunwald E, Kloner RA (1982): The "stunned myocardium": prolonged, postischemic ventricular dysfunction. *Circulation* 66(6):1146–1149.

Buchwald A, Klein HH, Lindert S, Pich S, Nebendahl K, Wiegand V, Kreuzer H (1989a): Effect of intracoronary superoxide dismutase on regional function in stunned myocardium. *J Cardiovasc Pharmacol* 13:258–264.

Buchwald A, Klein HH, Lindert S, Pich S, Oberschmidt R, Nebendahl K, Kreuzer H (1989b): Effect of α-tocopherol (vitamin E) in a porcine model of stunned myocardium. *J Cardiovasc Pharmacol* 14:46–52.

Burton KP, McCord JM, Ghai G (1984): Myocardial alterations due to free-radical generation. *Am J Physiol* 246:H776–H783.

Burton KP, Morris AC, Massey KD, Buja LM, Hagler HK (1990a): Free radicals alter ionic calcium levels and membrane phospholipids in cultured rat ventricular myocytes. *J Mol Cell Cardiol* 22:1035–1047.

Burton KP, Nazeran H, Hagler HK (1990b): Free radicals alter ionic calcium transients observed in isolated adult rat cardiac myocytes. *Circulation* 82(suppl III):452 (abstract).

Casale AS, Bulkley GB, Bulkley BH, Flaherty JT, Gott VL, Gardner TJ (1983): Oxygen free radical scavengers protect the arrested globally ischemic heart upon reperfusion. *Surg Forum* 34:313–316.

Charlat ML, O'Neill PG, Egan JM, Abernethy DR, Michael LH, Myers ML, Roberts R, Bolli R (1987): Evidence for a pathogenetic role of xanthine oxidase in the "stunned" myocardium. *Am J Physiol* 252:H566–H577.

Coretti MC, Koretsune Y, Chacko VP, Zweier JL, Marban E (1990): Intracellular calcium overload and glycolytic inhibition as consequences of exogenously-generated free radicals in rabbit hearts. *Circulation* 82(suppl III):700 abstract.

Dage RC, Anderson BA, Mao SJT, Koerner JE (1991): Probucol reduces myocardial dysfunction during reperfusion after short-term ischemia in rabbit heart. *J Cardiovasc Pharmacol* 17:158–165.

Davies KJA (1987): Protein damage and degradation by oxygen radicals. I. General aspects. *J Biol Chem* 262:9895–9901.

Davies MJ (1989): Direct detection of radical production in the ischaemic and reperfused myocardium: current status. *Free Rad Res Commun* 7:275–284.

DiGuiseppi J, Fridovich I (1982): Oxygen toxicity in *streptococcus sanguis*: The relative importance of superoxide and hydroxyl radicals. *J Biol Chem* 257:4046–4051.

Eddy LJ, Stewart JR, Jones HP, Engerson TD, McCord JM, Downey JM (1987): Free radical-producing enzyme, xanthine oxidase, is undetectable in human hearts. *Am J Physiol* 253:H709–H711.

Engler R (1987): Granulocytes and oxidative injury in myocardial ischemia and reperfusion. *Fed Proc* 46:2395–2396.

Engler R, Covell JW (1987): Granulocytes cause reperfusion ventricular dysfunction after 15-minute ischemia in the dog. *Circ Res* 61:20–28.

Engler RL, Dahlgren MD, Morris DD, Peterson MA, Schmid-Schonbein GW (1986): Role of leukocytes in response to acute myocardial ischemia and reflow in dogs. *Am J Physiol* 251:H314–H322.

Farber NE, Gross GJ (1990): Prostaglandin redirection by thromboxane synthetase inhibition: attenuation of myocardial stunning in canine heart. *Circulation* 81:369–380.

Farber NE, Vercellotti GM, Jacob HS, Pieper GM, Gross GJ (1988): Evidence for a role of iron-catalyzed oxidants in functional and metabolic stunning in the canine heart. *Circ Res* 63:351–360.

Ferrari R, Alfieri O, Curello S, Ceconi A, Cargnoni A, Marzollo P, Pardini A, Cardonna E, Visiolo O (1990): Occurrence of oxidative stress during reperfusion of the human heart. *Circulation* 81:201–211.

Forman MB, Puett DW, Cates CU, McCroskey DE, Beckman JK, Greene H, Virmani R (1988): Glutathione redox pathway and reperfusion injury. Effect of *N*-acetylcysteine on infarct size and ventricular function. *Circulation* 78:202–213.

Gardner TJ (1988): Oxygen radicals in cardiac surgery. *Free Radical Biol Med* 4:45–50.

Go LO, Murry CE, Richard VJ, Jennings RB, Weischedel GR, Reimer KA (1988): Myocardial neutrophil accumulation during reperfusion after reversible or irreversible ischemic injury. *Am J Physiol* 255:H1188–H1198.

Goldhaber JI, Ji S, Lamp ST, Weiss JN (1988): Effects of exogenous free radicals on electromechanical function and metabolism in isolated rabbit and guinea pig ventricle: implications for ischemia and reperfusion injury. *J Clin Invest* 83:1800–1909.

Grinwald PM (1982): Calcium uptake during postischemic reperfusion in the isolated rat heart: influence of extracellular sodium. *J Mol Cell Cardiol* 14:359–365.

Gross GJ, Farber NE, Hardman HF, Warltier DC (1986): Beneficial actions of superoxide dismutase and catalase in stunned myocardium of dogs. *Am J Physiol* 250:H372–H377.

Grum CM, Gallagher KP, Kirsh MM, Shlafer M (1989): Absence of detectable xanthine oxidase in human myocardium. *J Mol Cell Cardiol* 21:263–267.

Halliwell B, Gutteridge JMC (1984): Oxygen toxicity, oxygen radicals, transition metals and disease. *Biochem J* 219:1–14.

Hearse DJ, Humphrey SM, Bullock GR (1978): The oxygen paradox and the calcium paradox: two facets of the same problem? *J Mol Cell Cardiol* 10:641–668.

Hearse DJ (in press): Stunning: a radical re-view. *Cardiovasc Drugs Ther.*

Holzgrefe HH, Gibson JK (1988): Enhanced function recovery in the stunned canine myocardium by pretreatment with oxypurinol. *J Am Coll Cardiol* 11:208A (abstract).

Homans DC, Sublett E, Asinger R, Peterson D, Bache RJ (1988): SOD + catalase does not attenuate regional left ventricular dysfunction following exercise induced ischemia. *J Am Coll Cardiol* 78:II–76 (abstract).

Huizer T, de Jong JW, Nelson JA, Czarnecki W, Serruys PW, Bonnier JJRM, Troquay R (1989): Urate production by human heart. *J Mol Cell Cardiol* 21:691–695.

Illes RW, Silverman NA, Krukenkamp IB, del Nido PJ, Levitsky S (1989): Amelioration of postischemic stunning by desferrioxamine-blood cardioplegia. *Circulation* 80(suppl III):III-30–III-35.

Jackson CV, Mickelson JK, Pope TK, Rao PS, Lucchesi BR (1986): O_2 free radical-mediated myocardial and vascular dysfunction. *Am J Physiol* 251: H1225–H1231.

Jarasch ED, Bruder G, Heid HW (1986): Significance of xanthine oxidase in capillary endothelial cells. *Acta Physiol Scan* 548(suppl):39–46.

Jeremy RW, Becker LC (1989): Neutrophil depletion does not prevent myocardial dysfunction after brief coronary occlusion. *J Am Coll Cardiol* 13:1155–1163.

Jeroudi MO, Triana FJ, Patel BS, Bolli R (1990): Effect of superoxide dismutase and catalase, given separately, on myocardial "stunning." *Am J Physiol* 259:H889–H901.

Johnson DL, Horneffer PJ, Dinatale JM Jr., Gott VL, Gardner TJ (1987): Free radical scavengers improve functional recovery of stunned myocardium in a model of surgical coronary revascularization. *Surgery* 102:334–340.

Juneau CF, Ito BR, del Balzo U, Engler RL (1991): Severe neutrophil depletion by leukocyte filters or cytoxic drugs does not improve the recovery of contractile function in stunned myocardium. *Circulation* 84(4):II–655.

Kaneko M, Beamish RE, Dhalla NS (1989a): Depression of heart sarcolemmal Ca^{2+}-pump activity by oxygen free radicals. *Am J Physiol* 256:H368–H374.

Kaneko M, Elimban V, Dhalla NS (1989b): Mechanism for depression of heart sarcolemmal Ca^{2+} pump by oxygen free radicals. *Am J Physiol* 257:H804–H811.

Kerber RE, Shasby DM, Seabold J, Kieso R, Fox-Eastham K (1989): Does reduction of leukocyte accumulation in reperfused myocardium affect stunning?

Circulation 80(suppl II):II–410 (abstract).

Kim M-S, Akera T (1987): O_2 free radicals: cause of ischemia-reperfusion injury to cardiac Na^+-K^+-ATPase. *Am J Physiol* 252:H252–H257.

Klein HH, Pich S, Lindert S, Nebendahl K, Niedmann P, Kreuzer H (1989): Combined treatment with vitamins E and C in experimental myocardial infarction in pigs. *Am Heart J* 118:667–673.

Klein HH, Pich S, Schuff–Werner P, Niedmann P, Blattmann U, Nebendahl K (in press): Trolox, a water-soluble vitamin E analogue, accelerates functional recovery but does not reduce infarct size in regionally ischemic, reperfused porcine hearts. *Am Heart J.*

Koerner JE, Anderson BA, Dage RC (1991): Protection against postischemic myocardial dysfunction in anesthetized rabbits with scavengers of oxygen-derived free radicals: superoxide dismutase plus catalase, *N*-2-mercaotopropionyl glycine and captopril. *J Cardiovasc Pharmacol* 17:185–191.

Kramer JH, Mak IT, Weglicki WB (1984): Differential sensitivity of canine cardiac sarcolemmal and microsomal enzymes to inhibition by free radical-induced lipid peroxidation. *Circ Res* 55:120–124.

Krause SM, Jacobus WE, Becker LC (1989): Alterations in cardiac sarcoplasmic reticulum calcium transport in the postischemic "stunned" myocardium. *Circ Res* 65:526–530.

Kusuoka H, Marban E (in press). Cellular mechanisms of myocardial stunning. *Annu Rev Physiol.*

Kusuoka H, Porterfield JK, Weisman HF, Weisfeldt ML, Marban E (1987): Pathophysiology and pathogenesis of stunned myocardium. Depressed Ca^{2+} activation of contractions as a consequence of reperfusion-induced cellular calcium overload in ferret hearts. *J Clin Invest* 79:950–961.

Leiboff MS, Arroyo CM, Schaer GL, Mergner GW, Kramer JH, Miller DL, Visner MS, Weglicki WB (1988): Free radical generation in an *in vivo* model of regional myocardial stunning. *FASEB J* 2:A818 (abstract).

Lesnefsky EJ, Dauber IM, Horwitz LD (1991): Myocardial sulfhydryl pool alterations occur during reperfusion after brief and prolonged myocardial ischemia *in vivo.* *Circ Res* 68:605–613.

Lucchesi BR, Mullane KM (1986): Leukocytes and ischemia-induced myocardial injury. *Annu Rev Pharmacol Toxicol* 26:201–224.

Marban E, Kitakaze M, Koretsune Y, Yue DT, Chacko VP, Pike MM (1990): Quantification of $[Ca^{2+}]_i$ in perfused hearts: critical evaluation of the 5F-BAPTA and nuclear magnetic resonance method as applied to the study of ischemia and reperfusion. *Circ Res* 66:1255–1267.

McCord JM (1985): Oxygen-derived free radicals in postischemic tissue injury. *N Engl J Med* 312:159–163.

Menasche P, Grousset C, Gauduel Y, Piwnica A (1986): A comparative study of free radical scavengers in cardioplegic solutions. Improved protection with peroxidase. *J Thorac Cardiovasc Surg* 92:264–271.

Miki S, Ashraf M, Salka S, Sperelakis N (1988): Myocardial dysfunction and ultrastructural alterations mediated by oxygen metabolites. *J Mol Cell Cardiol* 20:1009–1024.

Murphy JG, Smith TW, Marsh JD (1988): Mechanisms of reoxygenation-induced calcium overload in cultured chick embryo heart cells. *Am J Physiol* 254: H1133–H1141.

Murry CE, Richard VJ, Jennings RB, Reimer KA (1989): Free radicals do not cause myocardial stunning after four 5 minute coronary occlusions. *Circulation* 80(suppl II):II-296 (abstract).

Muxfeldt M, Schaper W (1987): The activity of xanthine oxidase in hearts of pigs, guinea pigs, rats, and humans. *Basic Res Cardiol* 82:486–492.

Myers ML, Bolli R, Lekich RF, Hartley CJ, Roberts R (1985): Enhancement of recovery of myocardial function by oxygen free-radical scavengers after reversible regional ischemia. *Circulation* 72:915–921.

Myers ML, Bolli R, Lekich RF, Hartley CJ, Michael LH, Roberts R (1986a): *N*-2-mercaptopropionylglycine improves recovery of myocardial function following reversible regional ischemia. *J Am Coll Cardiol* 8:1161–1168.

Myers CL, Weiss SJ, Kirsh MM, Shepard BM, Shlafer M (1986b): Effects of supplementing hypothermic crystalloid cardioplegic solutions with catalase, superoxide dismutase, allopurinol, or deferoxamine on functional recovery of globally ischemic and reperfused isolated hearts. *J Thorac Cardiovasc Surg* 91:281–289.

Nejima J, Knight DR, Fallon JT, Uemura N, Manders WT, Canfield DR, Cohen MV, Vatner SF (1989): Superoxide dismutase reduces reperfusion arrhythmias but fails to salvage regional myocardial function or myocardium at risk in conscious dogs. *Circulation* 79:143–153.

O'Neill PG, Charlat ML, Kim H-S, Pocius J, Michael LH, Hartley CJ, Roberts R, Bolli R (1987): Lipoxygenase inhibitor nafazatrom fails to attenuate post-ischemic ventricular dysfunction. *Cardiovasc Res* 21:755–760.

O'Neill PG, Charlat ML, Michael LH, Roberts R, Bolli R (1989): Influence of neutrophil depletion on myocardial function and flow after reversible ischemia. *Am J Physiol* 256:H341–H351.

Patel BS, Jeroudi MO, O'Neill PG, Roberts R, Bolli R (1990): Effect of human recombinant superoxide dismutase on canine myocardial infarction. *Am J Physiol* 258:H369–H380.

Przyklenk K, Kloner RA (1986): Superoxide dismutase plus catalase improve contractile function in the canine model of the "stunned" myocardium. *Circ Res* 58:148–156.

Przyklenk K, Kloner RA (1989): "Reperfusion injury" by oxygen-derived free radicals? Effect of superoxide dismutase plus catalase, given at the time of reperfusion, on myocardial infarct size, contractile function, coronary microvasculature, and regional myocardial blood flow. *Circ Res* 64:86–96.

Przyklenk K, Whittaker P, Kloner RA (1990): *In vivo* infusion of oxygen free radical substrates causes myocardial systolic, but not diastolic dysfunction. *Am Heart J* 119:807–815.

Puett DW, Forman MB, Cates CU, Wilson BH, Handle KR, Friesinger GC, Virmani R (1987): Oxypurinol limits myocardial stunning but does not reduce infarct size after reperfusion. *Circulation* 76:678–686.

Reeves JP, Bailey CA, Hale CC (1986): Redox modification of sodium-calcium exchange activity in cardiac sarcolemmal vesicles. *J Biol Chem* 261:4948–4955.

Renlund DG, Gerstenblith G, Lakatta EG, Jacobus WE, Kallman CH, Weisfeldt ML (1984): Perfusate sodium during ischemia modifies postischemic functional and metabolic recovery in the rabbit heart. *J Mol Cell Cardiol* 16:795–801.

Romaschin AD, Rebeyka I, Wilson GJ, Mickle DAG (1987): Conjugated dienes in ischemic and reperfused myocardium: an *in vivo* chemical signature of oxygen free radical mediated injury. *J Mol Cell Cardiol* 19:289–302.

Rowe GT, Manson NH, Caplan M, Hess ML (1983): Hydrogen peroxide and hydroxyl radical mediation of activated leukocyte depression of cardiac sarcoplasmic reticulum: participation of the cyclooxygenase pathway. *Circ Res* 53:584–591.

Schott RJ, Nao BS, McClanahan TB, Simpson PJ, Stirling MC, Todd RF III, Gallagher KP (1989): $F(ab')_2$ Fragments of anti-mo1 (904) monoclonal antibodies do not prevent myocardial stunning. *Circ Res* 65:1112–1124.

Schrier GM, Hess ML (1988): Quantitative identification of superoxide anion as a negative inotropic species. *Am J Physiol* 24:H138–H143.

Sekili S, Li XY, Zughayb M, Sun JZ, Bolli R (1991): Evidence for a major pathogenetic role of hydroxyl radical in myocardial "stunning" in the conscious dog. *Circulation* 84:II-656.

Shattock MJ, Manning AS, Hearse DJ (1982): Effects of hydrogen peroxide on cardiac function and postischemic functional recovery in the isolated "working" rat heart. *Pharmacology* 24:118–122.

Shea MJ, Simpson PJ, Wernsm SW, Buda AJ, Alborzy-Khaghany AS, Hoff PT, Lucchesi BR (1987): Effect of neutrophil depletion on recovery of "stunned" myocardium. *Clin Res* 35:327A (abstract).

Shlafer M, Kane PF, Kirsh MM (1982a): Superoxide dismutase plus catalase enhances the efficacy of hypothermic cardioplegia to protect the globally ischemic, reperfused heart. *J Thorac Cardiovasc Surg* 83:830–839.

Shlafer M, Kane PF, Wiggins VY, Kirsh MM (1982b): Possible role for cytotoxic oxygen metabolites in the pathogenesis of cardiac ischemic injury. *Circulation* 66(suppl I):I-85–I-92.

Stewart JR, Blackwell WH, Crute SL, Loughlin V, Greenfield LJ, Hess ML (1983): Inhibition of surgically induced ischemia/reperfusion injury by oxygen free radical scavengers. *J Thorac Cardiovasc Surg* 86:262–272.

Tani M (1990): Effects of anti-free radical agents on Na^+, Ca^{2+}, and function in reperfused rat hearts. *Am J Physiol 259 (Heart Circ Physiol* 28):H137–H143.

Thompson JA, Hess ML (1986): The oxygen free radical system: A fundamental mechanism in the production of myocardial necrosis. *Prog Cardiovasc Dis* 28:449–462.

Triana JF, Jamaluddin U, Li XY, Bolli R (1990): Oxygen free radicals cause myocardial stunning after repetitive ischemia. *Circulation* 82:III-36 (abstract).

Triana JF, Li XY, Jamaluddin U, Thornby JI, Bolli R (1991): Postischemic myocardial "stunning": identification of major differences between the open-chest and the conscious dog and evaluation of the oxy-radical hypothesis in the latter model. *Circ Res* in press.

Weisel RD, Mickle DAG, Finkle CD, Tumiati LC, Madonik MM, Ivanov J, Burton GW, Ingold KU (1989): Myocardial free-radical injury after cardioplegia. *Circulation* 80(suppl III):III-14–III-18.

Werns S, Ventura A, Li GC, Lucchesi BR (1989): Amflutizole, a xanthine oxidase inhibitor, does not attenuate myocardial stunning in the canine heart. *Circulation* 80(suppl II):II-295 (abstract).

Westlin W, Mullane KM (1988): Does captopril attenuate reperfusion-induced myocardial dysfunction by scavenging free radicals? *Circulation* 77(suppl I):I-30–I-39.

Westlin W, Mullane KM (1989): Alleviation of myocardial stunning by leukocyte and platelet depletion. *Circulation* 80:1828–1836.

Ytrehus K, Gunnes S, Myklebust R, Mjos OD (1987): Protection by superoxide dismutase and catalase in the isolated rat heart reperfused after prolonged cardioplegia: a combined study of metabolic, functional and morphometric ultrastructural variables. *Cardiovasc Res* 21:492–499.

Ytrehus K, Myklebust R, Mjos OD (1986): Influence of oxygen radicals generated by xanthine oxidase in the isolated perfused rat heart. *Cardiovasc Res* 20:597–603.

Zhu WX, Myers ML, Hartley CJ, Roberts R, Bolli R (1986): Validation of a single crystal for measurement of transmural and epicardial thickening. *Am J Physiol* 251:H1045–H1055.

Zughaib M, Sekili S, Li XY, Triana JF, McCay PB, Bolli R (1991): Detection of free radical generation in the "stunned" myocardium in the conscious dog using spin trapping techniques. *FASEB J* 5:A704.

Chapter 13

Oxygen Free Radicals in the Pathophysiology of Myocardial Ischemia/Reperfusion

James N. Weiss, Joshua I. Goldhaber, and Sen Ji

In the setting of an acute myocardial infarction, there is a limited time window after the onset of coronary artery occlusion after which reperfusion is associated with irreversible cardiac injury. Cellular morphology by light and electron microscopy may be only mildly abnormal immediately before reperfusion, but becomes markedly distorted with sarcolemmal disruption, swollen mitochondria, and contraction band necrosis within minutes of reperfusion (Jennings and Ganote, 1974). It has been traditionally held that the major component of this irreversible injury occurs during the ischemic period itself, and that reperfusion only unmasks the latent damage (Braunwald and Kloner, 1985). However, it has become apparent that the conditions of reperfusion can significantly influence ultimate recovery of function (Buckberg, 1986; Kloner et al., 1989), consistent with the possibility that the act of reperfusion may convert potentially reversible ischemic injury into irreversible damage. In addition to this type of irreversible reperfusion injury, reperfusion after shorter durations of ischemia is associated with reversible myocardial dysfunction, or stunning, in which contractile and metabolic abnormalities may persist for hours to weeks before eventually recovering (Braunwald and Kloner, 1982). It has been shown that modifying the conditions of reperfusion can also reduce the severity of stunning (Kitakaze et al., 1988). Reperfusion can also produce lethal arrhythmias, the causes of which are not well understood (Manning and Hearse, 1984) but appear to involve predominantly nonreentrant mechanisms (Pogwizd and Corr, 1987).

Oxygen Free Radicals in Tissue Damage
Merrill Tarr and Fred Samson, Editors

With the growing appreciation of the cytotoxic actions of oxygen free radicals (OFR) and the observation that conditions favoring OFR accumulation may occur during ischemia at low oxygen tensions and especially during reperfusion with the sudden reexposure to high oxygen tensions, OFRs have become an attractive candidate to explain a component of reperfusion abnormalities in heart. Defining the role of OFRs in myocardial stunning, irreversible injury, and reperfusion arrhythmias is of more than academic interest as methods of achieving reperfusion have assumed increasing importance in the clinical management of acute myocardial infarction.

Evidence that OFRs play a role in reperfusion abnormalities in heart comes from three groups of observations: OFR levels are transiently elevated during reperfusion and possibly during ischemia, inhibition of OFR accumulation during ischemia/reperfusion has been shown to reduce the severity of reperfusion abnormalities under some conditions, and exposure of the heart to OFR-generating systems has deleterious effects on cardiac function resembling reperfusion abnormalities. However, many controversies remain and the hypothesis that OFRs play a major role in reperfusion abnormalities is not universally accepted. Key issues that need further clarification are: a) what are the important pathways of OFR generation during ischemia and reperfusion in heart, b) which species of OFRs cause tissue injury, c) how does the site of generation of OFRs (e.g., extracellular vs. intracellular) affect their ability to injure cellular organelles, d) which organelles are the primary targets of injury by OFRs, e) is there a specific time window during ischemia/reperfusion when the heart is maximally susceptible to OFR-induced injury vs. injury from other mechanisms, f) how do OFR scavengers confined to the extracellular space (e.g., superoxide dismutase) protect cytoplasmic organelles from injury, g) what is the relative importance of OFR-induced injury to the vasculature vs. the myocardium, and h) how do the levels and duration of exposure to OFRs occurring during ischemia/reperfusion relate to those causing myocardial dysfunction, injury, and arrhythmias during exposure to exogenous OFR-generating systems? This chapter briefly reviews some of these issues. The major focus is on the specific mechanisms by which OFRs alter cardiac function.

OFR Production During Ischemia and Reperfusion

Several techniques have been used to detect OFR production during myocardial ischemia and reperfusion. Electron paramagnetic resonance

(EPR) is the most common technique, and has been used to measure OFR levels directly, in freeze-clamped tissue samples from ischemic and reperfused heart. Although there has been some controversy over the possibility that tissue preparation methods may lead to artifacts (Baker et al., 1988; Zweier et al., 1989), Zweier et al. (1987) found that OFR tissue concentration in isolated rabbit hearts increased from a control level of 0.9 ± 0.2 μM to 2.8 ± 0.4 μM after 10 min of ischemia and to 6.8 ± 0.3 μM after reperfusion. OFR levels remained elevated for more than 60 min. EPR can also be used in conjunction with spin trap agents such as α-phenyl-N-tertbutylnitrone (PBN), which react with the otherwise short-lived free radicals, allowing the production of stable PBN spin adducts to be monitored continuously and nondestructively in venous effluent. These studies have documented increased OFR production during reperfusion, but have not consistently provided evidence of OFR accumulation during the ischemic period itself (Arroyo et al., 1987; Bolli et al., 1988; Garlick, 1987). Tissue chemiluminescence, either unenhanced or enhanced with substances that emit light when oxidized, is another nondestructive technique that has recently been developed to monitor tissue free radical production. Studies with this technique have detected increased OFR production during reperfusion, but not during the period of ischemia (Henry et al., 1990). Finally, indirect methods, such as monitoring malondialdehyde and conjugated diene formation as an index of OFR-initiated lipid peroxidation, have been used to infer the presence of cytotoxic OFR levels.

A limitation of all of these methods is their inability to localize sites of origin of OFRs. Both intracellular and extracellular pathways of OFR exist in the myocardium. Normally about 5% of myocardial O_2 metabolism involves the univalent reduction of molecular oxygen resulting in the formation of superoxide anion radicals ($\cdot O_2^-$) (Thompson and Hess, 1986). To deal with this high level of oxidative stress, an endogenous system of free radical scavengers is present in heart and other tissues. Superoxide dismutase (SOD) dismutes $\cdot O_2^-$ to form hydrogen peroxide, which is then converted to H_2O and O_2 by peroxidases such as glutathione peroxidase, or by catalases in some noncardiac tissues. The endogenous free radical scavenger system is desgined to prevent the formation of the hydroxyl free radical $\cdot OH$, which is generally believed to be much more directly cytotoxic than either $\cdot O_2^-$ or H_2O_2. $\cdot OH$ is generated from $\cdot O_2^-$ and H_2O_2 by a metal-catalyzed Haber-Weiss reaction in which $\cdot O_2^-$ reduces a transition metal ion (M) such as Fe^{3+} and the reduced metal then reacts with H_2O_2 by a Fenton reaction to form $\cdot OH$:

$$\cdot O_2^- + M^{n+} \rightarrow O_2 + M^{(n-1)+}$$

$$M^{(n-1)+} + H_2O_2 \rightarrow M^{n+} + \cdot OH + \cdot OH^-$$

Because $\cdot OH$ is so highly reactive, it is likely to react within a very close radius, on a molecular scale, to its site of formation near the transition metal. Thus, proteins containing iron, copper, or other transition metals (such as various mitochondrial enzymes) and lipid membranes in their vicinity may be selectively susceptible to damage by $\cdot OH$. Once $\cdot OH$ is formed, the heart has no direct scavenger mechanism to prevent it from causing cellular injury.

During ischemia and reperfusion, elevated OFR levels may result both from an increase in OFR production and the loss of endogenous free radical scavenging activity (Ferrari et al., 1983; Shlafer et al., 1987). OFR production may increase significantly by several mechanisms. In addition to the basal level of $\cdot O_2^-$ generation by mitochondria mentioned above, disruption of the mitochondrial electron transport system can result in autoxidation of flavoprotein and ubisemiquinone to form $\cdot O_2^-$ radicals. Elevated intracellular Ca^{2+} during ischemia may activate phospholipase C and stimulate arachidonic acid metabolism, whose intermediates also generate OFRs (Kuehl et al., 1980). Conversion of xanthine dehydrogenase to xanthine oxidase has also been documented during myocardial ischemia (Chambers et al., 1985), possibly caused by elevated cystolic Ca^{2+} (McCord, 1985). Xanthine oxidase catalyzes the conversion of xanthine or hypoxanthine and molecular oxygen to H_2O_2, $\cdot O_2^-$, and urea; this reaction is facilitated by the accumulation of hypoxanthine, an end-product of adenosine triphosphate (ATP) breakdown, in ischemic myocardium. However, whereas xanthine oxidase activity is high in the canine and rat heart, it is very low in the hearts of many other mammals including humans (de Jong et al., 1990). Whether a small pool located in endothelial cells may serve as a significant source of OFR production in human hearts remains an open question. Endothelial cells also produce free radicals such as NO$\cdot$ (a major component of endothelial-derived relaxing factor) that can react with $\cdot O_2^-$ to generate $\cdot OH$ in a metal-independent fashion (Beckman et al., 1990). In the extracellular compartment, autoxidation of catecholamines is another pathway for OFR production, and in the inflammatory response to ischemia, leukocytes generate large amounts of $\cdot O_2^-$ and H_2O_2 (Weiss, 1989). Thus, multiple mechanisms are likely to contribute to increased production of

OFRs during ischemia and reperfusion, and may be sufficient to exceed intrinsic antioxidant defenses of the myocardium that have been concurrently compromised by ischemia (Ferrari et al. 1983; Shlafer et al., 1987).

Evidence that Reducing OFR Accumulation During Ischemia/Reperfusion Reduces Reperfusion Abnormalities

The most compelling but perhaps also most controversial evidence that OFRs play a causative role in reperfusion abnormalities comes from studies in which the effects of reducing OFR accumulation during ischemia/reperfusion on postischemic cardiac function have been examined. Two approaches have been used to reduce accumulation of OFR during ischemia and reperfusion: increasing the antioxidant capacity of the myocardium with exogenous OFR scavengers, and inhibiting OFR production. Examples in the former group include SOD, polyethylene glycol-SOD (to prolong its half-life), catalase, dimethylthiourea, and N-(2-mercaptopropionyl)-glycine or mannitol (to scavenge ·OH) and α-tocopherol. Examples in the latter group are allopurinol and oxypurinol (inhibitors of xanthine oxidase), desferrioxamine (an iron chelator to prevent metal-catalyzed ·OH production), inhibitors of complement activation (to reduce chemoattraction and activation of neutrophils), and leukocyte depletion. Various studies have focused on different forms of reperfusion abnormalities: stunning, irreversible injury, and reperfusion arrhythmias. Several recent reviews have discussed these data in detail (e.g., Bolli, 1988, 1990; Kloner et al., 1989). In general, most studies have found the OFR scavengers and inhibitors had a protective effect during ischemic/reperfusion, although negative results have been obtained in enough studies to leave the issue controversial. In evaluating these discrepant results, however, a number of factors must be kept in mind. It is highly likely that experimental factors such as the species of animal, duration of ischemia, cardiac workload, composition of the coronary perfusate, and the timing of the delivery of OFR scavengers or inhibitors during ischemia/reperfusion may be critical determinants of the contribution that OFR, as opposed to other factors, make to reperfusion abnormalities. Even seemingly minor experimental differences could have significant effects on the outcome of experiments testing whether OFR scavengers or inhibitors have cardioprotective effects. In many studies, direct documentation that a given regimen of OFR scavengers or inhibitors reduced OFR accumulation was not obtained. Even if it was documented that OFR accumulation was reduced, the reduction may not necessarily have occurred in the critical pathway causing cytotoxicity. Some OFR scavengers or inhibitors may primarily affect OFR accumulation originating from only one pathway (e.g., allopurinol or leukocyte depletion) or in one compartment (e.g., SOD and catalase remain extracellular). On the other hand, the possibility that OFR scavengers or inhibitors may have nonspecific cardioprotective effects unrelated to their effects on OFR accumulation must also be considered. The time after reperfusion at which recovery of cardiac function is assessed may also be very important. Several studies have shown that early beneficial effects of OFR scavengers or inhibitors on myocardial stunning, irreversible injury, or reperfusion arrhythmias were not maintained at later time points (Buchwald et al., 1987; Tanaka et al., 1990; Yamada et al., 1990). The importance of OFR scavengers or inhibitors in preventing injury to the vasculature vs. myocardium is also not settled. It has been shown that reperfusion causes abnormalities in vascular function ranging from defective EDRF production (probably the vascular endothelial equivalent of stunning) to irreversible injury (no reflow). Both abnormalities have been shown to be reduced by OFR scavengers (Lefer et al., 1990; Przyklenk and Kloner, 1989). The extent to which vascular abnormalities contribute to the ultimate postischemic cardiac dysfunction and injury is not resolved.

Mechanisms of OFR-Induced Cardiac Injury

General mechanisms of OFR-induced injury. OFR-induced injury results from the interaction of highly reactive OFR species with fatty acids, causing lipid peroxidation of membranes, and proteins, causing oxidation of amino acids and sulfhydryl groups and destruction of polypeptide chains. As noted above, ·OH is thought to be the most important species responsible for cytotoxicity, based on its known chemical properties and the general but not universal observation that ·OH scavengers and inhibitors that have no effect on $\cdot O_2^-$ or H_2O_2 accumulation have approximately similar cardioprotective effects as OFR scavengers and inhibitors that reduce accumulation of all three OFR species. Because of the extremely high reactivity of ·OH, ·OH-induced injury is likely to be site specific, localized initially to organelles in which transition metals such as iron or copper are present to catalyze its formation. However, as destruction of these organelles causes release of transition metals into the cytoplasm, more generalized injury may result. In contrast, $\cdot O_2^-$ and H_2O_2 have longer lifetimes, and even if generated extracellularly, may readily permeate cell membranes, directly in the case of H_2O_2 and via anion channels in the case of $\cdot O_2^-$ (Henry et al., 1990). This may explain the ability of extracellular generated OFRs (e.g., from leukocytes) to cause intracellular injury to myocytes, as well as the effectiveness of OFR scavengers confined to the extracellular space, such as SOD and catalase, to reduce injury to intracellular organelles. By neutralizing extracellular $\cdot O_2^-$ and H_2O_2, these scavengers create a gradient-favoring efflux of intracelllar $\cdot O_2^-$ and H_2O_2 to the extracellular space, reducing their intracellular concentrations below the antioxidant threshold. This idea is supported by recent studies in which OFR accumulation was monitored with enhanced chemiluminescent in isolated Langendorff-perfused rat hearts subjected to ischemia and reperfusion (Henry et al., 1989, 1990). These investigators found that SOD attenuated the increase in chemiluminescence during reperfusion under control conditions, but not in the presence of the anion channel blocker DIDS. They interpreted these results as indicating that the chemiluminescence signal arose predominantly from the intracellular OFRs and that the ability of SOD to scavenge $\cdot O_2^-$ required its influx to the extracellular compartment via anion channels. It is intriguing to speculate that manipulating anion channels might provide another method of modulating OFR-induced myocardial injury. In principle, blocking cardiac anion channels during ischemia and reperfusion could have either a cardioprotectivce effect, by inhibiting influx of extracellularly generated $\cdot O_2^-$ or a deleterious effect, by inhibiting efflux of intracellularly generated $\cdot O_2^-$. It is interesting to note that Cl^- channels in heart are highly regulated by beta receptor stimulation, which is enhanced during ischemia and early reperfusion. Whether a component of the cardioprotective effects of beta receptor blockade during ischemia/reperfusion may be related to reduced activation of Cl^- channels is an interesting but unanswered question. Beta blockers have been shown to have a protective effect against OFR-induced injury (Mak et al., 1990), although the mechanism may be related to a direct inhibitory effect on lipid peroxidation by OFRs (Mak and Weglicki, 1988).

Intracellular Ca^{2+} and the pathophysiology of reperfusion abnormalities. To understand the potential link between OFR accumulation and reperfusion abnormalities, it is helpful to review briefly some aspects of the pathophysiology of these abnormalities. Myocardial stunning is characterized by two features: reversible depression of contractile performance and depletion of tissue high energy phosphates (for review see Bolli, 1990). The cause of metabolic depression is unknown, but is presumed to reflect direct injury to the cellular metabolic machinery and possibly loss of adenine nucleotide precursors. The depressed contractile function does not directly result from the metabolic defect, since the contractile response of stunned myocardium to beta adrenergic stimulation is similar to that of normal myocardium despite the depressed levels of tissue energy phosphates (Becker et al., 1986; Bolli et al., 1985). Rather, a primary defect in excitation-contraction coupling has been identified in which intracellular Ca^{2+} transients are normal but myofilament Ca^{2+} sensitivity and maximal Ca^{2+}-activated force are reduced (Kusuoka et al., 1987, 1990). It has been hypothesized that this defect may result from elevated intracellular Ca^{2+} levels during early ischemia and reperfusion, since contractile dysfunction indistinguishable from stunning could be produced by inducing transient intracellular Ca^{2+} overload with elevated $[Ca^{2+}]_o$ in the absence of ischemia/reperfusion. In addition, reducing $[Ca^{2+}]_o$ at the time of reperfusion attenuated stunning (Kitakaze et al., 1988). Inability to regulate intracellular Ca^{2+} levels is also a hallmark of irreversible reperfusion injury, in which massive intracellular Ca^{2+} overload is the final common pathway leading to cell death. The role of intracellular Ca^{2+} overload in reperfusion arrhythmias is not well understood. However, mapping studies indicate that the mechanism of reperfusion arrhythmias is typically nonreentrant in nature (Pogwizd and Corr, 1987), suggesting that triggered activity or other forms of automaticity may be commonly involved. The ability of elevated intracellular Ca^{2+} to promote triggered activity and other forms of automaticity is well known, and could be an important factor predisposing the heart to the development of reperfusion arrhythmias. Thus, a defect in the intracellular Ca^{2+} regulation is a common theme, although undoubtedly not the sole etiologic factor, in the various forms of reperfusion abnormalities in heart.

OFRs and intracellular Ca^{2+} regulation. Evidence that OFRs interfere with intracellular Ca^{2+} regulation in heart is now substantial. In isolated intact heart preparations, exposure to OFR-generating systems stimulated many aspects of irreversible reperfusion injury, including persistent cellular K^+ loss, APD shortening leading to inexcitability, loss of systolic force development, a progressive and irreversible rise in diastolic tension (contracture), and depressed metabolic function (Goldhaber et al., 1989). Recently, Corretti et al. (1991) have directly demonstrated using ^{19}F-BAPTA NMR spectroscopy that OFR-induced contractile and metabolic changes were accompanied by a prominent rise in intracellular Ca^{2+} that preceded major depletion of high energy phosphates. Importantly, these investigators also established with EPR spectroscopy that the levels of OFRs produced by their OFR-generating system (a 4-min exposure to H_2O_2 and Fe^{3+}-chelate) were comparable to those occurring during ischemia/reperfusion in the same animal model. In these studies, exposure to exogenous OFR-generating systems caused irreversible injury analogous to irreversible reperfusion injury. Whether briefer, less intense exposure to OFRs would have caused reversible cardiac depression analogous to stunning is unknown. However, OFR-generating systems have been shown to depress myofibrillar ATPase activity in some studies (Ventura et al., 1985), which could partially account for the abnormal myofilament Ca^{2+} responsiveness of stunned myocardium. Similar results documentaing intracellular Ca^{2+} overload in response to exogenous OFR-generating systems have been obtained in isolated myocytes (Burton et al., 1990; Josephson et al., 1991). As diastolic Ca^{2+} overload developed in isolated myocytes, spontaneous intracellular Ca^{2+} oscillations and aftercontractions occurred (Josephson et al., 1991). It is likely that similar changes in intracellular Ca^{2+} were the cause of OFR-induced afterdeopolarizations and triggered activity observed in isolated myocytes in other studies (Barrington et al., 1988). Furthermore, OFR-generating systems have been documented to cause similar arrhythmias in intact heart preparations (Hearse et al., 1989; Manning et al., 1988; Nakaya et al., 1987; Pallandi et al., 1987). These results are all consistent with the hypothesis that intracellular Ca^{2+} overload is an important factor underlying the development of OFR-induced arrhythmias, although direct effects of OFRs on arrhythmogenic ionic currents may also be important. OFRs have been reported to affect several ionic currents, causing accelerated rundown of the Ca^{2+} current (Goldhaber et al., 1989; Tarr and Valenzeno, 1991), activation of the ATP-sensitive K^+ current (Goldhaber et al., 1989), depression of other K^+ currents including the delayed outward rectifier I_K (Tarr and Valenzeno, 1991) and inward rectifier I_{K1} (Coetzee and Opie, 1988), and alterations in the Na^+ current (Bhatnagar et al., 1990; Tarr and Valenzeno, 1991) in isolated myocytes.

What are the mechanisms by which OFR-generating systems cause abnormal cellular Ca^{2+} regulation in heart? OFR-induced intracellular Ca^{2+} overload does not appear to result from a nonspecific increase in membrane leakiness, since myocytes remained impermeable to macromolecules when the initial increase in intracellular Ca^{2+} occurred (Josephson et al., 1991). Increased Ca^{2+} entry via L-type Ca^{2+} channels is also not responsible, since the L-type Ca^{2+} current was depressed or unchanged by OFR-generating systems (Bhatnagar et al., 1990; Goldhaber et al., 1989). Although Ca^{2+} channel blockers have been reported to have a protective effect against OFR-induced intracellular Ca^{2+} overload and injury in isolated myocyctes (Josephson et al., 1990), this could be attributed to a reduction in the overall Ca^{2+} burden to the heart rather than specific inhibition of the Ca^{2+} current. Considerable evidence points to impaired function of the sarcoplasmic reticulum. In heart, the sarcoplasmic reticulum (SR) is the intracellular Ca^{2+} storage and release site for most of the Ca^{2+} which activates contraction. Ca^{2+} release from the SR occurs through Ca^{2+} release channels that are triggered to open during excitation by the influx of extracellular Ca^{2+} predominantly through sarcolemmal L-type Ca^{2+} channels. Ca^{2+} is then resequestered by a Ca^{2+}-ATPase located in the SR membrane and stored in the SR for re-release with the next beat. The SR Ca^{2+}-ATPase is quantitatively the major mechanism by which Ca^{2+} is removed from the cytoplasm after each individual beat. Since the capacity of the SR to store Ca^{2+} is limited, however, over the course of several beats other processes such as Na^+-Ca^{2+} exchange, the sarcolemmal Ca^{2+} pump and mitochondrial Ca^{2+} sequestration may have important effects on intracellular Ca^{2+}. It has been shown in isolated SR vesicles that Ca^{2+} uptake is depressed after exposure to OFR-generated systems because of depression of the SR Ca^{2+}-ATPase (Rowe et al., 1983) and/or increased passive Ca^{2+} permeability (Okabe et al., 1988) involving phospholamban- (Morris et al., 1990) and calmodulin-dependent processes (Okabe et al., 1989a). This is consistent with the finding in intact patch-clamped isolated myocytes loaded with FURA-2 that caffeine-induced intracellular Ca^{2+} transients, an index of SR Ca^{2+} content, were depressed after exposure to H_2O_2 (Goldhaber et al., 1991). The Na^+-Ca^{2+} exchange current during the caffeine-induced intracellular Ca^{2+} transients remained intact. However, OFR-generated systems have been reported to have both depressant (Dixon et al., 1990; Reeves et al., 1986) and stimulatory effects (Okabe et al., 1989b; Shi et al., 1989) on Na^+-Ca^{2+} exchange in isolated sarcolemmal vesicles. Na^+-K^+-ATPase has been reported to be depressed by

OFR-generating systems (Kim and Akera, 1987; Kukreja et al., 1990), and an increase in intracellular Na^+ by this mechanism could indirectly inhibit the ability of Na^+-Ca^{2+} exchange to extrude Ca^{2+} from the cytoplasm. OFR-generated systems have also been reported to impair the sarcolemmal Ca^{2+} ATPase (Kaneko et al., 1989), although it is unlikely that this would have any major effect on intracellular Ca^{2+} regulation, since current evidence suggests that this mechanism of Ca^{2+} removal is probably minor in mammalian ventricle.

The effects of OFR-generating systems on cellular metabolism deserve special mention in relation to intracellular Ca^{2+} regulation. OFRs have been shown to impair the function of isolated mitochondria (Malis and Bonventre, 1986; Richter and Frei, 1988) and to depress anaerobic glycosis by inhibiting glyceraldehyde 3-phosphate dehydrogenase (Hyslop et al., 1987). In isolated myocytes in which intracellular Ca^{2+} overload was prevented by buffering the cystol with high concentrations of EGTA, OFR-generating systems were shown to inhibit directly both glycolysis and oxidative phosphorylation (Goldhaber et al., 1989). Since intracellular Ca^{2+} overload can be a consequence of metabolic inhibition, the possibility must be considered that OFR-induced metabolic inhibition contributes to intracellular Ca^{2+} overload, in addition to any direct effects of OFRs on intracellular Ca^{2+} regulatory processes. This remains an open question at present, but it is interesting that in the NMR study by Corretti et al. (1991), intracellular Ca^{2+} overload in isolated rabbit hearts exposed to H_2O_2 + Fe^{3+} was preceded by a prominent increase in sugar phosphate compounds indicative of glycolytic inhibition. It has been previously speculated that glycosis may play a special role in intracellular Ca^{2+} regulation in heart and other tissues (Lynch and Paul, 1983; Paul et al., 1989; Weiss and Hiltbrand, 1985), and this will be an interesting area for future investigation.

It is important to state that OFR-induced impairment of intracellular Ca^{2+}-regulation is unlikely to be the sole factor responsible for intracellular Ca^{2+} overload during ischemia and reperfusion for several reasons. First, a significant rise in diastolic Ca^{2+} has been documented to occur during early ischemia (Lee et al., 1987; Marban et al., 1987; Steenbergen et al., 1987), and whether OFR accumulation occurs at all during this period is still controversial. Second, intracellular Na^+ accumulation during ischemia and/or initial reperfusion (Pike et al., 1990) may be a major factor leading to intracellular Ca^{2+} loading by promoting Ca^{2+} entry via Na-Ca exchange. The causes of intracellular Na^+ accumulation under these conditions are controversial, but roles for Na^+-H^+ exchange and/or Na^+-K^+ pump inhibition have been proposed.

Summary and Conclusions

The three forms of reperfusion abnormalities in heart are reversible myocardial stunning, irreversible injury, and reperfusion arrhythmmias. In addition to the pathological events occurring during the period of ischemia, evidence suggests that the act of reperfusion itself may contribute independently to the pathogenesis of these abnormalities. During ischemia, cellular antioxidant defenses are compromised and conditions favoring increased OFR production upon reperfusion occur. As a result, during initial reperfusion marked OFR accumulation occurs and may persist for hours. Whether significant OFR accumulation occurs during the period of ischemia itself remains controversial. The observation that OFR scavengers and inhibitors can ameliorate myocardial stunning, irreversible injury, and reperfusion arrhythmias under some conditions suggests that OFR accumulation can contribute significantly to these reperfusion abnormalities. From a pathophysiological standpoint, intracellular Ca^{2+} overload has been implicated to play a role in all three forms of reperfusion abnormalities. Likewise, OFR-generating systems have been shown to cause intracellular Ca^{2+} overload and to induce irreversible tissue injury and arrhythmias resembling those seen in postischemic hearts. The mechanism of OFR-induced impairment of cellular Ca^{2+} regulation is not completely understood, but may predominantly involve injury to the sarcoplasmic reticulum, although other Ca^{2+} regulatory processes may also be affected. OFR-generating systems also cause direct metabolic injury by inhibiting glycosis and oxidative phosphorylation, which may contribute to the prolonged metabolic defect in stunned myocardium and the lack of recovery of metabolic function in irreversibly damaged myocardium. A role of OFR-induced metabolic injury in contributing to intracellular Ca^{2+} overload is also possible. These many observations indicate that OFRs are likely to be an important etiologic factor in myocardial stunning, irreversible injury, and reperfusion arrhythmias. However, definitive proof of this hypothesis is still lacking.

Acknowledgments. Supported by NIH grants RO1 HL36729, RO1 HL 44880, and Research Career Development Award K04 HL01890 (to J.N.W.), by American Heart Association, Greater Los Angeles Affiliate Senior Fellowship 912 F1-1 (to J.I.G.) and Clinician-Scientist Award 921 Cs-2 (to J.I.G.), and by the Laubisch Cardiovascular Research Fund, the Kawata Endowment and the Flintridge Foundation.

References

Arroyo CM, Cramer JH, Dickens BF, Weglicki WB (1987): Identification of free radicals in myocardial ischemia/reperfusion by spin trapping with nitrone DMPO. *FEBS Lett* 221:101–104.

Baker JE, Felix CC, Olinger GN, Kalyanaraman B (1988): Myocardial ischemia and reperfusion: Direct evidence for free radical generation electron spin resonance spectroscopy. *Proc Natl Acad Sci USA* 85:2786–2789.

Barrington PL, Meier CF Jr., Weglicki WB (1988): Abnormal electrical activity induced by free radical generating systems in isolated cardiocytes. *J Mol Cell Cardiol* 20:1163–1178.

Becker LC, Levine JH, DiPaula AF, Guarnieri T, Aversano T (1986): Reversal of dysfunction in postischemic stunned myocardium by epinephrine and post-extrasystolic potentiation. *J Am Coll Cardiol* 7:580–589.

Beckman JS, Beckman TW, Chen J, Marshall PA, Freeman BA (1990): Apparent hydroxyl radical production by peroxynitrite: implications for endothelial injury from nitric oxide and superoxide. *Proc Natl Acad Sci USA* 87:1620–1624.

Bhatnagar A, Srivastava SK, Szabo G (1990): Oxidative stress alters specific membrane currents in isolated cardiac myocytes. *Circ Res* 67:525–549.

Bolli R (1988): Oxygen-derived free radicals and postischemic myocardial dysfunction (stunned myocardium). *J Am Coll Cardiol* 12:239–249.

Bolli R (1990): Mechanism of myocardial "stunning." *Circulation* 82:723–738.

Bolli R, Patel BS, Jeroudi MO, Lai EK, McCay PB (1987): Demonstration of free radical generation in "stunned" myocardium of intact dogs with the use of the spin trap α-phenyl-*n-tert*butyl nitrone. *J Clin Invest* 82:476–485.

Bolli R, Zhu WX, Myers ML, Hartley CJ, Roberts R (1985): Beta-adrenergic stimulation reverses postischemic myocardial dysfunction without producing subsequent functional deterioration. *Am J Cardiol* 56:964–968.

Braunwald E, Kloner RA (1982): The stunned myocardium: prolonged, post-ischemic ventricular dysfunction. *Circulation* 66:1146–1149.

Braunwald E, Kloner RA (1985): Myocardial reperfusion: a double-edged sword? *J Clin Invest* 76:1713–1719.

Buckberg GD (1986): Studies of controlled reperfusion after ischemia. I: when is cardiac muscle irreversibly damaged? *J Thorac Cardiovasc Surg* 92:483–487.

Buchwald A, Klein HH, Nebendahl K, Pich S, Kruezer H (1987): Effect of intracoronary superoxide dismutase on regional myocardial function after brief periods of ischemia. *Circulation* 76(suppl IV):IV-229 Abstract.

Burton KP, Morris AC, Massey KD, Buja LM, Hagler HK (1990): Free radicals alter ionic calcium levels and membrane phospholipids in culture rat ventricular myocytes. *J Mol Cell Cardiol* 22:1035–1047.

Chambers DE, Parks DA, Patterson G, et al. (1985): Xanthine oxidase as a source of free radical damage in myocardial ischemia. *J Mol Cell Cardiol* 17:145–152.

Coetzee WA, Opie LH (1988): Electrophysiological effects of free oxygen radicals on guinea pig ventricular myocytes. *J Mol Cell Cardiol* 21(suppl V):S.17.

Coretti MC, Koretsune Y, Kusuoka H, Chacko VP, Zweier JL, Marban E (1991): Glycolytic inhibition and calcium overload as consequences of exogenously-generated free radicals in rabbit hearts. *J Clin Invest* 88:1014–1025.

Dixon IM, Kaneko M, Hata T, Panagia V, Dhalla NS (1990): Alterations in cardiac membrane Ca^{2+} transport during oxidative stress. *Mol Cell Biochem* 99:125–133.

Ferrari R, Ceconi C, Curello S, Guarnieri, Calderera CM, Albertini A, Visioli O (1985): Oxygen-mediated myocardial damage during ischaemia and reperfusion: Role of the cellular defenses against oxygen toxicity. *J Mol Cell Cardiol* 17:937–945.

Garlick PB, Davies MJ, Hearse DJ, Slater TF (1987): Direct detection of free radicals in the reperfused rat heart using electron spin resonance spectroscopy. *Circ Res* 61:757–760.

Goldhaber JI, Ji S, Lamp ST, Weiss JN (1989): Effect of exogenous free radicals in electromechanical function and metabolism in isolated rabbit and guinea pig ventricle. *J Clin Invest* 83:1800–1809.

Goldhaber JI, Liu E, Weiss JN (1991): Effects of H_2O_2 on $[Ca^{2+}]_i$ and excitation-contraction coupling in isolated cardiac myocytes. *Circulation* 84(Suppl II):II-549.

Hearse DJ, Kusama Y, Bernier M (1989): Rapid electrophysiological changes leading to arrhythmias in the aerobic rat heart. *Circ Res* 65:146–153.

Henry TD, Archer SL, Nelson DP, Weir EK, From AHL (1989): Effect of anion channel blockade on oxygen radical induced chemiluminescence in postischemic myocardium. *Circulation* 80(Suppl. II):II-31.

Henry TD, Archer SL, Nelson D, Weir EK, From AHL (1990): Enhanced chemiluminescence as a measure of oxygen-derived free radical generation during ischemia and reperfusion. *Circ Res* 67:1453–1461.

Hyslop PA, Hinshaw DB, Halsey WA Jr., Schraufstatter IU, Sauerheber RD, Spraggs RG, Jakcson JH, Cochrane CG (1988): Mechanisms of oxidant-mediated cell injury. *J Biol Chem* 1665–1675.

Jennings RB, Ganote CE (1974): Structural changes in myocardium during acute ischemia. *Circ Res* 34/35(Suppl. III):III-156–III-171.

de Jong JWD, Meer PVD, Nieukoop AS, Huizer T, Stroeve RJ, Bos E (1990): Xanthine oxidoreductase activity in perfused hearts of various species, including humans. *Circ Res* 67:770–773.

Josephson RA, Silverman HS, Lakatta EG, Stern MD, Zweier JL (1991): Study of the mechanisms of hydrogen peroxide and hydroxyl free radical-induced cellular injury and calcium overload in cardiac myocytes. *J Biol Chem* 266:2354–2361.

Kaneko M, Beamish RE, Dhalla NS (1989): Depression of heart sarcolemmal Ca^{2+}-pump activity by oxygen free radicals. *Am J Physiol* 256:H368–H374.

Keuhl EA, Humes JL, Ham EA (1980): Inflammation: the role of peroxidase-derived products. *Adv Prostagland Thromb Res* 6:77–86.

Kim M-S, Akera T (1987): O_2 free radicals: cause of ischemia-reperfusion injury to cardiac Na^+-K^+-ATPase. *Am J Physiol* 252:H252–H257.

Kitakaze M, Weisman HF, Marban E (1988): Contractile dysfunction and ATP depletion after transient calcium overload in perfused ferret hearts. *Circulation* 77:685–695.

Kloner RA, Przyklenk K, Whittaker P (1989): Deleterious effects of oxygen radical in ischemia/reperfusion: resolved and unresolved issues. *Circulation* 80:1115–1127.

Kukreja RC, Weaver AB, Hess ML (1990): Sarcolemmal Na^+-K^+-ATPase: inactivation by neutrophil-derived free radicals and oxidants. *Am J Physiol* 259:H1330–H1336.

Kusuoka H, Koretsune Y, Chacko VP, Weisfeldt ML, Marban E (1989): Excitation-contraction coupling in postischemic myocardium. *Circ Res* 66:1268–1276.

Kusuoka H, Porterfield JK, Weisman HF, Weisfeldt ML, Marban E (1987): Pathophysiology and pathogenesis of stunned myocardium. *J Clin Invest* 79:950–961.

Lee H-C, Smith N, Mohabir R, Clusin WT (1987): Cystolic calcium transients from the beating mammalian heart. *Physiol Sci* 84:7793–7797.

Lefer DJ, Ma X, Johnson G, Lefer AM (1990): Endothelium preservation as a major mechanism of cardioprotection by superoxide dismutase in myocardial ischemia-reperfusion injury. *Circulation* 82(suppl III):III-170.

Lynch RM, Paul RJ (1983): Compartmentation of glycolytic and glycogenolytic metabolism in vascular smooth muscle. *Science* 222:1344–1346.

Mak IT, Kramer JH, Freedman AM, Tse SYH, Weglicki WB (1990): Oxygen radical-mediated injury of myocytes-protection. *J Mol Cell Cardiol* 22:687–695.

Mak IT, Weglicki WB (1988): Protection by β-blocking agents against free radical-mediated sarcolemmal lipid peroxidation. *Circ Res* 63:262–266.

Malis CD, Bonventre JV (1986): Mechanism of calcium potentiation of oxygen free radical injury renal mitochondria. *J Biol Chem* 261:14201–14208.

Manning A, Bernier M, Crome R, Little S, Hearse D (1988): Reperfusion-induced arrhythmias: a study of the role of xanthine oxidase-derived free radicals in the rat heart. *J Mol Cell Cardiol* 20:35–45.

Manning AS, Hearse DJ (1984): Reperfusion-induced arrythmias: mechanisms and prevention. *J Mol Cell Cardiol* 16:497–518.

Marban E, Kitakaze M, Kusuoka H, Porterfield JK, Yue DT, Chacko VP (1987): Intracellular free calcium concentration measured with ^{19}F NMR spectroscopy in intact ferret hearts. *Physiol Sci* 84:6005–6009.

McCord JM (1985): Oxygen-derived free radicals in postischemic tissue injury. *N Engl J Med* 312:159–163.

Morris TE, Prakash XV, Sulakhe PV (1990): Free radicals-induced rapid dysfunction of sarcoplasmic reticulum calcium pump in rat cardiomyocytes. *Circulation* 82:III-264.

Nakaya H, Tohse N, Kanno M (1987): Electrophysiological derangements induced by lipid peroxidation in cardiac tissue. *Am J Physiol* 253:H1089–H1097.

Okabe E, Odajima C, Taga R, Kukreja RC, Hess ML, Ito H (1988): The effect of oxygen free radicals on calcium permeability and calcium loading at steady state in cardiac sarcoplasmic reticulum. *Mol Pharmacol* 34:388–394.

Okabe E, Sugihara M, Tanaka K, Sasaki H, Ito H (1989a): Calmodulin and free oxygen radicals interaction with steady state-state calcium accumulation and passive calcium permeability of cardiac sarcoplasmic reticulum. *J Pharmacol Exper Ther* 250:286–292.

Okabe E, Fujimaki R, Murayama M, Ito H (1989b): Possible mechanism responsible for mechanical dysfunction of ischemic myocardium: a role of oxygen free radicals. *Jpn Circ J* 53:1132–1137.

Pallandi RT, Perry MA, Campbell TJ (1987): Proarrhythmic effects of an oxygen-derived free radical generating system on action potentials recorded from guinea pig ventricular myocardium: A possible cause of reperfusion-induced arrhythmias. *Circ Res* 61:50–54.

Paul RJ, Hardin CD, Raeymaekers L, Wuytack F, Casteels R (1989): Preferential suport of Ca^{2+} uptake in smooth muscle plasma membrane vesicles by an endogenous glycolytic cascade. *FASEB J* 3:2298–2301.

Pike MM, Kitakaze M, Marban E (1990): ^{23}Na-NMR measurements of intracellular sodium in intact perfused ferret hearts during ischemia and reperfusion. *Am J Physiol* 259:H1767–H1773.

Pogwizd SM, Corr PB (1987): Electrophysiologic mechanisms underlying arrhythmias due to a reperfusion of ischemic myocardium. *Circulation* 76:404–426.

Przyklenk K, Kloner RA (1989): "Reperfusion injury" by oxygen-derived free radicals? Effect of superoxide dismutase + catalase, given at time of reperfusion, on myocardial infarct size, contractile function, coronary microvasculature and regional myocardial blood flow. *Circ Res* 64:86–96.

Reeves JP, Bailey CA, Hale CC (1986): Redox modification of sodium-calcium exchange activity in cardiac sarcolemmal vesicles. *J Biol Chem* 261:4948–4955.

Richter C, Frei B (1987): Ca^{2+} release from mitochondria induced by prooxidants. *Free Rad Biol Med* 4:365–375.

Rowe GT, Manson NH, Caplan M, Hess ML (1983): Hydrogen peroxide and hydroxyl radical mediation of activated leukocyte depression of cardiac sarcoplasmic reticulum. *Circ Res* 53:584–591.

Shi ZQ, Davison AJ, Tibbits GF (1989): Effects of active oxygen generated by DDT/Fe^{2+} on cardiac Na^{+}/Ca^{2+} exchange and membrane permeability to Ca^{2+}. *J Mol Cell Cardiol* 21:1009–1016.

Shlafer M, Myers CL, Adkins S (1987): Miochondrial hydrogen peroxide generation and activities of glutathione peroxidase and superoxide dismutase following global ischemia. *J Mol Cell Cardiol* 19:1195–1206.

Steenbergen C, Murphy E, Levy L, London RE (1987): Elevation in cystolic free calcium contraction early in mitochondrial ischemia in perfused rat heart. *Circ Res* 60:700–707.

Tanaka M, Stoler RC, FitzHarris GP, Jennings RB, Reimer KA (1990): Evidence against the "early protection-delayed death" hypothesis of superoxide dismutase therapy in experimental myocardial infarction. *Circ Res* 67:636–644.

Tarr M, Valenzeno DP (1991): Modification of cardiac ionic currents by photosensitizer-generated reactive oxygen. *J Molec Cell Cardiol* 23:639–649.

Thompson JA, Hess ML (1986): The oxygen free radical system: a fundamental mechanism in the production of myocardial necrosis. *Prog Cardiovasc Dis* 28:449–462.

Ventura C, Guarnieri C, Caldarera CM (1985): Inhibitory effect of superoxide radicals on cardiac myofibrillar ATPase activity. *J Biochem* 34:267–274.

Weiss JN, Hiltbrand B (1985): Functional compartmentation of glycolytic versus oxidative metabolism in isolated rabbit heart. *J Clin Invest* 75:436–447.

Weiss SJ (1989): Tissue destruction by neutrophils. *N Eng J Med* 320:365–376.

Yamada M, Hearse DJ, Curtis MJ (1990): Reperfusion and readmission of oxygen. *Circ Res* 67:1211–1224.

Zweier JL, Flaherty JT, Weisfeldt ML (1987): Direct measurement of free radical generation following reperfusion of ischemic myocardium. *Proc Natl Acad Sci USA* 84:1404–1407.

Zweier JL, Kuppusamy P, Williams R, Rayburn BK, Smith D, Weisfeldt ML, Flaherty JT (1989): Measurement and characterization of postischemic free radical generation in the isolated perfused heart. *J Biol Chem* 264:18890–18895.

Chapter 14

Reactive Oxygen-Induced Modifications of Cardiac Electrophysiology: A Comparison of the Effects of Rose Bengal and other Reactive Oxygen Generators

Merrill Tarr and Dennis P. Valenzeno

Recently we and other investigators have used the reaction of light with the photosensitizer Rose Bengal (RB) as a tool to investigate the effects of reactive oxygen species (ROS) on cardiac tissue. These studies have provided new and useful information regarding ROS effects on the action potential and ionic currents in cardiac tissue. The favorable comparison (see below) of the cardiac electrophysiological modifications produced by photoactivation of RB to those produced by other ROS generators supports the conclusion that photosensitizers provide a biologically relevant method to evaluate ROS-induced cellular damage. What follows is a summary of the electrophysiological modifications observed by us and other investigators studying the effects that photoactivation of RB has on whole heart and isolated cardiac cells. These results are compared to those that others have obtained using different methods to generate ROS such as (a) dihydroxyfumaric acid (DHF), (b) xanthine oxidase (XO), (c) hydrogen peroxide (H_2O_2), and (d) organic hydroperoxides such as *tert*-butyl hydroperoxide (TBH) and cumene hydroperoxide (CH). The relevance of these studies to electrophysiological modifications in ischemic-reperfusion injury is also presented.

Both DHF and the reaction of XO with xanthine, hypoxanthine, or purine have often been used to generate the superoxide radical. However,

Oxygen Free Radicals in Tissue Damage
Merrill Tarr and Fred Samson, Editors

these methods will also produce the highly reactive hydroxyl radical since the dismutation of superoxide yields H_2O_2, which in the presence of iron yields the hydroxyl radical via the Fenton reaction. Hydroxyl radicals will also be formed after the use of H_2O_2 alone. However, DHF, XO, and H_2O_2 have been used extensively, since superoxide, H_2O_2, and the hydroxyl radical are thought to be important ROS biologically. Organic hydroperoxides such as CH and TBH initiate peroxidation of unsaturated membrane lipids and have been used on the assumption that ROS-induced cellular damage involves peroxidation of membrane lipids. By comparison, the photosensitizer RB is primarily a singlet oxygen generator, although it does produce some superoxide. Unfortunately, emphasis on superoxide-initiated reactions in many types of ROS-induced damage and singlet oxygen in photomodification-induced damage has left the impression that photomodification studies may produce results that are not relevant to other forms of ROS-induced membrane damage. We hope the following discussion will change this impression.

Arrhythmia Induction in Whole Heart

It is now recognized that the reintroduction of oxygen to ischemic or hypoxic cardiac tissue can produce arrhythmias within minutes. It is likely that a variety of events combine to increase the heart's susceptibility to arrhythmias during the early phase of reperfusion. However, the coincidence of the timing of a burst of ROS within the first minute of reperfusion with the onset of arrhythmias at these early times provides circumstantial evidence for ROS involvement in reperfusion-induced arrhythmias. The obvious question arises: Can ROS, by themselves, induce arrhythmias in nonischemic tissue?

Hearse et al. (1989) demonstrated that photomodification with RB (250 nM) produced rapid changes in the electrocardiogram (ECG) of isolated, aerobic rat hearts. In their study, rat hearts were perfused aerobically for 10 min without RB and for 5 min with RB. During this time no electrophysiological changes were observed. The hearts were then illuminated uniformly with green light via 200 fiberoptic cables and profound electrophysiological changes occurred within seconds. The initial changes (within 12 s) consisted of inversion of the terminal portion of the T wave and an increase in Q-T interval. This was followed by production of ventricular premature beats (within 2 min), ventricular tachycardia (within 3 min), and complete atrioventricular block (within

6 min). These results indicate that ROS can rapidly induce electrophysiological changes and arrhythmias in intact heart even in the absence of ischemia or reperfusion.

Nakaya et al. (1987) perfused isolated guinea pig hearts with the organic hydroperoxides CH or TBH. Both of these hydroperoxides altered electrical activity of the heart as assessed with bipolar electrograms (a localized ECG measurement). Conduction disturbances manifested by conduction delays and fractionation of the electrograms occurred after about 25 min of exposure to the hydroperoxides. Arrhythmias also occurred and ventricular fibrillation was induced by ventricular pacing. These changes in electrical activity occurred concomitantly with an increase in malondialdehyde (MDA) content of the myocardium, suggesting the involvement of lipid peroxidation.

Recently, Lesnefsky et al. (1991) reported that ventricular arrhythmias could be induced by a single extrastimulus during continuous pacing of intact rabbit ventricles perfused with H_2O_2. Extrastimulus-induced ventricular arrhythmias occurred in four of six hearts perfused with 10^{-5}M H_2O_2 compared to two of eight hearts exposed to 5×10^{-6}M H_2O_2 and zero of six hearts exposed to 10^{-6}M H_2O_2. No spontaneous arrhythmias occurred in any of these hearts.

A direct comparison of the arrhymthias induced by exposing healthy hearts to ROS to those associated with reperfusion of ischemic heart tissue is difficult, since ischemia itself alters cardiac electrical activity in a manner that may alter the nature of the arrhythmia. However, it is well known that ventricular tachycardia, accelerated idioventricular rhythms, ventricular premature beats, and rate-dependent fibrillation occur within the first minutes of reperfusion after an ischemic episode (Murdock et al. 1980; Pogwizd and Corr, 1987).

Action Potential Modification

Arrhythmia induction, by its very nature, must in some manner be related to alterations in the electrophysiological properties of the heart. One aspect of this is the waveform of the action potential. The question arises: Do ROS modify the action potential in a manner that could be arrhythmogenic?

We have investigated the effects that photomodification with RB has on the action potential of isolated frog atrials cells, and have found that they depend on (a) the location (extracellular or intracellular) of the photosensitizer, and (b) the intensity and duration of ROS exposure (Tarr and Valenzeno, 1989). Continuous ROS production using green light at an intensity of 1.5 mW/cm^2 and an extracellular RB concentration of 0.125 μM produced an initial increase in action potential duration (APD). However, after about 3 min of ROS exposure APD began to decrease rapidly and the action potential became spikelike at about 6 min. In contrast, when RB was applied only intracellularly via the patch pipette solution containing 0.125 μM RB, similar illumination produced only a decrease in APD. In both cases, however, the amplitude of the action potential decreased with time during ROS exposure. In a more recent study, Tarr and Valenzeno (1991) used a higher concentration of RB (0.5 μM) extracellularly, higher illumination intensity (6.5 mW/cm^2), and intermittent rather than continuous illumination with green light to produce ROS. The results of that study demonstrated that even very brief (< 10 s) ROS exposures can produce profound effects on action potential amplitude and duration, and also demonstrated that whether an increase or decrease in APD follows ROS exposure depends on both the intensity and duration of ROS production. For example, 4 s of high intensity illumination (6.5 mW/cm^2) applied immediately after a normal action potential caused the duration of the next action potential (stimulus interval of 20 s) to be increased by more than 50% and the duration remained increased during subsequent action potentials. In contrast, 12 s of such illumination, again applied immediately after a normal action potential, caused the duration of the next action potential to be reduced by about 20% (see Fig. 6 of Tarr and Valenzeno, 1991), and the durations of subsequent action potentials were further reduced. Thus, under appropriate circumstances, ROS can produce a decrease in APD without first increasing action potential duration. By comparison, with lower intensity illumination (1.5 mW/cm^2) and RB concentration, minutes of illumination were required before the APD decreased. Although an ROS-induced increase in APD might be considered to be antiarrhythmogenic, a dependency of action potential modification on the duration and intensity of ROS exposure could result in inhomogeneities in action potential waveform in reperfused cardiac tissue that could be highly arrhythmogenic. Regions exposed to high levels of ROS may have decreased APD compared to those exposed to lower ROS levels. Such inhomogeneities in APD may contribute to reentrant arrhythmias that can occur upon reperfusion of the ischemic myocardium.

Matsuura and Shattock (1991a,b) have recently assessed the effects of photomodification with RB (10–100 nM) on the action potential of isolated rabbit ventricular cells. They also found a marked initial increase in APD followed by a decrease in APD after prolonged exposure to ROS.

An initial increase in APD followed by a subsequent decrease has also been reported using other methods to generate ROS. Barrington and coworkers (Barrington, 1990; Barrington et al. 1988) reported that the XO reaction, DHF, or H_2O_2 produced similar three-stage changes in the action potential of isolated mammalian (canine or feline) ventricular myocytes. Initially (stage 1), both the action potential amplitude and duration increased while the resting potential of the cell remained unchanged. After long exposure (3–8 min) to these agents, the second stage began and was associated with the development of both early and delayed afterdepolarizations. The third stage began with still longer exposure. Excitability was lost during this stage, but in some cells the APD decreased dramatically before the loss of excitability. Recently, Beresewicz and Horackova (1991) reported results similar to those of Barrington and coworkers. They found that 30 μM H_2O_2 produced an initial increase in APD. In contrast to Barrington's results where the increase in APD was occasionally followed by a decrease before loss of excitability, Beresewicz and Horackova consistently observed a decrease in APD in rat and guinea pig ventricular myocytes with prolonged H_2O_2 exposure. However, cell-to-cell variability in the time course of the APD changes occurred and in a few cases only decreases in APD were observed. Interestingly, the sodium channel blocker tetrodotoxin (10 μM) prevented the H_2O_2-induced increase in APD whereas the calcium channel blocker nifedipine (3 μM) did not, suggesting that delayed sodium inactivation may contribute to the increased APD. The H_2O_2-induced action potential modifications were prevented by intracellular application of the iron chelator desferroxamine, suggesting that the extracellulary applied H_2O_2 crossed the cell membrane and caused hydroxyl radical generation intracellularly by iron-catalyzed Fenton chemistry.

Hydrogen peroxide–induced increases in APD followed by decreases have also been reported to occur in multicellular cardiac preparations (Firek and Beresewicz, 1989; Hayashi et al. 1989). Increases in APD have also been observed with the XO reaction in guinea pig myocytes (Coetzee and Opie, 1988) and with DHF, also in guinea pig myocytes (Cerbai et al., 1991; Coetzee and Opie, 1988).

While ROS-induced increases in APD have been reported by several investigators as discussed above, other investigators have reported only decreases in APD after ROS generation by DHF (Coetzee et al., 1990), the XO reaction (Coetzee et al., 1990; Goldhaber et al., 1989; Tsushima and Moffat, 1990), and H_2O_2 (Coetzee et al., 1990; Goldhaber et al., 1989). Nakaya et al. (1987) have also reported only a decrease in APD after exposure of cardiac tissue to the organic hydroperoxides CH and TBH. By comparison, in some preparations a complete lack of effect of the XO reaction on action potential duration has been reported (Pallandi et al., 1987; Tsushima and Moffat, 1990).

A combination of factors may explain why in some cases DHF, XO and/or H_2O_2 increase APD, whereas in other cases only decreases or no alteration in APD occurred. Tsushima and Moffat (1990) reported recently that whereas purine (5.75 mM) + XO (0.025 U/ml) failed to alter APD in canine papillary muscle, this combination reduced APD in canine Purkinje fibers in a frequency-dependent manner. At short cycle length (300 ms) a decrease in APD occurred within 10 min of exposure to purine + XO. However, at a longer cycle length (500 ms) the APD did not decrease significantly until 40 min, and still longer times were required at a cycle length of 800 ms. Thus, the type of cardiac preparation as well as frequency of stimualtion can affect the experimental results. Pallandi et al. (1987) used a cycle length of 1000 ms in their study on guinea pig ventricular myocardium and failed to observe any effects on APD after 30 min of exposure to purine (2.3 mM) plus XO (0.02–0.04 U/ml), a result consistent with those of Tsushima and Moffat (1990). However, Goldhaber et al. (1989) found a marked reduction in APD within 17 min of exposure of the isolated rabbit intraventricular septum to xanthine (1 mM) + XO (0.01 U/ml) stimulated at an 800-ms cycle length.

As our work with RB photomodification has demonstrated, the intensity of ROS exposure can affect the outcome regarding ROS-induced changes in APD. High intensity exposure produces a rapid decline in APD whereas a lower intensity exposure produces an increase in APD followed by a decrease. Comparable concentration effects have been reported for H_2O_2. Barrington (1990) stated that a low H_2O_2 concentration (0.001%) caused a sequential increase in APD followed by afterdepolarizations. By comparison, higher H_2O_2 concentrations (0.01%) rapidly decreased APD and depolarized the cell with few afterdepolarizations. Goldahaber et al. (1989) reported that long exposure (up to 60 min) to low H_2O_2 concentrations (0.1–0.01 mM) were required to reduce APD in rabbit intraventricular septa. By comparison, 1mM H_2O_2 reduced APD within a few minutes. Nevertheless, Hayashi et al. (1989) reported that the relatively high H_2O_2 concentration of 10 mM produced an initial increase in APD in intact right ventricles of guinea pig hearts. Thus, there is no apparent simple relationship between H_2O_2 concentration and the H_2O_2-induced changes in APD.

Although variability does exist in the ROS-induced changes in APD, this variability is not related in any obvious manner to the method used to generate the ROS. In fact, many different ROS generators produce similar results in a variety of cardiac preparations. Perhaps the basis for the variability will become clear when more is known regarding the dependency of ROS effects on (a) species, (b) preparation (intact heart vs. isolated cell), (c) ROS concentration, and (d) heart rate or stimulus frequency.

Caution must be exercised in using the action potential modifications induced in healthy cardiac tissue after exposure to ROS to predict the nature of action potential modifications that occur during reperfusion of ischemic or hypoxic cardiac tissue. Action potential modifications resulting from metabolic modifications of ionic currents during the ischemic episode may affect the outcome of ROS-induced modifications during reperfusion. Furthermore, the status of cell or tissue metabolic state as well as the antioxidants available to the preparation may affect the outcome of ROS exposure. Nevertheless, it should be noted that a rapid increase in action potential duration occurs during reoxygenation of hypoxic cardiac tissue (Hayashi et al., 1987).

Oscillatory Membrane Potentials and Oscillatory Currents

Oscillatory membrane potentials are thought to play a role in arrhythmic activity related to cellular calcium overload as may occur during reperfusion of ischemic and/or hypoxic cardiac tissue. These potentials are of two types. Early afterdepolarizations (EAD) are positive-going potential fluctuations occurring during the repolarization phase of the action potential. Delayed afterdepolarizations (DAD) occur after repolarization has been completed. When the afterdepolarizations are large enough to reach the threshold potential for activation of a regenerative inward current, the resultant action potentials are referred to as "triggered." Oscillatory inward currents elicited by a rise in internal free ionized calcium are thought to produce these afterdepolarizations. Several mechanisms can underlie both the increase in intracellular calcium as well as the oscillatory inward current. Increases in intracellular calcium during ischemic-reperfusion injury may result from (a) release of calcium from intracellular calcium stores, (b) inhibition of calcium efflux, or (c) increased calcium influx. Oscillatory inward currents elicited by the rise in intracellular calcium have been shown to be related to (a) a calcium-activated nonspecific cation channel, and (b) an electrogenic Na-Ca exchange mechanism.

In our work with single frog atrial cells, we have not observed the induction of either afterdepolarizations or oscillatory inward currents by RB-generated ROS. By comparison, Matsuura and Shattock (1991a,b) found that photoactivation of RB produced afterdepolarizations, triggered action potentials, and oscillatory inward currents in isolated rabbit ventricular cells. The oscillatory current remained inward even at positive membrane potentials (+30 mV) but was markedly suppressed by replacing extracellular sodium with lithium. Since the reversal potential of the calcium-activated nonselective cation current is near 0 mV and the Na-Ca exchanger does not transport lithium, it appears that the oscillatory inward current induced by RB-generated ROS is due to electrogenic Na-Ca exchange. The fact that RB-generated ROS produce afterdepolarizations and oscillatory inward currents in the mammalian cardiac cell but fail to do so in frog atrial cells should not be construed as an inconsistency. It is well recognized that not only does the frog cardiac cell have less sarcoplasmic reticulum than the mammalian cardiac cell, but calcium influx from the extracellular fluid plays a dominant role in increasing intracellular calcium in the frog cell, whereas release of calcium from the sarcoplasmic reticulum is more important in the mammalian cell. Calcium release from sarcoplasmic reticulum, especially the calcium-induced calcium release prevalent in mammalian but not frog cardiac cell, may be important in producing the increase in intracellular calcium which, in turn, elicits an oscillatory inward Na-Ca exchange current.

Afterdepolarizations and triggered action potentials are produced in mammalian cardiac preparations by ROS generators other than RB. Both EAD, DAD, and triggered activity have been reported after exposure to DHF (Barrington et al., 1988; Cerbai et al., 1991), X + XO (Barrington et al., 1988), and H_2O_2 (Beresewicz and Horackova, 1991; Hayashi et al., 1989). Delayed afterpotentials and triggered activity have also been reported after exposure to the organic hydroperoxides CH and TBH (Nakaya et al., 1987). Hayashi et al. (1989) and Beresewicz and Horackova (1991) demonstrated that the H_2O_2-induced DAD and triggered activity were prevented by pretreatment with ryanodine, indicating calcium release from the sarcoplasmic reticulum was responsible for inducing the DAD. However, ryanodine did not block the H_2O_2-induced EAD (Beresewicz and Horackova, 1991). Cerbai et al. (1991) demonstrated that chelation of the intracellular calcium prevented the DHF-induced EAD in isolated guinea pig ventricular cells, and Beresewicz and Horackova (1991) found that tetrodotoxin prevented H_2O_2-induced EAD. Thus, the ROS-induced afterdepolarizations are related to increases in intracellular calcium. DAD appear to be related to release of calcium from the sarcoplasmic reticulum, whereas EAD are not. The tetrodotoxin sensitivity of EAD observed by Beresewicz and Horackova (1991) suggest involvement of the Na-Ca exchanger. An increase in intracellular sodium, perhaps due to delayed sodium inactivation, would decrease calcium efflux via the Na-Ca exchanger and thereby increase intracellular calcium. Whether or not afterdepolarizations are observed after ROS exposure may depend not only on the mechanisms normally used by the cell to regulate intracellular calcium but also on the calcium load carried by the cell before ROS exposure.

The afterdepolarization and triggered activity after exposure of normal cardiac tissue to ROS are similar to those occurring during reoxygenation of hypoxic cardiac tissue. For example, Hayashi et al. (1987) demonstrated that reoxygenation of either hypoxic guinea pig papillary muscle or isolated guinea pig ventricular myocytes resulted in aftercontractions, DAD, and automaticity. These were prevented by pretreatment of the muscle or cells with ryanodine. Similar results were obtained by Thandroyen et al. (1988) during reperfusion of the ischemic rat heart. That oscillations in intracellular calcium levels are important in the production of afterdepolarizations during reperfusion has been confirmed using intracellular calcium probes (Allshire et al., 1987; Smith and Allen, 1988).

Sodium Current

The inward sodium current (I_{Na}) produces the upstroke phase (i.e., phase 0) of atrial, ventricular, and Purkinje cardiac action potentials. Alterations in I_{Na} will therefore affect cardiac excitability and conduction velocity. Our work with frog atrial cells has demonstrated that photoactivation of RB not only suppresses I_{Na}, but it also slows I_{Na} inactivation and shifts the I_{Na}–voltage relationship in the depolarizing direction (Tarr and Valenzeno, 1991). The shift in the current–voltage (I/V) relationship results in a greater I_{Na} suppression at negative than at positive membrane potentials. Bhatnagar et al. (1990) reported somewhat similar effects on I_{Na} in frog ventricular cells during exposure to TBH. The current was suppressed and its rate of inactivation slowed but, in contrast to our results with RB, TBH did not shift the I_{Na}–voltage relationship. Beresewicz and Horackova (1991) presented data they interpreted as an H_2O_2-induced slowing of I_{Na} inactivation without alteration in the magnitude of I_{Na}. We know of no studies in which the effects of DHF or XO on I_{Na} in cardiac tissue have been investigated.

Calcium Current

The inward calcium current (I_{Ca}) plays an important role in controlling the magnitude as well as the duration of the cardiac action poential. Decreases in I_{Ca} would be expected to decrease the magnitude and duration of the action potential whereas an increase in I_{Ca} would have the opposite effect. Since increases as well as decreases in action potential magnitude and duration after ROS exposure have been reported, variability in ROS effects on I_{Ca} may contribute to these effects.

In contrast to the rather limited number of studies regarding ROS effects on cardiac I_{Na}, there have been several studies on I_{Ca}. Our work with photoactivation of RB has demonstrated that ROS suppress I_{Ca} in frog atrial cells without affecting either the kinetics of I_{Ca} or the I/V relationship (Tarr and Valenzeno, 1991). Matsuura and Shattock (1991a,b) also reported that I_{Ca} in rabbit ventricular cells is decreased after photoactivation of RB.

Studies using other ROS generators have given variable results in that both decreases and increases in I_{Ca} have been reported. Goldhaber et al. (1989) reported that both the XO reaction and H_2O_2 reduced I_{Ca} in single guinea pig ventricular cells. They also reported that H_2O_2 did not affect either the kinetics or voltage dependency of I_{Ca}. However, Beresewicz and Horackova (1991) reported that H_2O_2 did not affect I_{Ca} in the same preparation. In a preliminary report Coetzee and Opie (1988) reported that DHF as well as the XO reaction increased I_{Ca} in isolated guinea pig ventricular cells. By comparison, Cerbai et al. (1991) reported that DHF reduced I_{Ca} in the same preparation. Cerbai et al. (1991) also reported that DHF did not affect either the kinetics or voltage dependency of I_{Ca}. A biphasic effect on I_{Ca} has been reported by Sato et al. (1989) for the effect of TBH in rabbit sinoatrial (SA) node. An initial, slight increase in I_{Ca} was followed by a marked reduction in I_{Ca}. Again, there was no effect of TBH on either the kinetics or voltage dependency of I_{Ca}. However, Bhatnagar et al. (1990) reported that TBH did not affect I_{Ca} in isolated frog ventricular cells even after prolonged exposure that completely suppressed I_{Na}. The reasons for these highly variable effects on I_{Ca} remain to be determined, but it seems clear that in some cases H_2O_2, the XO reaction, DHF, and TBH affect I_{Ca} in a manner similar to that by photoactivation of RB.

Delayed Rectifier Potassium Current

The delayed rectifier potassium current (I_K) plays an important role in repolarizing the cell during the action potential. A decrease in I_K would produce an increased APD, whereas an increase would decrease APD. Our results (Tarr and Valenzeno, 1991; Valenzeno and Tarr, 1991b) demonstrate that photoactivation of RB suppresses I_K in frog atrial cells without altering the kinetics of I_K activation. The effects on the I_K–voltage relationship appear to depend on the duration or intensity of ROS exposure. Short duration or low intensity exposure to ROS does not affect the I/V relationship (Tarr and Valenzeno, 1991). However, a shift in the I/V relationship in the depolarizing direction can occur with long duration ROS exposure (*unpublished observations*).

Reductions in I_K have also been reported after exposure of cardiac muscle to other ROS generators. Cerbai et al. (1991) reported that DHF reduced I_K in guinea pig ventricular cells with no apparent shift in the I/V relationship. Sato et al. (1989) stated that TBH reduced I_K in the rabbit SA node but did not show any representative data. Although Beresewicz and Horackova (1991) did not attempt to distinguish H_2O_2 effects on different potassium currents, they did observe that H_2O_2 reduced outward currents during depolarizing voltage pulses to positive membrane potentials. Such a result would be consistent with an H_2O_2-induced reduction of I_K.

Inward Rectifier Potassium Current

The inward-rectifying potassium current (I_{K1}) plays an important role in controlling the resting potential of cardiac muscle. Alterations in this current, therefore, could produce changes in resting potential. We have found that photoactivation of RB suppresses I_{K1} (Tarr and Valenzeno, 1991) in single frog atrial cells. Matsuura and Shattock (1991b) also reported a reduction in I_{K1} in rabbit ventricular cells after photoactivation of RB.

Other ROS generators have also been reported to reduce I_{K1}. Coetzee and Opie (1988) in a preliminary study found that DHF as well as the XO reaction reduced I_{K1} in guinea pig ventricular myocytes. However, in the same preparation Cerbai et al. (1991) did not observe an I_{K1} effect of DHF. Bhatnagar et al. (1990) also observed no alteration in I_{K1} after exposure of frog ventricular cells to TBH. The results of Beresewicz and Horackova (1991) suggest that H_2O_2 reduces I_{K1}.

Time-Independent Currents

Currents in addition to I_{Na}, I_{Ca}, I_K, and I_{K1} can play important roles in modulating the electrophysiological properties of cardiac tissue. Our work has demonstrated that photoactivation of RB activates a current in the frog atrial cell that we have designated as I_{leak} (Tarr and Valenzeno, 1991; Valenzeno and Tarr, 1991b). This is a time-independent current in that I_{leak} does not alter its magnitude during a sustained voltage clamp pulse. Longer ROS exposure is required to activate I_{leak} than to suppress I_{Ca} and I_K. Accordingly, the reduction in action potential amplitude and increase in APD that occurs early during photoactivation of RB are related primarily to ROS-induced suppression of I_{Ca} and I_K respectively. The marked shortening of the action potential after longer ROS exposure is related to the ROS-induced increase in I_{leak}. The rapid depolarization of the cell's resting potential after prolonged ROS exposure is also related to activation of I_{leak}.

We recently completed an extensive study designed to characterize the properties of this ROS-induced I_{leak}. The details of these results will be the subject of a future paper. However, a brief synopsis of the properties of I_{leak} is in order: (a) The I_{leak}-voltage relationship is fairly linear over a potential range of –110 mV to +50 mV; (b) The reversal potential (E_R) of I_{leak} shifts with time during ROS exposure from an initial value of –70 mV to a final value near 0 mV; (c) A significant increase (more than 30-fold) in conductance accompanies the depolarizing shift in E_R; (d) Activation of I_{leak} by ROS is not affected by the presence or absence of extracellular calcium; (e) Currents carried by both sodium and potassium ions contribute to I_{leak} throughout ROS exposure; (f) An increase in the sodium contribution to I_{leak} with time during ROS exposure accounts for the depolarizing shift in E_R; (g) Cesium produces complex effects on

I_{leak} and its use either extracellularly or intracellularly markedly affects both the E_R of I_{leak} as well as I_{leak} conductance.

Taken together, our results demonstrate that with time ROS enhance membrane permeability such that after prolonged ROS exposure there is a significant current flow through a relatively nonselective pathway. However, the ROS-induced activation of I_{leak} is not dependent on activation of the calcium-dependent nonselective cation channel since ROS activate I_{leak} in calcium-free as well as in calcium containing media. The fact that current carried by sodium contributes to I_{leak} throughout ROS exposure is significant, since sodium influx via this pathway could cause intracellular sodium to increase with time during continuous ROS exposure. Since an increase in intracellular sodium reduces calcium efflux via the Na-Ca exchanger, activation of I_{leak} by ROS may contribute to ROS-induced cell contracture and eventually cell death related to cellular calcium overload.

Recently, Matsuura and Shattock (1991b) reported that photoactivation of RB produces an increase in membrane conductance in isolated rabbit ventricular myocytes. The voltage dependency of the current related to this ROS-induced conductance showed a slight outward-going rectification and an E_R near 0 mV. The slope conductance of this current increased from an initial value of 0.5 nS to 12 nS after 7 min of ROS exposure. Matsuura and Shattock concluded that this ROS-induced current is the nonselective cation current activated by a rise in intracellular free ionized calcium. This conclusion was based primarily on the fact that the I/V relationship rotated with time during ROS exposure about an E_R near 0 mV; that is, the E_R expected for the nonselective cation current. However, the experimental protocol used by Matsuura and Shattock may have affected their results. They used 130 mM cesium chloride in the patch pipette solution as well as 5.4 mM cesium chloride in the extracellular fluid. We have observed that high internal cesium in combination with cesium (20 mM) in the extracellular fluid causes the ROS-induced I_{leak} in the frog atrial cell to have an E_R near 0 mV early as well as late during ROS exposure—a result similar to that reported by Matsuura and Shattock. However, as stated previously, we have found that with internal potassium rather than cesium the E_R of the ROS-induced current is near –70 mV early during ROS exposure. Since the calcium-activated nonselective cation channel is equally permeable to potassium and cesium (Ehara et al., 1988), substitution of cesium for potassium in the intracellular fluid should not have affected E_R if the ROS-induced current is related to current flow through the calcium-activated nonselective cation channel.

We know of no reports where the use of ROS generators other than photosensitizers have activated a time-independent current having the properties of the I_{leak} induced by photoactivation of RB. Goldhaber et al. (1989) reported that the XO reaction and H_2O_2 activated the ATP-sensitive potassium current. Although this current is a time-independent current, it clearly has properties different from I_{leak} activated by photoactivation of RB.

Different ROS Generators May Produce A Similar ROS Family

As the comparison of the electrophysiological modifications produced by photoactivation of RB to those produced by DHF, XO, H_2O_2, and organic hydroperoxides indicates, a variety of ROS generators can produce similar effects. What is the basis for these similarities? There are several possibilities. One may be that the ROS-induced modifications are not highly dependent on the type of ROS; that is, singlet oxygen, superoxide, hydroxyl radical, and organic peroxyl radicals may produce similar modifications of membrane ionic channels. Another may be that the modifications produced by a variety of ROS involve a common pathway such as peroxidation of unsaturated membrane lipids. A third possibility is that a similar family of ROS is produced regardless of the reaction used to initiate ROS production. In the case of ROS production by RB, the latter could occur simply because RB produces both singlet oxygen and superoxide. However, as discussed below, reactions exist by which superoxide-initiated reactions as well as H_2O_2-initiated reactions may result in the production of singlet oxygen.

Mechanisms by which singlet oxygen can be generated under biological conditions have been reviewed recently by us (Valenzeno and Tarr, 1991a) and by Cadenas (1989). The reader is referred to those reviews for a more extensive discussion of the various reaction schema. Suffice it to say that superoxide may be oxidized to singlet oxygen after the protonation of superoxide to the highly reactive perhydroxyl radical. Since this protonation reaction has a pK of 4.8, it is tempting to suggest that formation of the perhydroxyl radical will not occur to any significant extent at physiological pH. However, protons can be concentrated in specific cellular environments or near anionic surfaces. For example, the lysosomal membrane contains a H^+ pump that pumps H^+ into the lysosome, thereby maintaining the lumen at a pH of about 5. Also, the polyanionic surface of cell membranes creates an unusual microenvironment with pH in the so-called Gouy-Chapman-Stern layer at the surface of phospholipid membranes being 2 to 3 units less than in the bulk fluid phase (Siesjo et al., 1989). Thus, superoxide formed near the outside surface of cell membranes may be rapidly protonated to the perhydroxyl radical. Superoxide produced inside the cell can be transported by either an anion channel or an anion exchange to the outside surface where it also could be protonated. Thus, superoxide produced either extracellularly or intracellularly may result in the formation of singlet oxygen near the external surface of cell membranes or within acidic intracellular compartments.

Just as the reaction of two perhydroxyl radicals can yield singlet oxygen, the reaction of two lipid peroxyl radicals can also produce singlet oxygen via the Russell mechanism. Since lipid peroxyl radicals are formed by ROS-induced lipid peroxidation, such peroxidation may result in the formation of singlet oxygen.

It is also possible for singlet oxygen to be reduced to superoxide if an electron donor having an appropriate oxidation potential is available. Saito and coworkers have studied the efficiency of superoxide generation in the reaction of electron-rich compounds with singlet oxygen (Inoue et al., 1984; Inoue et al., 1985; Saito and Matsuura, 1984; Saito et al., 1983). As expected, the efficiency of superoxide formation depended on the oxidation potential of the electron donor. Substrates with oxidation potentials more than 0.5 V produced little, if any, superoxide, since the reduction potential of singlet oxygen is 0.65 V. It is unlikely, therefore, that singlet oxygen formed by RB at or near the external surface of the cell membrane, as in our experiments, would be reduced to superoxide, since the electron donors of appropriate oxidation potentials would not be available. However, the formation of superoxide from singlet oxygen can be of importance biologically, since there are a number of electron-rich biologically relevant intracellular substrates with oxidation potentials less than 0.5 V. For example, both NADH and 5-hydroxytrytophan produce superoxide when they react with singlet oxygen (Inoue et al., 1984).

Summary

The results that we and other investigators have obtained on cardiac cells with RB-generated ROS are representative of those obtained by other investigators using other methods to generate ROS. The sequence of an initial ROS-induced increase in APD followed by a marked reduction in APD and depolarization of the cell's resting potential is produced by RB-generated ROS (Matsuura and Shattock, 1991a,b; Tarr and Valenzeno, 1989) as well as by the XO reaction, DHF, and H_2O_2 (Barrington, 1990; Barrington et al., 1988; Beresewicz and Horackova, 1991). Early and delayed afterdepolarizations and triggered action potentials are produced in mammalian cardiac cells by RB-generated ROS (Matsuura and Shattock, 1991a,b) as well as by DHF (Barrington et al., 1988; Cerbai et al., 1991), XO (Barrington et al., 1988), and H_2O_2 (Beresewicz and Horackova, 1991; Hayashi et al., 1989). Organic hydroperoxides also produce DADs and triggered action potentials (Nakaya et al., 1987). Suppression of I_{Na} is produced by RB-generated ROS (Tarr and Valenzeno, 1991) as well as by organic hydroperoxide (Bhatnagar et al., 1990) and H_2O_2 (Beresewicz and Horackova, 1991). Reduction in I_{Ca} is produced by RB-generated ROS (Tarr and Valenzeno, 1991) as well as by XO, H_2O_2 DHF, and organic hydroperoxides (Cerbai et al., 1991; Goldhaber et al., 1989; Sato et al., 1989). A reduction in I_K is produced by RB-generated ROS (Tarr and Valenzeno, 1991; Valenzeno and Tarr, 1991b) and a DHF-induced reduction in I_K has also been reported (Cerbai et al., 1991). Studies on I_{leak} induced by ROS generators other than RB have not been reported, although Goldhaber et al. (1989) reported that XO and H_2O_2 activate the ATP-sensitive potassium current ($I_{K(ATP)}$). The RB-induced I_{leak}, however, has different properties than $I_{K(ATP)}$.

Acknowledgment. Supported by NIH grant RO1 HL43008 and by an American Heart Association Grant-in-Aid (890714).

References

Allshire A, Piper M, Cuthbertson KSR, Cobbold PH (1987): Cystolic free Ca^{2+} in single rat heart cells during anoxia and reoxygenation. *Biochem J* 244:381–385.

Barrington PL (1990): Effects of free radicals on the electrophysiological function of cardiac membranes. *Free Rad Biol and Med* 9:355–365.

Barrington PL, Meier CF Jr., Weglicki WB (1988): Abnormal electrical activity induced by free radical generating systems in isolated cardiocytes. *J Mol Cell Cardiol* 20:1163–1178.

Beresewicz A, Horackova M (1991): Alterations in electrical and contractile behavior of isolated cardiomyocytes by hydrogen peroxide: possible ionic mechanisms. *J Mol Cell Cardiol* 23:899–918.

Bhatnagar A, Srivastava SK, Szabo G (1990): Oxidative stress alters specific membrane currents in isolated cardiac myocytes. *Circ Res* 67:535–549.

Cadenas E (1989): Biochemistry of oxygen toxicity. *Annu Rev Biochem* 58:79–110.

Cerbai E, Ambrosio G, Porciatti F, Chiariello M, Giotti A, Mugelli A (1991): Cellular electrophysiological basis for oxygen radical-induced arrhythmias. *Circulation* 84:1773–1782.

Coetzee WA, Opie LH (1988): Electrophysiological effect of free oxygen radicals on guinea pig ventricular myocytes. *J Mol Cell Cardiol* 20(suppl V), S.17.

Coetzee WA, Owen P, Dennis SC, Saman S, Opie LH (1990): Reperfusion damage: free radicals mediate delayed membrane changes rather than early ventricular arrhythmias. *Cardiovasc Res* 24:156–164.

Ehara T, Noma A, Ono K (1988): Calcium-activated non-selective cation channel in ventricular cells isolated from adult guinea-pig hearts. *J Physiol* 403:117–133.

Firek I, Beresewicz A (1989): Electrophysiological effects of H_2O_2 on guinea pig ventricular muscle. Permissive role of iron. *J Mol Cell Cardiol* 21(suppl IV), S. 39.

Goldhaber JI, Ji S, Lamp ST, Weiss JN (1989): Effects of exogenous free radicals on electromechanical function and metabolism in isolated rabbit and guinea pig ventricle. *J Clin Invest* 83:1800–1809.

Hayashi H, Miyata H, Watanabe H, Kobayashi A, Yamazaki N (1989): Effects of hydrogen peroxide on action potentials and intracellular Ca^{2+} concentration of guinea pig heart. *Cardiovasc Res* 23:767–773.

Hayashi H, Ponnambalam C, McDonald TF (1987): Arrhythmic activity in reoxygenated guinea pig papillary muscles and ventricular cells. *Circ Res* 61:124–133.

Hearse DJ, Kusama Y, Bernier M (1989): Rapid electrophysiological changes leading to arrhythmias in the aerobic rat heart. *Circ Res* 65:146–153.

Inoue K, Matsuura T, Saito I (1984): A convenient method for detecting the superoxide ion from singlet oxygen reactions of biological systems: superoxide formation from hydrogenated nicotinamide adenine dinucleotide and 5-hydroxytryptophan. *J Photochem* 25:511–518.

Inoue K, Matsuura T, Saito I (1985): Importance of single-electron transfer in singlet oxygen reaction in aqueous solution. Oxidation of electron-rich thioanisoles. *Tetrahedron* 41:2177–2181.

Lesnefsky EJ, Williams GR, Rubinstein JD, Hogue TS, Horwitz LD, Reiter MJ (1991): Hydrogen peroxide decreases effective refractory period in the isolated heart. *Free Rad Biol Med* 11:529–535.

Matsuura H, Shattock MJ (1991a): Membrane potential fluctuations and transient inward currents induced by reactive oxygen intermediates in isolated rabbit ventricular cells. *Circ Res* 68:319–329.

Matsuura H, Shattock MJ (1991b): Effects of oxidant stress on steady-state background currents in isolated ventricular myocytes. *Am J Physiol* 261 (*Heart Circ Physiol 30*):H1358–H1365.

Murdock DK, Loeb JM, Euler DE, Randall WC (1980): Electrophysiology of coronary reperfusion. *Circulation* 61:175–182.

Nakaya H, Tohse N, Kanno M (1987): Electrophysiological derangements induced by lipid peroxidation in cardiac tissue. *Am J Physiol* 253 (*Heart Circ Physiol 22*):H1089–H1097.

Pallandi RT, Perry MA, Campbell TJ (1987): Proarrhythmic effects of an oxygen-derived free radical generating system on action potentials recorded from guinea pig ventricular myocardium: a possible cause of reperfusion-induced arrhythmias. *Circ Res* 61:50–54.

Pogwizd SM, Corr PB (1987): Electrophysiologic mechanisms underlying arrhythmias due to reperfusion of ischemic myocardium. *Circulation* 76:404–426.

Saito I, Matsuura T (1984): Formation of superoxide ion via one-electron-transfer from organic electron donors to singlet oxygen. In: *Oxygen radicals in chemistry and biology*, Bors W, Saran M, Tait D, eds. Berlin: Walter de Gruyter, pp 535–538 .

Saito I, Matsuura T, Inoue K (1983): Formation of superoxide ion via one-electron transfer from electron donors to singlet oxygen. *J Am Chem Soc* 105:3200–3206.

Sato N, Nishimura M, Tanaka H, Homma N, Watanabe Y (1989): Augmentation and subsequent attenuation of Ca^{2+} current due to lipid peroxidation of the membrane caused by t-butyl hydroperoxide in the rabbit sinoatrial node. *Br J Pharmacol* 98:721–723.

Siesjo BK, Agardh CD, Bengtsson F (1989): Free radicals and brain damage. *Cerebrovasc Brain Metab Rev* 1:165–211.

Smith GL, Allen DG (1988): Effects of metabolic blockade on intracellular calcium concentration in isolated ferret ventricular muscle. *Circ Res* 62:1223–1236.

Tarr M, Valenzeno DP (1989): Modification of cardiac action potential by photosensitizer-generated reactive oxygen. *J Mol Cell Cardiol* 21:539–543.

Tarr M, Valenzeno DP (1991): Modification of cardiac ionic currents by photosensitizer-generated reactive oxygen. *J Mol Cell Cardiol* 23:639–649.

Thandroyen FT, McCarthy J, Burton KP, Opie LH (1988): Ryanodine and caffeine prevent ventricular arrhythmias during acute myocardial ischaemia and reperfusion in rat heart. *Circ Res* 62:306–314.

Tsushima RG, Moffat MP (1990): Differential effects of purine/xanthine oxidase on the electrophysiologic characteristics of ventricular tissues. *J Cardiovasc Pharm* 16:50–58.

Valenzeno DP, Tarr M (1991a): Membrane photomodification and its use to study reactive oxygen effects. In: *Photochemistry and photophysics*, Rabek JF, ed. Boca Raton, Fl: CRC Press Inc.

Valenzeno DP, Tarr M (1991b): Membrane photomodification of cardiac myocytes: Potassium and leakage current. *Photochem Photobiol* 53:195–201.

Index

Related Birkhäuser Titles

Free Radicals and Aging
Ingrid Emerit and Britton Chance, Editors
ISBN 0-8176-2744-8, 1992, 440 Pages

Lipid Soluable Antioxidants: Biochemistry and Clinical Applications
A.S.H. Ong and L. Packer, Editors
ISBN 0-8176-2667-0, 1992, 656 Pages

Molecular Biology of Receptors that Couple to G-Proteins
Mark R. Brann, Editor
ISBN 0-8176-3465-7, 1992, 256 Pages

Emerging Strategies in Neuroprotection
Paul J. Marangos and Harbans Lal, Editors
ISBN 0-8176-3544-0, 1992, 320 Pages

Formation and Regeneration of Nerve Connections
Sansar C. Sharma and James W. Fawcett, Editors
ISBN 0-8176-3563-7, 1992, 320 Pages

Genetically Defined Animal Models of Neurobehavioral Dysfunctions
Peter Driscoll, Editor
ISBN 0-8176-3460-6, 1992, 328 Pages

Cholinergic Basis for Alzheimer Therapy
Robert Becker and Ezio Giacobini, Editors
ISBN 0-8176-3566-1, 1991, 494 Pages